AF333831

INDUSTRIAL CHEMISTRY

New Applications, Processes and Systems

INDUSTRIAL CHEMISTRY
New Applications, Processes and Systems

Harold H. Trimm, PhD, RSO

*Chairman, Chemistry Department, Broome Community College;
Adjunct Analytical Professor, Binghamton University,
Binghamton, New York, U.S.A.*

William Hunter III

Researcher, National Science Foundation, U.S.A.

Apple Academic Press

Industrial Chemistry: New Applications, Processes and Systems

© Copyright 2011*
Apple Academic Press Inc.

First Published in the Canada, 2011
Apple Academic Press Inc.
3333 Mistwell Crescent
Oakville, ON L6L 0A2
Tel. : (888) 241-2035
Fax: (866) 222-9549
E-mail: info@appleacademicpress.com
www.appleacademicpress.com

The full-color tables, figures, diagrams, and images in this book may be viewed at www.appleacademicpress.com

ISBN 978-1-926692-83-8

Harold H. Trimm, PhD, RSO
William Hunter III

Printed at Thomson Press (I) Ltd.
9 8 7 6 5 4 3 2 1

Cover Design: Psqua

Library and Archives Canada Cataloguing in Publication Data
CIP Data on file with the Library and Archives Canada

CONTENTS

INTRODUCTION

Industrial chemistry is the practical application of chemistry to industrial processes. It is closely allied to the field of chemical engineering, where reactions are run on a large scale and the economics of the process are being constantly evaluated. In industry, the main focus is to make a profit. Industrial chemists take readily available raw materials and turn them into finished products for the consumer; the lower the price of the raw material, the better the economics of the process. Common raw materials include salts, limestone, and petroleum. Industrial chemists use their knowledge of chemistry to turn these substances into consumer products such as synthetic fibers, dyes, pigments, agricultural chemicals, drugs, packaging materials, plastics, and polymers. One of the largest applications of industrial chemistry is the refining of crude oil into various products such as liquefied petroleum gas, naphtha, gasoline, kerosene, jet fuel, diesel, heating oil, greases, waxes, coke, and asphalt.

In oil refineries, crude oil is separated into various consumer products by the process of distillation. The crude oil is heated and the vapors rise through a fractionating tower to a point corresponding to their boiling point. There is a great deal of chemistry and research occurring at the cracking and reforming units, which can break down heavier molecules or reassemble smaller molecules to tailor the products formed to consumer demand.

The green revolution now allows farmers to produce as much as ten times the crop yield as before. Much of this increase in the world's food supply is due to the production of chemical fertilizers by the agrochemical industry. The Haber reaction allows agrochemists to turn the raw materials nitrogen and hydrogen into ammonia, which is then turned into fertilizer. The process is very energy intensive, and there is intense research to improve the process.

Plastics are usually produced by disassembling fossil fuels into simple molecules (monomers), which are then reassembled into long polymers. The properties of the plastic produced can be controlled by which monomers are selected, the length of the polymer chain, and how the chains interact (crosslinking). Plastics have become an integral part of everyday life. They are used for packaging as well as major structural components of consumer products.

The majority of chemists are hired by industry. Some of the areas where chemists are employed include agriculture, biotechnology, education, chemical sales, consulting, environmental, food and flavor, forensics, geochemistry, hazardous waste, health, pharmaceuticals, petroleum, polymer, paper, research, and water treatment.

Industrial chemists are constantly working on ways to produce consumer goods with less cost and less waste. There is an increased focus on green chemistry, which is the design of chemical products and processes that reduce or eliminate the use or generation of hazardous substances. Some of the fields that are studied by industrial chemists include agricultural, organic, pharmaceutical, soap, beauty aids, rubber, resin, paint, polymer, dye, pigment, inorganic, and household and office products. By mass, the largest industrial chemistry products are petroleum, chloralkali, sulfuric acid, ammonia, and phosphoric acid. Economists use the production of sulfuric acid as a means of estimating country's manufacturing capabilities.

— Harold H. Trimm, PhD, RSO

Trace Determination of Linear Alkylbenzene Sulfonates: Application in Artificially Polluted Soil — Carrots System

Caroline Sablayrolles, Mireille Montréjaud-Vignoles,
Jérôme Silvestre and Michel Treilhou

ABSTRACT

Surfactants are widely used in household and industrial products. The risk of incorporation of linear alkylbenzene sulfonates (LAS) from biosolids, waste-water, and fertilizers land application to the food chain is being assessed at present by the European Union. In the present work, a complete analytical method for LAS trace determination has been developed and successfully applied to LAS (C10–C13) uptake in carrot plants used as model. These carrots were grown in soil with the trace organics compounds added directly into the plant containers in pure substances form. LAS trace determination ($\mu g\ kg^{-1}$

dry matter) in carrots samples was achieved by Soxtec apparatus and high-performance liquid chromatography-fluorescence detection. The methodology developed provides LAS determination at low detection limits (5 µg kg⁻¹ dry matter) for carrot sample (2 g dry matter) with good recoveries rate (>90%). Transfer of LAS has been followed into the various parts of the carrot plant. LAS are generally found in the carrot leaves and percentage transfer remains very low (0.02%).

Introduction

Linear alkylbenzene sulfonates (LASs) are synthetic anionic surfactants which were introduced in the 1960s as more biodegradable replacements for highly branched alkyl benzene sulfonates [1, 2]. LASs are nonvolatile compounds produced by alkylation and sulfonation of benzene [3]. LASs are a mixture of homologues and phenyl positional isomers, each containing an aromatic ring sulfonated at the para position and attached to a linear alkyl chain at any position except the terminal one (Figure 1). The product is generally used in detergents and cleaning products in the form of the sodium salt for domestic and industrial uses [2–4]. Commercially available products are very complex mixtures containing homologues with alkyl chains ranging from 10 to 13 carbon units (C10–C13). It corresponds to a compromise between cleaning capacity, on the one hand and biodegrading and toxicity, on the other hand. LASs have been extensively used for over 30 years with an established global consumption of 2 millions tons per year [5]. The properties of LASs differ greatly depending on the alkyl chain length and position on benzene sulfonate group. It has been found that longer LASs homologues have higher octanol/water partition coefficient (Kow) values [6, 7]. In fact, the homologues with long chain have a greater capacity of adsorption on the solids and a greater insolubility in the presence of calcium or magnesium [8]. In general, a decrease in alkyl chain length is accompanied by a decrease in toxicity [5]. Dermal contact is the first source of human exposure to LASs. Minor amounts of LASs may be ingested in drinking water, on utensils, and food. The daily intake of LASs via these media (exposure from direct and indirect skin contact as well as from inhalation and from oral route in drinking water and dishware) can be estimated to be about 4 µg/kg body weight [9]. Occupational exposure to LASs may occur during the formulation of various products, but no chronic effects in humans have been noticed. In great concentration (500–2000 mg kg⁻¹), LASs could have a long-term effect [8]. Indeed, their dispersing capacity could induce the release of others compounds present in soil [10]. Generated scrubbing could involve the biodisponibility of these new compounds [2].

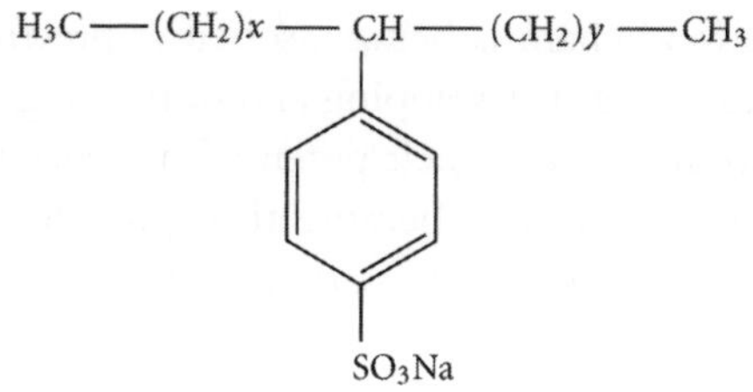

$$H_3C \text{---} (CH_2)x \text{---} CH \text{---} (CH_2)y \text{---} CH_3$$

SO_3Na

Figure 1. General chemical structure of linear alkylbenzene sulfonate (LASs), where x and y corresponds with the number of CH^2 on each side of the benzene sulphonate group (7 • x + y • 10).

After use, LASs are discharged into wastewater treatment plants and dispersed into the environment through effluent discharge into surface waters and sludge disposal on lands. Moreover, LASs can be introduced directly into the grounds: their emulsifying and dispersing properties make them essential in the formulations of fertilizers and pesticides [11]. They are thus present in many compartments of the environment (sediments, aquatic environments, grounds…). LASs have been detected in raw sewage with a concentration range of $1-15$ mgL^{-1} [9], in sludge with concentrations between $3-15$ g kg^{-1} of dry matter [5, 8, 9], in surface waters at $2-47$ µgL^{-1} concentration range [9], and in soil at concentrations below 1 mg kg^{-1} [9, 10]. LASs can be degraded under aerobic conditions however are persistent under anaerobic conditions [9, 12]. Moreover, Lara-Martín and colleagues have shown that this surfactant can be degraded in sulfate-reducing environments such as marine sediments [13].

The determination of LASs in environmental samples is usually performed using liquid chromatographic methods with UV detection [4, 14, 15], fluorescence detection [16], or mass spectrometric detection [15, 17, 18] which allows the identification and determination of LASs isomers and homologues. There are a more limited number of gas chromatography methods [19, 20] which can be due to the low volatility of these compounds, being necessary the use of derivatisation reactions of the sulfonate group to obtain more volatile compounds. Capillary electrophoresis with UV detection has been also used for the determination of the sum, homologues and isomers of LASs in household products and wastewater samples [21]. Methods for the quantification of LASs in soil [18], in sewage sludge [18, 19], in sediment [18, 22], in biological organisms [17, 23], or in water [14, 15, 20, 24] can be reported. However, these methods cannot be directly applied to plant analysis. Specific purification steps were needed. The main problem for analysis of organic pollutants in plants comes from the complexity of the matrix. Plants have a particular tissue structure, which depend on the species and the age, and are highly rich in pigments, essential oil, fatty acids, or alcohols.

The risk of incorporation of LASs from biosolids, wastewater, and fertilizers land application to the food chain is being assessed at present by the European Union. The present study aims at developing and optimizing a method for LASs quantitative determination to detect their potential presence in food plants Carrots (Daucus carota L.) were studied because they provide a good maximizing uptake model for research. Indeed, carrots are root crops with high lipid content [25] and therefore LASs (amphiphilic nature) should easily be dissolved in these lipids.

Experimental

Standards and Reagents

All chemicals used were analytical quality. Methanol, acetonitrile, and HPLC water were purchased from VWR Merck (France). Sodium perchlorate ($NaClO_4$) and sodium dodecylsulfate (SDS) were obtained by VWR Prolabo (France). The cellulose extraction thimbles of 20 mm × 80 mm size were purchased from Schleicher & Schuell (France). Fontainebleau sand (particle size 150–300 μm) to control boiling and powdered Florisil (Florisil PR particle size 60–100 mech- magnesium silicate) to adsorb grease were added to the sample in the cellulose extraction cartridge. The HPLC separation was performed with an Inertsil ODS3 column (C_{18}) of 25 cm long × 0.46 cm internal diameter and 5 μm particle size purchased from Supelco (France).

Condea Chimie SARL supplied the Marlon ARL which is a commercial surfactant powder containing 80% of C10–C13 LASs. This commercial homologue LASs mixture has the following homologue mass distribution: C10 (14.3%), C11 (35.7%), C12 (30.8%), and C13 (19.2%). Stock standard solution (1 g L^{-1}) was prepared by dissolving 312.5 mg Marlon ARL in 250 mL methanol / SDS aqueous solution at 5 10^{-3} mol L^{-1} (50/50, v/v).

Samples Collection

In order to test the different stages of the analytical protocol, carrots (Daucus carrota var. Amsterdam A.B.K. Bejo) were cultivated under greenhouse conditions (mean temperature 24°C, 14-hour light, 50% relative humidity). Fifteen days after germination, seedlings were transplanted into glass pots containing soil only (control pots) and soil artificially polluted with LASs (Marlon ARL): 1 g of Marlon ARL were mixed to 2 kg dry matter of soil. Four pots (2 L volume; 15 10^{-2} m high; 13 10^{-2} m diameter) per treatment, each containing 7 carrots, were used. A soil sample was collected from the upper 20 cm of the field horizon from a French agricultural research station and was sieved at 5 mm. All plants were watered with nutrient solution (KNO_3 5, KH_2PO_4 2, $Ca(NO_3)_2$ 5, $MgSO_4$ 1.5 mmol L^{-1} for macronutrients, and Fe 15, Mn 0.49, Cu 0.06, Zn 0.11, Mo 0.01, B 0.26 mg L^{-1} for micronutrient) to 2/3 of field capacity in order to allow roots oxygenation. Carrots were harvested after 3 months and divided into leaves, peel, and core. Each part of the plants was carefully washed three times with demineralised water.

Sample Extraction

LASs determination in environmental samples was carried out according to a protocol of several determinative steps, that is, pretreatment, extraction, and analysis (Figure 2). Prior to extraction, samples were lyophilized and grounded up using a household grinder. This method allows recuperation of a maximum amount of trace organics.

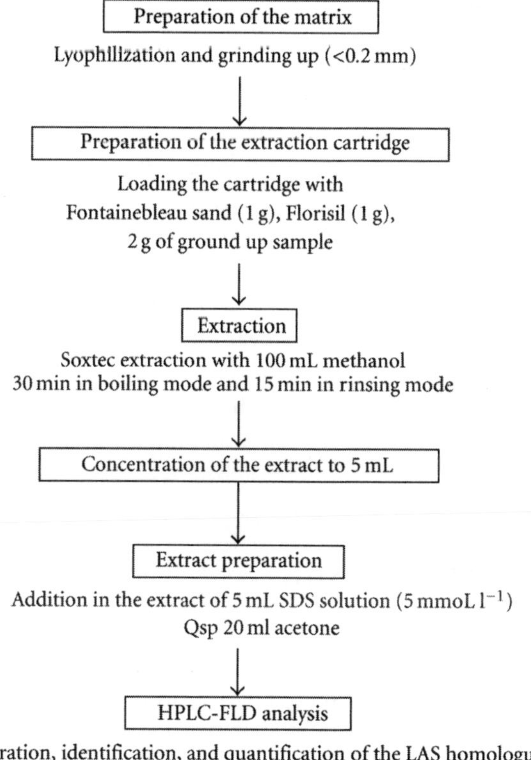

Figure 2. Description of the different treatment stages for LASs quantitative determination.

The solid/liquid extraction was carried out with a Soxtec System HT2 (Tecator, France). This apparatus is a semiautomated apparatus working on the Soxhlet principle, while allowing extractions which are more rapid, economical (better solvent recuperation), and safe (dissociation of extraction and heating units). It is composed of two parts: an oil bath plus a unit with two plates heated by the oil, and above, systems for fixing the cartridge and for cooling. Two grams of lyophilized sample were extracted with 100 mL of methanol for 45 minutes. The sample was placed in a cellulose cartridge which was capped with cotton wool immersed in methanol (boiling mode) for 30 minutes to give a rapid, total contact. Next, the cartridge was lifted up above the still boiling solvent (rinsing mode) allowing the condensing solvent to rinse the sample. Then, a rotary evaporator (Rotavapor, Büchi) and 30C° temperature controlled bath was used to concentrate the solvent down to 10 mL. This extract is concentrated to 5 mL under a stream of nitrogen. 5 mL SDS solution (5 mmol L^{-1}) and 10 mL acetone were added into the extract.

Sample Analysis

P200 (Spectra Physics) pump linked to a Hewlett Packard 1100 fluorimetric detection computer-controlled with data acquisition and processing using Normasoft (ICS) software, was employed. The system was equipped with an Inertsil ODS3 column and a guard column 2 cm long, 4.6 mm in diameter with a particle size of 5 μm (Supelco, France). The volume injected (20 μL) was reproducible and flow was 1.5 mL min^{-1} with the elution lasting for 35 minutes. The mobile phase was acetonitrile (A) and a premixed water/acetonitrile (75/25, v/v) solution containing sodium perchlorate (10 g L^{-1}) (B) gradient programme (15%A–85%B at start, 2 minutes hold, 11 minutes linear gradient to 40%A–60%B, 1 minutes hold, 8 minutes linear gradient to 70%A–30%B and 5 minutes hold, 5 minutes linear gradient to 15%A–85%B and 3 v hold) and carried out at room temperature. The fluorescence detector operated at excitation-emission wavelengths of 225–305 nm.

Results and Discussion

Extraction Time

Lyophilised carrots sample were spiked with LASs mixture (500 μg kg^{-1} dry matter) just prior to extraction to test the effectiveness of the extraction procedure. This level has been chosen in order to be in coherence with the experimental culture. Three extraction times (45, 90, and 180 minutes) have been tested. Three replicates have been made for each extraction procedure. Recovery rate % and standard deviation for extraction times were found such as: 45 minutes : 91 ± 4; 90 minutes : 91 ± 9; 180 minutes : 89 ± 8. The results showed that there is no significant difference between the 3 extraction times. Thus, the total extraction time has been set at 45 minutes (30 minutes in boiling mode and 15 minutes in rinsing mode).

Separation, Calibration Graphs, and Limits of Detection

The homologues separation has been set up using reversed phase high-performance liquid chromatography coupled with fluorescence detector (HPLC-FLD) which has proved to be a successful method for determination of LASs in water and solid samples [14, 22, 23, 26, 27]. LASs separation by homologues is illustrated in Figure 3. This chromatograph corresponds to the analysis of a sample.

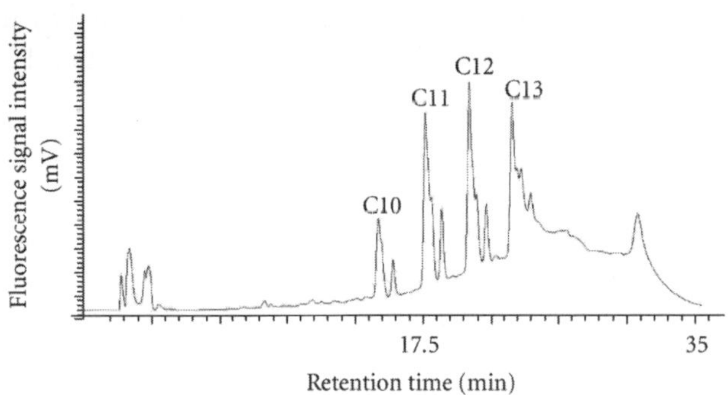

Figure 3. HPLC-FLD chromatogram obtained for carrot sample spiked with LASs (50 µgg⁻¹ dry matter). Chromatographic condition: Inertsil ODS3 colum of 250 × 4.6 mm (5 µg), flow rate = 1.5 mL min⁻¹, fluorescence detection (ex: 225 nm–em: 305 nm), 20 µL of injection volume.

Calibration curves (equation: $y=3.642 \cdot 10^4 x + 4.307 \cdot 10^6$) were generated using linear regression analysis and over the established concentration range (5–100 µgmL⁻¹) gave good fits ($r^2 > 0.990$). LASs were identified by the retention time compared to the qualitative standard. The quantitative calculations are made from the peak area. LASs were determined as the sum of homologous C10 to C13 LASs.

The precision of the method is given as the repeatability expressed as the relative standard deviation (R.S.D.%) and is an evaluation of the overall extraction-analysis procedure [28]. It is calculated from 5 replicates of 5 carrot samples. The R.S.D. was found to be 5% for the concentration level 30 mg LASs kg⁻¹ dry matter.

The recoveries rate were found in the range of 91±4% for carrot samples. The blank values, obtained from an empty sampler, were <5 µg kg⁻¹ dry matter.

The quantification limit has been based on 10 standard deviations (S.D.) of 10 replicates samples and was determined to be 25 µgkg⁻¹ of dry matter. The detection limit, based on 3 standard deviations (S.D.) of 10 replicates samples, was determined to be 5 µgkg⁻¹ of dry matter. This limit of detection is 100 times lower than those obtained by Gron et al. [27], and Mortensen et al. [26], for carrots analysis.

Application

This method has been applied to carrots samples exposed to LASs with objective of evaluating potential bioconcentration of this surfactant.

Initially carrots cultures on sand–LASs pure substances mixtures were carried out but carrots did not develop. Indeed, LASs induce a modification of the sand capillarity properties: water do not go up when it is introduced by the lower part of the farming system and water fall down when it is introduced by the higher part of the substrate. Thus, carrots cultures on soil with LASs pure substances were set up.

Statistical data processing were carried out after a variance analysis by the test with multiple degree of Newman-Keuls at P<.05 (Statistical Software, Sigma Stat 2.00). The same letter in a column means as there is no significant difference with P<.05. On the other hand, a different letter means that there is a significant difference between treatments.

Table 1 (line 1 and 2) presents the number of repetitions per treatment and the number of carrots per pot. In a general way, 7 carrots per pot were transplanted. With the stage of harvest, there do not remain inevitably these 7 carrots because of climatic and watering conditions. However, there is no significant difference between treatments regarding the number of carrots per pot.

Table 1. Number of carrots per pot; dry matter production; LASs concentration in soil, and LASs concentration in carrots. Mean values of four replications followed by the same letter in a column are not significantly different at P<.05, ± SE, standard error in variance analysis.

Line			Control carrots	LASs carrots
1	Number of pots	—	6	6
2	Number of carrots per pot	carrot	7.00 ±0.50[b]	6.25 ± 0.48[b]
		Peel	82.5 ± 4.8[a]	85.0 ±23.6[a]
3	Dry matter production (mg per plant, mean ± S.E.)	Core	188.0 ±9.5[a]	197.0 ± 54.2[a]
		Leaves	210.0 ±15.7[a]	233.0 ± 19.7[a]
4	Total content of LASs in soil (initial) (mg kg⁻¹ dry matter, mean ± SE)	Soil	50 ± 4	500 ± 20
		Peel	0.47 ± 0.04	0.63 ± 0.06
5	Total content of LASs in carrot plants (mg kg⁻¹ dry matter, mean ± SE)	Core	0.26 ± 0.03	0.31 ± 0.03
		Leaves	0.06 ± 0.01	0.26 ± 0.02

Dry matter production is presented in Table 1 (line 3). Newman-Keuls tests showed that there were no significant differences in growth between the carrots on soil only (control) and those on soil with LASs pure substances. It is thus clear that even in very great quantity, LASs do not inhibit the development of carrot under our experimental conditions. This result is in agreement with [27].

Average levels of LASs in core, peel, and leaves of carrot plants are presented in Table 1 (line 4). HPLC-FLD apparatus gives concentration results in mg L^{-1}. Knowing sample mass and taking into account all analytical steps, results can be expressed in terms of mg kg^{-1} of dry matter.

The percentage of transfer of LASs in carrot was calculated by submitting the ratio of the mass of LASs found in carrot on the mass of LASs present initially in the pot. Only 0.02% of LASs initially spiked in soil have been uptake by carrot plants. Percentage repartition of LASs have been calculated by taking into account LASs flux in each carrot compartment and LASs total flux transferred in the three compartments. Mean of 12% LASs transferred are found in peel, 23% in core, and 66% in leaves. LASs are compounds likely to be transferred from the ground towards plant [27, 29, 30]. LASs properties responsible for this behavior are a great solubility in water without micelle formation and weak affinity for organic matter. Thus, LASs can be transferred towards the plants by water absorption by the roots. Indeed, LASs are surface-active anion which has the characteristic to have absorbent groups (sodium sulfonate) and hydrophobic groups (alkyl chain) (2). Thus, carrots are able to accumulate hydrophobic compounds [31] but the uptake observed in this experimentation remains very low.

Conclusions

An analytical protocol for LAS determination in carrots using Soxtec extraction (methanol, 30 minutes) and HPLC-FLD quantification has been developed. The methodology developed provides good recoveries rate (91 ± 4%), good precision (5%) and low detection limits (5 µg kg^{-1} dry matter) for carrot sample (2 g dry matter). This method has been applied to the study of LAS bioavailability in carrots cultivated on soil enriched with pure trace organics. LAS have been traced in the various parts of the plant. LAS are generally found in carrot leaves. Plant analyses of LAS in carrots showed a weak plant uptake with LAS added as spike solution.

Acknowledgements

This research has been supported by Anjou Recherche (Veolia Environnement, Paris). Authors acknowledge Sabrina Guili, Pauline Pinel and Ecole Nationale de Formation Agronomique's staff for technical support.

References

1. R. A. Rapaport and W. S. Eckhoff, "Monitoring linear alkyl benzene sulfonate in the environment: 1973–1986," Environmental Toxicology and Chemistry, vol. 9, no. 10, pp. 1245–1257, 1990.

2. J. Jensen, "Fate and effects of linear alkylbenzene sulphonates (LAS) in the terrestrial environment," The Science of the Total Environment, vol. 226, no. 2-3, pp. 93–111, 1999.

3. G. Thoumelin, Les Tensio-Actifs dans les Eaux Douces et Marines: Analyse, Comportement, Ecotoxicologie, Reperes Ocean, Lille, France, 1995.

4. A. Marcomini, F. Filipuzzi, and W. Giger, "Aromatic surfactants in laundry detergents and hard-surface cleaners: linear alkylbenzenesulphonates and alkylphenol polyethoxylates," Chemosphere, vol. 17, no. 5, pp. 853–863, 1988.

5. W. De Wolf and T. Feijtel, "Terrestrial risk assessment for linear alkyl benzene sulfonate (LAS) in sludge-amended soils," Chemosphere, vol. 36, no. 6, pp. 1319–1343, 1998.

6. K. Jones and J. Stevens, Organic Contaminants in Sewage Sludge Applied to Agricultural Land, Water Industry Research, London, UK, 2002.

7. G.-G. Ying, "Fate, behavior and effects of surfactants and their degradation products in the environment," Environment International, vol. 32, no. 3, pp. 417–431, 2006.

8. D. Schowanek, H. David, R. Francaviglia, et al., "Probabilistic risk assessment for linear alkylbenzene sulfonate (LAS) in sewage sludge used on agricultural soil," Regulatory Toxicology and Pharmacology, vol. 49, no. 3, pp. 245–259, 2007.

9. HERA, "Human and environmental risk assessment of ingredients of household cleaning products," Report for LAS., 2004, http://www.heraproject.com/.

10. S. D. Haigh, "A review of the interaction of surfactants with organic contaminants in soil," in Science of the Total Environment, vol. 185, no. 1–3, pp. 161–170, 1996.

11. L. Carlsen, M.-B. Metzon, and J. Kjelsmark, "Linear alkylbenzene sulfonates (LAS) in the terrestrial environment," The Science of the Total Environment, vol. 290, no. 1–3, pp. 225–230, 2002.

12. M. T. García, E. Campos, I. Ribosa, A. Latorre, and J. Sànchez-Leal, "Anaerobic digestion of linear alkyl benzene sulfonates: biodegradation kinetics and metabolite analysis," Chemosphere, vol. 60, no. 11, pp. 1636–1643, 2005.

13. P. A. Lara-Martín, A. Gómez-Parra, T. Köchling, J. L. Sanz, R. Amils, and E. González-Mazo, "Anaerobic degradation of linear alkylbenzene sulfonates in coastal marine sediments," Environmental Science and Technology, vol. 41, no. 10, pp. 3573–3579, 2007.

14. A. Di Corcia, M. Marchetti, R. Samperi, and A. Marcomini, "Liquid chromatographic determination of linear alkylbenzenesulfonates in aqueous environmental samples," Analytical Chemistry, vol. 63, no. 11, pp. 1179–1182, 1991.

15. S. Wangkarn, P. Soisungnoen, M. Rayanakorn, and K. Grudpan, "Determination of linear alkylbenzene sulfonates in water samples by liquid chromatography-UV detection and confirmation by liquid chromatography-mass spectrometry," Talanta, vol. 67, no. 4, pp. 686–695, 2005.

16. M. L. Sanchez-Martinez, M. P. Aguilar-Caballos, S. A. Eremin, and A. Gomez-Hens, "Determination of linear alkylbenzenesulfonates in water samples by immunoaffinity chromatography with fluorescence detection," Analytica Chimica Acta, vol. 553, no. 1-2, pp. 93–98, 2005.

17. J. Tolls, R. Samperi, and A. Di Corcia, "Bioaccumulation of LAS in feral fish studied by a novel LC-MS/MS method," Environmental Science and Technology, vol. 37, no. 2, pp. 314–320, 2003.

18. P. Eichhorn, O. Lopez, and D. Barcelo, "Application of liquid chromatography-electrospray-tandem mass spectrometry for the identification and characterisation of linear alkylbenzene sulfonates and sulfophenyl carboxylates in sludge-amended soils," Journal of Chromatography A, vol. 1067, no. 1-2, pp. 171–179, 2005.

19. J. McEvoy and W. Giger, "Determination of linear alkylbenzenesulfonates in sewage sludge by high-resolution gas chromatography/mass spectrometry," Environmental Science and Technology, vol. 20, no. 4, pp. 376–383, 1986.

20. R. Alzaga, A. Peña, L. Ortiz, and J. M. Bayona, "Determination of linear alkylbenzensulfonates in aqueous matrices by ion-pair solid-phase microextraction-in-port derivatization-gas chromatography-mass spectrometry," in Journal of Chromatography A, vol. 999, no. 1-2, pp. 51–60, 2003.

21. W.-H. Ding and C.-H. Liu, "Analysis of linear alkylbenzenesulfonates by capillary zone electrophoresis with large-volume sample stacking," Journal of Chromatography A, vol. 929, no. 1-2, pp. 143–150, 2001.

22. S. Morales-Muñoz, J. L. Luque-García, and M. D. Luque de Castro, "Screening method for linear alkylbenzene sulfonates in sediments based on water Soxhlet extraction assisted by focused microwaves with on-line preconcentration/

derivatization/detection," Journal of Chromatography A, vol. 1026, no. 1-2, pp. 41–46, 2004.

23. M. Saez, V. M. León, A. Gómez-Parra, and E. González-Mazo, "Extraction and isolation of linear alkylbenzene sulfonates and their intermediate metabolites from various marine organisms," Journal of Chromatography A, vol. 889, no. 1-2, pp. 99–104, 2000.

24. T. Reemtsma, "Methods of analysis of polar aromatic sulfonates from aquatic environments," Journal of Chromatography A, vol. 733, no. 1-2, pp. 473–489, 1996.

25. S. R. Wild and K. C. Jones, "Polynuclear aromatic hydrocarbon uptake by carrots grown in sludge-amended soil," Journal of Environmental Quality, vol. 21, no. 2, pp. 217–225, 1992.

26. G. K. Mortensen, H. Egsgaard, P. Ambus, E. S. Jensen, and C. Grøn, "Influence of plant growth on degradation of linear alkylbenzene sulfonate in sludge-amended soil," Journal of Environmental Quality, vol. 30, no. 4, pp. 1266–1270, 2001.

27. C. Grøn, F. Laturnus, G. K. Mortensen, et al., "Plant uptake of LAS and DEHP from sludge amended soil," in Persistent, Bioaccumulative, and Toxic Chemicals—I. Fate and Exposure, vol. 772 of ACS Symposium Series, pp. 99–111, American Chemical Society, Washington, DC, USA, 2000.

28. French normalisation, XPT90-210 Water Quality: Protocole d'Évaluation d'une Méthode Alternative d'Analyse Physico-Chimique Quantitative par Rapport à une Méthode de Référence, AFNOR Press, Paris, France, 1999.

29. S. A. Klein, D. Jenkins, and P. H. Mc Gauthey, "The fate of alkylbenzene sulfonate in soils and plants—I: reduction by soils," Journal Water Pollution Control Federation, vol. 35, pp. 636–654, 1963.

30. N. Litz, H. W. Doering, M. Thiele, and H.-P. Blume, "The behavior of linear alkylbenzenesulfonate in different soils: a comparison between field and laboratory studies," Ecotoxicology and Environmental Safety, vol. 14, no. 2, pp. 103–116, 1987.

31. S. O. Petersen, K. Henriksen, G. K. Mortensen, et al., "Recycling of sewage sludge and household compost to arable land: fate and effects of organic contaminants, and impact on soil fertility," Soil and Tillage Research, vol. 72, no. 2, pp. 139–152, 2003.

Are Biogenic Emissions a Significant Source of Summertime Atmospheric Toluene in the Rural Northeastern United States?

M. L. White, R. S. Russo, Y. Zhou, J. L. Ambrose,
K. Haase, E. K. Frinak, R. K. Varner, O. W. Wingenter,
H. Mao, R. Talbot and B. C. Sive

ABSTRACT

Summertime atmospheric toluene enhancements at Thompson Farm in the rural northeastern United States were unexpected and resulted in a toluene/benzene seasonal pattern that was distinctly different from that of other anthropogenic volatile organic compounds. Consequently, three hydrocarbon sources were investigated for potential contributions to the enhancements

*during 2004–2006. These included: (1) increased warm season fuel evapo-
ration coupled with changes in reformulated gasoline (RFG) content to meet
US EPA summertime volatility standards, (2) local industrial emissions and
(3) local vegetative emissions. The contribution of fuel evaporation emission
to summer toluene mixing ratios was estimated to range from 16 to 30 pptv
d^{-1}, and did not fully account for the observed enhancements (20– 50 pptv)
in 2004–2006. Static chamber measurements of alfalfa, a crop at Thompson
Farm, and dynamic branch enclosure measurements of loblolly pine trees in
North Carolina suggested vegetative emissions of 5 and 12 pptv d^{-1} for crops
and coniferous trees, respectively. Toluene emission rates from alfalfa are potentially much larger as these plants were only sampled at the end of the growing season. Measured biogenic fluxes were on the same order of magnitude as
the influence from gasoline evaporation and industrial sources (regional industrial emissions estimated at 7 pptv d^{-1}) and indicated that local vegetative emissions make a significant contribution to summertime toluene enhancements. Additional studies are needed to characterize the variability and
factors controlling toluene emissions from alfalfa and other vegetation types
throughout the growing season.*

Introduction

Toluene is a ubiquitous aromatic volatile organic compound (VOC) in the tropo-
sphere (Dewulf and Van Langenhove, 1997; Singh et al., 1985) that has been classified by the United States Environmental Protection Agency (US EPA) as an air
toxic for its detrimental effects on human central nervous system function with
acute exposure (Goldhaber et al., 1995). Its oxidation in the presence of nitrogen
oxides (NO_x) can lead to tropospheric ozone (O_3) formation, a secondary pollutant and respiratory irritant (Wang et al., 1998). Low volatility oxidation products
can also partition into particulate matter becoming a significant component of
fine aerosol mass (Schauer et al., 2002). Of particular importance to air quality in
rural environments, recent studies indicate that secondary organic aerosol (SOA)
formation from aromatic precursors, including toluene, is substantially faster under low NO_x conditions than under high NO_x conditions (Ng et al., 2007).

Toluene sources are understood to be primarily anthropogenic and include
combustion, fuel evaporation, solvent usage, and industrial processes (Singh and
Zimmerman, 1992). It is often assumed that these sources are concentrated in
urban areas and have emission rates that are consistently proportional to benzene,
another widely distributed aromatic VOC and air toxic with similar anthropogenic sources. These assumptions, along with several others regarding sinks and
air mass mixing, allow the use of toluene/benzene ratios in characterizing air mass

photochemical age at non-urban locations (e.g. Warneke et al., 2007; deGouw et al., 2005; Gelencser et al., 1997).

However, the discovery of elevated warm season toluene mixing ratios at a rural site in northern New England brings these assumptions into question. In the long-term data set of daily VOC measurements made at Thompson Farm in coastal New Hampshire, toluene followed a significantly different seasonal pattern than benzene and other common anthropogenic tracers which usually reached their minimum levels during summer. The presence of elevated toluene mixing ratios from late spring to early fall at this rural location, even in well-processed air masses, could reflect several influences including a seasonal cycle in urban anthropogenic emissions of toluene and/or an unidentified local warm season source.

Any seasonal cycle in urban anthropogenic toluene emissions most likely reflects changes in reformulated gasoline (RFG) hydrocarbon content to meet US EPA mandated VOC volatility requirements from 1 June to 15 September of each year (Romanow, 2008). To fulfill both these lower fuel Reid Vapor Pressure (RVP) requirements and fuel octane grades in summer, refineries often replace more volatile high octane compounds such as n-butane with heavier alkanes and aromatic compounds in their gasoline formulations (Gary and Handwerk, 2001; Lough et al., 2005). An analysis of individual hydrocarbon compound content in 2000–2001 summer and winter RFG samples from Milwaukee, Wisconsin, indicated that toluene exhibited the largest summertime increase of all hydrocarbons increasing from 1% (weight) of fuel in winter to approximately 10% in summer (Lough et al., 2005). While tunnel studies in the Milwaukee area indicated that these changes in fuel content did not significantly alter toluene emissions or toluene-to-benzene ratios from mobile source exhaust, they did affect the percent hydrocarbon composition of fuel headspace vapors with toluene, i-pentane, and 2,2,4-trimethylpentane exhibiting the largest percent increase for summertime fuels (Lough et al., 2005). The effect these seasonal gasoline content changes could have on fuel evaporation sources suggests a distinct yearly cycle in urban toluene emissions which must be considered in evaluating seasonal toluene variability in New England.

Local plants may also be seasonal toluene sources with particular significance in extensively vegetated rural areas such as northern New England. In a series of laboratory enclosure experiments, Heiden et al. (1999) showed that isotopically labeled $^{13}CO_2$ taken up by sunflowers was emitted as ^{13}C labeled toluene. In the same study, a combination of laboratory and field experiments indicated that toluene emission rates for sunflowers and pine trees increased with plant stress (i.e., pathogen attack, low nutrients, leaf wounding). Considering the significant influence regional biogenic VOC emissions have on tropospheric chemistry in coastal New England (Griffin et al., 2004; Mao et al., 2006; White et al., 2008), the potential contribution of vegetative toluene emissions in seasonal toluene enhancements at Thompson Farm warrants exploration.

Elevated toluene mixing ratios at rural locations have also been previously attributed to local industrial emissions. VOC measurements made as part of the Southern Oxidants Study in June 1995 revealed significant toluene enhancements at New Hendersonville, Tennessee, which correlated strongly with winds coming from the direction of a regional toluene emitting industrial facility (McClenny et al., 1998). Such observations highlight the impact that local anthropogenic sources can have on ambient toluene variability in rural locations.

In this paper we examine all three sources in more detail for their potential relative contribution to the summer toluene enhancements observed at Thompson Farm. In particular, we identify and quantify the contribution of seasonal changes in gasoline formulation to evaporative toluene emissions. We also present estimates of toluene emissions from alfalfa crops in New Hampshire and loblolly pine trees in North Carolina. Finally, we evaluate the impact of annual reported industrial releases of toluene in the local area around Thompson Farm.

Methods

Measurements from several studies were utilized in this paper, and a brief description of each follows. All data is presented in local time (LT) which is UTC-04:00 during daylight savings time and UTC-05:00 during the rest of the year. Seasons are defined as follows: winter is December–February, spring is March–May, summer is June–August, and autumn is September–November.

Thompson Farm Ambient VOC Measurements

Situated 25 km from the Gulf of Maine in rural Durham, NH, USA, the University of New Hampshire AIRMAP observation site at Thompson Farm (43.11°N, 70.95°W, 24m above sea level, a.s.l) is established on an active corn and alfalfa farm surrounded by mixed forest. Daily canister samples have been collected from 12 January 2004 to 31 May 2007 from the top of the 12m sampling tower and provide the most continuous record of interseasonal and interannual VOC variability for the site. Sample collection times ranged from 08:30 to 19:30 LT daily, with the majority of samples in all seasons obtained between 12:00 and 15:00. The air samples were collected in evacuated 2 L electropolished stainless steel canisters and analyzed at the University of New Hampshire on a three GC system equipped with two flame ionization detectors (FIDs), two electron capture detectors (ECDs), and a mass spectrometer (MS) for C_2-C_{10} nonmethane hydrocarbons

(NMHCs), C_1-C_2 halocarbons and C_1-C_5 alkyl nitrates (Sive et al., 2005; Zhou et al., 2005; Zhou et al., 2008).

All data from samples in the 90th percentile for ethyne (C_2H_2), nitrogen oxide (NO), total pentanes (i-pentane and n-pentane, and total butanes (i-butane and n-butane) for each season (winter, spring, summer, autumn) were excluded from the analyzed dataset to provide a representative picture of background mixing ratios independent of strong local vehicle exhaust and fuel evaporative influence. Data from samples with acetonitrile levels ≥ 150 pptv were also excluded to eliminate strong biomass burning influences (Lobert et al., 1990; Warneke et al., 2006). In total, both these exclusions affected approximately 25% of the dataset. While C_2H_2, butanes and pentanes were measured as part of the VOC analysis described above, NO was measured at the Thompson Farm observation tower using a chemiluminescent technique described by Mao and Talbot (2004a) and acetonitrile measurements were made using proton transfer reactionmass spectrometry (PTR-MS) (Talbot et al., 2005). Mean monoterpene mixing ratios from summer 2004–2006 and correlations between monoterpenes and toluene in 2006 were also calculated from PTR-MS measurements. Additionally, carbon monoxide (CO), used to calculate C_2H_2/CO ratios, was measured using infrared spectroscopy as described by Mao and Talbot (2004a). All mean measurements are presented as mean±standard error where the standard error of the mean was calculated as described by Taylor (1982).

Thompson Farm Vegetative Flux Measurements

Net toluene flux from alfalfa (Medicago sativa) was measured at Thompson Farm with a static chamber made of Lexan with an aluminum frame (61 cm×61 cm×46 cm) (Varner et al., 1999). Blank chamber tests conducted over a 10 year old concrete pad and a dirt road/bare soil indicated no VOC emission artifacts associated with the apparatus. During flux measurements, the chamber was set on an aluminum collar (61 cm×61 cm) placed in an alfalfa covered plot below the AIR-MAP observation tower a week before sampling. An ambient air sample was taken directly above the collar immediately prior to sampling by opening an evacuated 2 L electropolished stainless steel canister. The chamber was then placed into the collar lip and sealed with water. Three headspace samples were collected every 6 min in 2 L electropolished stainless steel canisters.

The collar was sampled twice on 25 September 2007. For the first flux measurement (before harvest), the vegetation within the chamber was intact. The chamber was then removed and the vegetation in the collar was clipped to within 2 inches of the ground. The harvested vegetation was left lying within the collar during the second flux measurement (after harvest) which was taken approximately

2 min later. The same collar was sampled again without disturbing vegetation re-growth on 5 October 2007. All collar measurements on 25 September and 5 October were made between 13:30 and 15:30 LT. Net fluxes (nmolm–2 d–1) were calculated as follows:

$$Flux = \frac{dC}{dt} \times \frac{V_C}{A_C} \tag{1}$$

where $\frac{dC}{dt}$ is the linear regression slope of the chamber headspace concentration (in nmolm^{-3}) versus time (d), V_c is the chamber volume (m^3), and A_c is the collar area (m^2). Flux errors were propagated as described by Taylor (1982) from the standard error of the linear regression slope and the estimated upper limit of uncertainty in the chamber volume (2%) and collar area (2%).

Duke Forest Vegetative Flux Measurements

In June 2005, VOC fluxes were measured from loblolly pine (Pinus taeda) and sweetgum (Liquidambar styraciflua) trees at the Duke Forest Free Atmospheric Carbon Enrichment (FACE) site (35°52′N, 79° 59′W, 163m a.s.l.) located in Chapel Hill in the central Piedmont region of North Carolina, USA (Sive et al., 2007). Flux measurements were collected every two hours over two 48-h sampling periods using dynamic branch enclosures made of large (36 L) clear Teflon bags supported by an external frame. A single branch from the tree sampled was carefully placed within the enclosure and exposed to a continuous flow of air from the canopy for 24 to 48 h prior to sampling. A mass flow meter monitored the rate of air flow into the bag (3–6 L min^{-1}) while a cold palladium catalyst was used for O_3 removal. Subsamples of air from the enclosure inlet and outlet were collected during each flux measurement and pressurized to 35 psig in 2 L electropolished stainless steel canisters. Canisters were then returned to UNH for analysis on the three GC system.

Vegetation fluxes (nmolm^{-2} LA d^{-1}) were calculated as follows:

$$Flux = [C_B - C_{PC}] \times flow \times LA^{-1} \tag{2}$$

where C is the number density in nmol m^{-3} of toluene in both the bag (B) and post-catalyst (PC) air samples, flow is the rate of air flow into the bag in m^3 d^{-1}, and LA represents the leaf area (m^2) of the branch enclosed. All loblolly pine fluxes presented in this paper were converted to nmolm^{-2} ground area d^{-1} by multiplying by the Ring 1 leaf area index (LAI) at the Duke Forest sampling site for June 2005 (7m^2 LAm^{-2} ground area) (FACTS-1, 2006). Flux errors were propagated as described by Taylor (1982) from the upper limit of measurement uncertainties for toluene (5%), leaf area (10%), and LAI (40%), and the standard error of the mean inlet flow rate during measurement. While measurements were made in both ambient and elevated (+200 ppmv) CO_2 environments, only the ambient measurements are reported for clarity. There was no significant difference in toluene fluxes measured in the two CO_2 regimes.

Warm Season Toluene Enhancements at Thompson Farm

The time series data for benzene and toluene, presented in Fig. 1 with 15 day moving averages, were constructed from the daily canister samples at Thompson Farm collected during 2004–2007 and filtered as described in the methods section. There was a distinct seasonal trend in benzene over the three year period with elevated mixing ratios in winter (Table 1; 140±2 pptv, 2004–2006 mean) most likely reflecting a combination of factors including less oxidation by hydroxyl radical (OH), seasonally reduced boundary layer heights, and increased emissions from wintertime combustion processes (Singh and Zimmerman, 1992; Cheng et al., 1997; Na and Kim, 2001). As OH levels and boundary layer heights increased in the spring, benzene mixing ratios decreased to a mean of 85±2 pptv (2004–2006). Minimum benzene values were observed in the summer months for all three years (49±2 pptv, 2004–2006 mean) and they began increasing again in autumn (75±2 pptv, 2004–2006, mean).

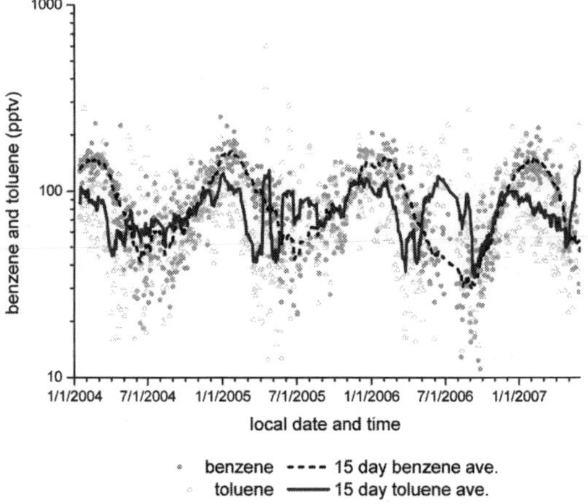

Figure 1. Benzene and toluene mixing ratios from daily canister measurements made 12 January 2004 to 31 May 2007. The 15 day moving averages of the individual measurements are displayed on the graph as solid and dashed lines.

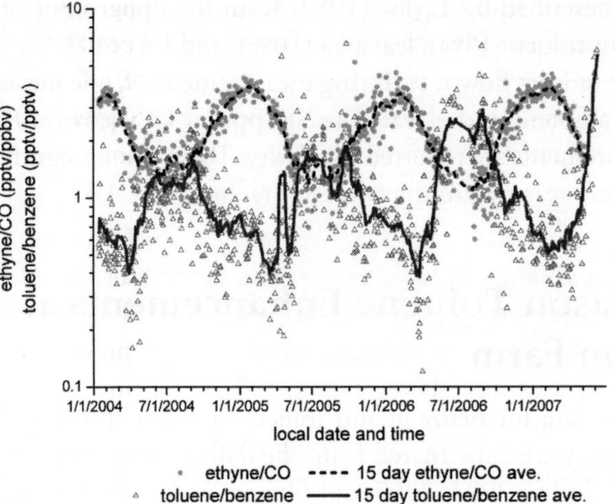

Figure 2. A comparison of C_2H_2/CO and toluene/benzene ratios as tracers of urban anthropogenic influence from the Thompson Farm daily canisters from 12 January 2004 through 31 May 2007. The 15 day moving averages of the daily ratios are displayed on the graph as solid and dashed lines.

Table 1. The seasonal mean mixing ratios ± standard errors for benzene, toluene, and the toluene/benzene ratio in the Thompson Farm daily cans from 12 January 2004 through 31 May 2006. The n values in parentheses are the number of samples included in the seasonal mean for that year after filtering the data set. The superscripts a, b, and c indicate means within each season that are significantly different according to an independent means t-test, $p < 0.05$, SPSS 15.0.1.1.

| | Mean (± st. err) | | |
Season and year	benzene (pptv)	toluene (pptv)	toluene/benzene (pptv/pptv)
winter			
2004 ($n=36$)	143±4[a]	89±7[ab]	0.63±0.05[a]
2004–5 ($n=57$)	144±5[a]	97±5[ab]	0.67±0.03[ab]
2005–6 ($n=50$)	141±4[a]	104±6[b]	0.73±0.03[b]
2006–7 ($n=60$)	133±4[a]	88±5[a]	0.66±0.04[ab]
spring			
2004 ($n=66$)	92±5[a]	64±5[a]	0.75±0.06[a]
2005 ($n=71$)	83±3[a]	60±10[a]	0.8±0.1[a]
2006 ($n=65$)	81±4[a]	64±5[a]	0.86±0.08[a]
2007 ($n=59$)	85±4[a]	72±6[a]	0.95±0.08[a]
summer			
2004 ($n=56$)	51±3[a]	63±5[a]	1.23±0.07[a]
2005 ($n=41$)	56±3[a]	85±9[b]	1.47±0.09[b]
2006 ($n=35$)	38±3[b]	100±10[b]	2.5±0.2[c]
fall			
2004 ($n=74$)	78±3[a]	79±7[a]	0.96±0.06[a]
2005 ($n=63$)	83±4[a]	100±10[a]	1.2±0.1[ab]
2006 ($n=64$)	61±4[b]	80±8[a]	1.4±0.1[b]

Similarly, toluene mixing ratios were elevated during winter with a seasonal mean of 95±3 pptv in all three years followed by a decrease to the springtime mean level of 56±4 pptv (Fig. 1 and Table 1). However, toluene increased again beginning in April or May and remained elevated into September during all three years, with mean values of 85±5 and 88±5 pptv for summer and autumn, respectively.

It should be noted that each successive year exhibited summertime toluene enhancements, with 2006 levels reaching a maximum mean of 100±10 pptv (Table 1). Subtracting the minimum toluene mixing ratios reached in April and May (42±3 pptv, 2004–2006 mean) from the daily summer means provides a rough estimate of the warm season enhancement levels (21±6, 43±9, and 50±10 pptv in 2004, 2005, and 2006, respectively).

These summertime toluene enhancements at Thompson Farm resulted in a toluene/benzene ratio seasonal pattern distinctly different from that of other anthropogenic VOC relationships. For example, the daily canister toluene/benzene ratio and C_2H_2/CO ratio are compared in Fig. 2 over the three year study period. Combustion sources of C_2H_2 and CO are largely concentrated in urban areas making them useful tracers of anthropogenic influence, particularly when biomass burning influences are minimal (Lobert et al., 1990; Warneke et al., 2006).

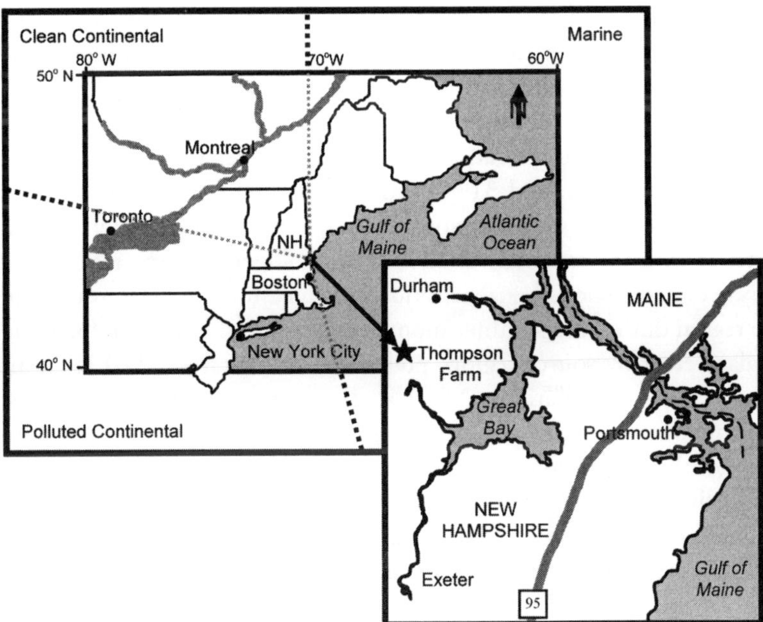

Figure 3. A map of the air mass source regions influencing Thompson Farm.

C_2H_2 also has a significantly faster rate of reaction with OH, its major sink, than CO making the ratio of the two compounds an indicator of air mass photochemical and mixing processes (Smyth et al., 1996). The C_2H_2/CO ratios observed at Thompson Farm reached their minimum values in summer reflecting higher levels of OH and increased air mass photochemical processing (Fig. 2). In contrast, the annual maximum toluene/benzene ratios at Thompson Farm occurred from June through September with the mean summer toluene-to-benzene ratio for all three years (1.9±0.1) significantly higher than all other months sampled (0.90±0.02; independent means t-test: p<0.001, SPSS v.15.0.1.1). Summer 2006 had the highest toluene/benzene ratios of all three years with a maximum (6.4 on 28 August 2006) nearly a factor of two larger than the emission ratios derived in urban plumes sampled directly in New England during NEAQS 2002 (3.7±0.3) (deGouw et al., 2005) and ICARTT 2004 (4.25) (Warneke et al., 2007).

A closer examination of the predominant source region impacting Thompson Farm during each season reveals that differences in air mass origin did not influence the summer toluene/benzene enhancements. Air mass source regions were classified as polluted continental, clean continental, and marine (Fig. 3) using 72 h back trajectories simulated using the NOAA HYSPLIT transport and dispersion model (Draxler and Rolph, 2003) for each daily sample collection time in the filtered data set.

During the summer months, the air masses sampled were relatively equally divided between the three different source regions with 37% coming from a clean continental area to the northwest, 33% from marine influenced areas to the northeast and east, and 30% from more polluted continental sources to the south and southwest. In contrast, the majority (54%) of air masses sampled during the winter came from clean continental source region, 15% from the clean marine influenced area, and 31% from the polluted continental source region. However, this shift in source region between summer and winter did not seem to correlate with the seasonal variation of the toluene/benzene ratios. More specifically, elevated toluene/benzene ratios were measured in air masses from every source region during the summer months (mean summer toluene/benzene ratios ± standard error by source region: polluted continental = 1.5±0.1, clean continental = 1.9±0.2, and marine = 2.0±0.4). These summer ratios were at least twice the toluene/benzene ratios observed in air masses from all directions during the winter months (mean winter toluene/benzene ratios ± standard error by source region: polluted continental = 0.73±0.05, clean continental = 0.61±0.02, and marine = 0.69±0.03). Because the daily canister data set was also filtered to remove the influence of fresh fuel evaporation and combustion sources on toluene mixing ratios, these anomalously high summertime tolueneto- benzene ratios must reflect additional toluene source influences in all source regions.

Evidence for Various Toluene Source Influences at Thompson Farm

To better characterize the influence of local and/or regional sources on toluene levels at Thompson Farm, the effect of photochemically processed urban fuel evaporation emissions on the observed toluene variability must be identified and quantified. A strong contribution from gasoline evaporation to the warm season toluene enhancements was indicated by concurrent spring and summer increases in both i-pentane and toluene mixing ratios (Fig. 4). The two compounds reached their minimum levels in April or May of each year followed by similar increases with the approach of summer fuel volatility requirements on 1 June. However, a closer examination of the relationship between toluene and i-pentane from April to May revealed distinctly that the springtime increases in toluene were independent of those in i-pentane, particularly in 2005 and 2006. For example, beginning on 15 April 2005, ambient toluene at Thompson Farm was significantly elevated (up to 600 pptv) for several days while ipentane levels continued to decrease. This episode was relatively short-lived, and toluene reached its springtime minimum on 10 May 2005 before rising concurrently with ipentane. In contrast, springtime toluene increases in 2006 followed a different pattern with a springtime minimum on 28 April 2006 several weeks earlier than the i-pentane minimum on 13 May 2006. Such variability provides further support for the influence of another toluene source(s) in addition to fuel evaporation on seasonal enhancements at Thompson Farm.

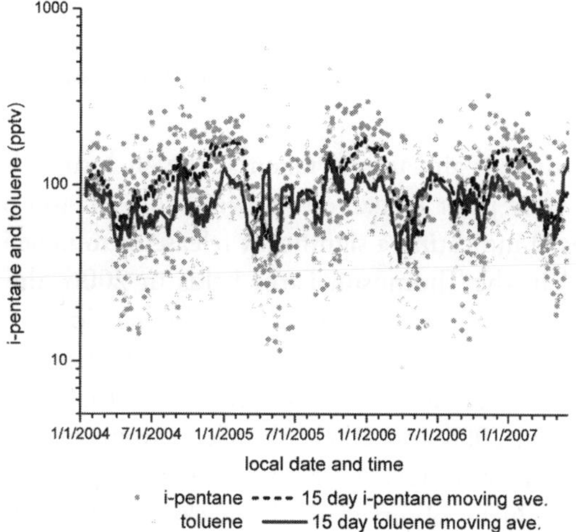

Figure 4. A comparison of i-pentane and toluene mixing ratios from the Thompson Farm daily canisters from 12 January 2004 through 31 May 2007. The 15 day moving averages of the individual measurements are displayed on the graph as solid and dashed lines.

The impact of these additional sources is also indicated in the scatter of the toluene/i-pentane correlation from June through August in 2004, 2005, and 2006 (Fig. 5). The implementation of summer fuel volatility requirements from 1 June to 15 September indicates that fuel evaporation emissions of toluene should have been at their greatest during this period. In Fig. 5, the background relationships between toluene and i-pentane were defined as the linear regression equations for daily canister data below the median values for i-pentane and toluene during each summer (2004–2006). The majority of the data in all years corresponded closely to the background relationship indicating that fuel evaporation was a major factor influencing summer toluene levels. In 2005 and 2006, there was also significant scatter with elevated toluene over a wide range of i-pentane levels (15–260 pptv). Elevated toluene mixing ratios were actually higher in 2005 (approximately 100–300 pptv) than in 2004 (50–200 pptv) despite a smaller range of i-pentane (15–150 pptv in summer2005) further suggesting a strong influence from additional toluene sources besides fuel evaporation. Compared to 2004 and 2005, the background relationship of toluene and i-pentane in summer 2006 was less well-defined implying that fuel evaporation was not as dominant a source to seasonal toluene enhancements that year. A higher background slope (0.7 ± 0.2 compared to 0.4 ± 0.1 and 0.5 ± 0.2 in 2004 and 2005, respectively) further implies input from an additional toluene source even in the cleanest air masses in 2006.

Potential toluene sources include the crop plants and coniferous trees surrounding Thompson Farm. Static enclosure flux measurements of alfalfa conducted at Thompson Farm in September 2007 revealed significant toluene emissions, particularly when the plants were harvested (Fig. 6 and Table 2). The alfalfa flux rates presented in Table 2 were calculated from the linear regression slopes shown in Fig. 6 as described in Sect. 2.2. Blank chamber tests made over a dirt road immediately prior to the alfalfa experiment showed that toluene increases were not chamber artifacts.

Furthermore, ambient air samples taken at the enclosure site increased from approximately 80 pptv prior to harvest at 14:40 to over 400 pptv immediately after harvest at 15:10 indicating a significant release of toluene in the area. Since alfalfa was planted in the Thompson Farm fields in 2006, the vegetative emissions presented in Fig. 6 could help explain the higher slope associated with the background toluene and i-pentane relationship that year (Fig. 5c). Unfortunately, there are no current toluene flux measurements from corn, the major crop planted at Thompson Farm in 2004 and 2005. Should future measurements prove that corn is indeed a toluene source, these emissions may have also influenced the scatter of elevated toluene observed in the background toluene and i-pentane relationship for 2005 (Fig. 5b). It should also be noted that the initial measurements of toluene flux rates from alfalfa presented in Table 2 were made at the end of September and beginning of October and emissions during the growing season could be significantly higher. For example, toluene fluxes measured from loblolly pine in North Carolina exhibited a strong temperature dependence (Fig. 7a) that, if applicable to alfalfa, indicate a substantial increase in flux rates during warmer seasons.

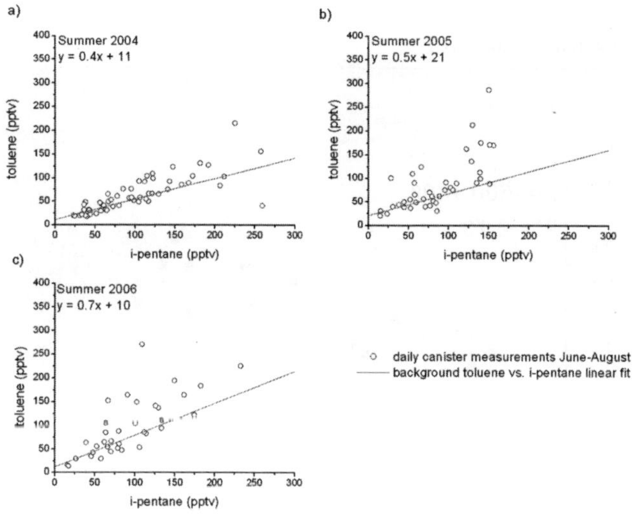

Figure 5. Toluene vs. i-pentane mixing ratios at Thompson Farm for June through August (a) 2004, (b) 2005, and (c) 2006. The background relationships were determined from the linear regressions of measurements below the median values for i-pentane (94, 79, and 81 pptv for 2004, 2005, and 2006, respectively) and toluene (54, 70, and 85 pptv for 2004, 2005, and 2006, respectively) for each summer.

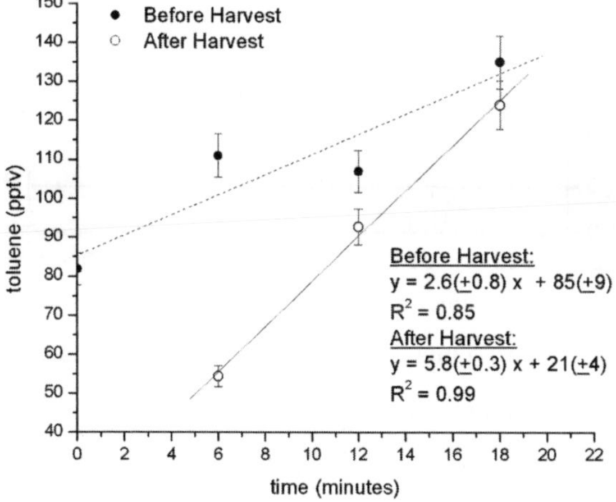

Figure 6. 25 September 2007 static chamber measurements from Thompson Farm of alfalfa toluene production before and after harvesting. Error bars represent the measurement uncertainty of the GC system.

Table 2. Vegetation toluene net flux rates from static and dynamic enclosure measurements. The loblolly pine net flux rate listed is the diurnally integrated flux rate from the 2 day sampling period. Flux errors were calculated as described in Sect. 2.2 and 2.3.

	Toluene flux nmol m^{-2}d^{-1}	r^2
9/25/07: Alfalfa before harvest	70±60	0.62
9/25/07: Alfalfa immediately after harvest	200±10	0.99
10/5/07: Alfalfa 2 weeks after harvest	80±50	0.31
6/4–6/6/05: Loblolly Pine	500±300	

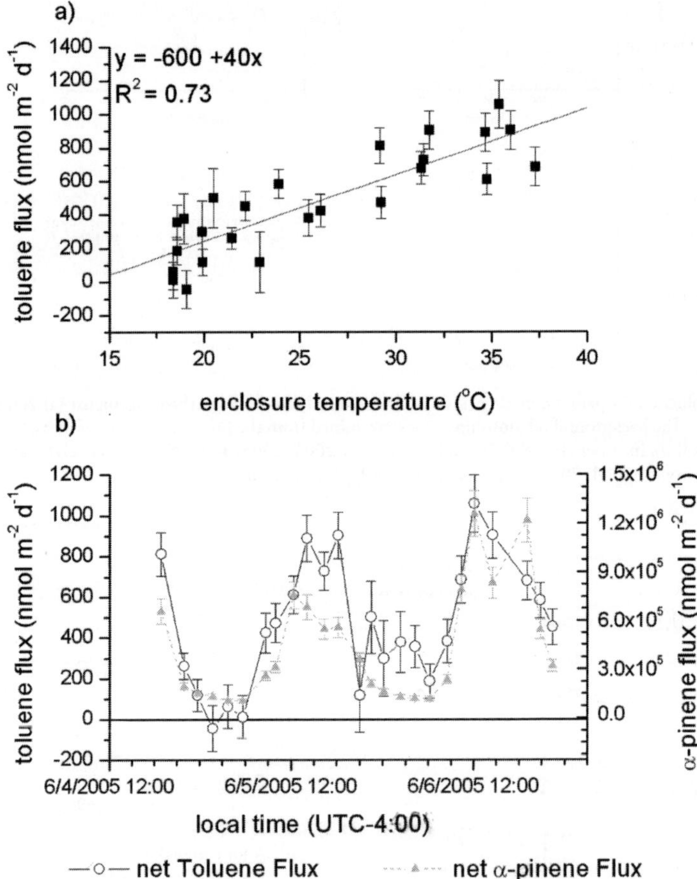

Figure 7. June 2005 dynamic branch enclosure measurements from Duke Forest in Chapel Hill, North Carolina of loblolly pine net toluene flux with a) net toluene flux vs. temperature and (b) the time series of net toluene and -pinene flux. Original fluxes were calculated as nmolm^{-2} leaf area d^{-1} and converted to nmolm^{-2} ground area d^{-1} by multiplying by the average leaf area index (LAI) at the Duke Forest sampling site for June 2005, 7m^2 leaf area 1m^{-2} ground area. Error bars represent the individual flux error propagated from the uncertainty of measurements used in flux calculation.

Assuming alfalfa emissions follow the loblolly pine temperature relationship, the average 9°C temperature difference between July and late September (2004–2007) at Thompson Farm would result in a summertime flux increase of approximately 360 nmolm^{-2} d^{-1} (or a total flux rate of approximately 430 nmolm^{-2} d^{-1} for unharvested alfalfa and 560 nmolm^{-2} d^{-1} after harvesting). Therefore, further study is warranted to quantify the temperature dependence of alfalfa flux and subsequently its seasonal cycle.

Our enclosure measurements of loblolly pine in North Carolina suggested comparable toluene flux magnitudes and diurnal emission patterns as those of Scots pines (Pinus sylvestris) sampled in Germany (Heiden et al., 1999). Loblolly pines also exhibited similar emission patterns between toluene and monoterpenes (Fig. 7b) that were consistent with correlations observed by Heiden et al. (1999) and suggest that biogenic toluene emission may be widespread for evergreen trees.

In contrast, negligible toluene production was evident in branch enclosure measurements of sweet gum (Liquidambar styraciflua), a deciduous tree species found in North Carolina. Furthermore, springtime emissions from local coniferous trees could explain the early toluene increase in May 2006 (Fig. 4). PTR-MS measurements of toluene and monoterpenes at Thompson Farm were more strongly correlated during the first two weeks of May (r^2=0.82) than at any other time during that year (January–April 2006 r^2=0.31, 14 May–August 2006, r^2=0.47, September– December r^2=0.41). This is also consistent with observations by Heiden et al. (1999) that toluene emissions from Scots pine were highest in spring.

In addition to the crops and trees surrounding Thompson Farm, local industry could also influence ambient toluene mixing ratios. According to the EPA's Toxic Release Inventory, there were two industrial facilities that released approximately 11 000 kg of toluene in 2005 located within a 20 km radius of Thompson Farm to the north and south (US Environmental Protection Agency, 2007). A wind direction analysis of measurements at Thompson Farm from June to August, when the seasonal enhancement was at its peak, revealed no distinct relationship between toluene and wind direction. With a lifetime on the order of days, the toluene lifetime is long enough for it to be dispersed and well-mixed with ambient air obscuring a directional relationship. However, an estimate of the daily ambient mixing ratio increase attributable to these local industrial emissions can be made by calculating the daily emission rate into the 20 km radius circle surrounding Thompson Farm (assuming a planetary boundary layer height of 1 km (Mao and Talbot, 2004b; Sive et al., 2007)). This rough approximation indicates that industrial emissions increase ambient toluene at Thompson Farm by 7 pptv d^{-1}. While significant, this value is still much less than warm season toluene enhancements (approximately 20– 50 pptv as estimated in Sect. 3). Moreover, industrial emissions cannot produce the seasonal toluene patterns observed as the facilities are in operation year round presumably with little seasonality in their source strength. All of this evidence together rules out local industry as the major source responsible for the summertime toluene enhancements.

Estimates of Source Contributions to Thompson Farm Summer Toluene Enhancements

In this section, the contributions to ambient summer toluene mixing ratios at Thompson Farm were quantified on a daily basis from the seasonal toluene sources, fuel evaporation and biogenic emissions (Table 3). Recognizing that there are assumptions associated with the calculations presented, these estimates provide an informative first estimate of the potential impact each toluene source could have on the seasonal enhancements observed.

Table 3. Estimates of warm season toluene source contributions and summer mean toluene enhancement over the 2004–2006 springtime minimum mean (42±3 pptv) at Thompson Farm. The errors given were propagated from the standard errors associated with the slopes and means used in calculation.

Year	Toluene from Fuel Evaporation (pptv d^{-1})	Toluene from Crop Plant Emissions (pptv d^{-1})	Toluene from Pine Tree Emissions (pptv d^{-1})	Summer Toluene Enhancement (pptv d^{-1})
2004	22±7	5±0.3	12±7	21 ±6
2005	16 ±6	5 ±0.3	12 ±7	43 ±9
2006	30 ±10	5 ±0.3	12 ±7	50 ±10

Fuel Evaporation

Contribution to the ambient toluene level from increased fuel evaporation was estimated by multiplying the slope of the background toluene-to-i-pentane relationship (given in Fig. 5a, b, and c) by the summertime i-pentane enhancements. It was assumed that the contribution from fuel evaporation to the ambient i-pentane level was minimal in winter and spring based on its temperature dependence. Therefore, we considered the minimum i-pentane mixing ratios from April and May (51±4 pptv, 2004–2006 mean) to be a background level, and the summertime i-pentane enhancement was estimated by subtracting this background value from the June–August i-pentane mean.

The fuel evaporation contribution for June–August 2004 (22±7 pptv) is consistent with the summer toluene enhancement above the springtime minimum for that year (21±6 pptv) and reflects the dominant fuel evaporation source influence indicated in the toluene versus i-pentane scatter plot for summer 2004 (Fig. 5a). In contrast, the estimates of toluene fuel evaporation contributions for summer 2005 and 2006 (16±6 and 30 ±10 pptv, respectively) cannot fully account for the toluene enhancements above springtime minimum in those years (43±9 and 50±10 pptv), further reinforcing the conclusion that additional toluene sources had important influences on the seasonal enhancements in those years. The 2006 summer fuel evaporation estimate may also include additional toluene source influences as the higher slope used to calculate the estimate reflected greater scatter in the toluene versus i-pentane background relationship in that year (Fig. 5c).

Biogenic

The potential toluene contributions from the crop plants and trees surrounding Thompson Farm were estimated by dividing the measured enclosure flux rates presented in Table 2 by a boundary layer height of 1 km (Mao and Talbot, 2004b; Sive et al., 2007). It should be noted that corn, rather than alfalfa, was planted at Thompson Farm in 2004 and 2005. Since corn has not been sampled for toluene emissions yet, the crop plant emission estimates presented for those years are based on the alfalfa toluene flux rates. Additionally, it was assumed that the diurnally integrated flux rates measured from loblolly pine in North Carolina are comparable to toluene emissions from the New England coniferous species surrounding Thompson Farm. The resulting estimates of biogenic toluene emissions (5±0.3 and 12±7 pptv d^{-1} for crops and coniferous trees, respectively) are on the same order of magnitude as industrial (7 pptv d^{-1}) and fuel evaporation emission (16–30 pptv d^{-1}) estimates. Should there be a temperature dependence of toluene emission from alfalfa and corn resembling the one presented in Fig. 7a, local vegetation could make summer contributions to the seasonal toluene enhancements at Thompson Farm greater by a factor of 3.

However, the apparent agreement noted in Sect. 5.1 between fuel evaporation estimates and summer toluene enhancements in 2004 suggest that biogenic emissions were overestimated for that summer. Significantly lower monoterpene levels in summer 2004 further indicate that regional biogenic emissions were reduced compared to the other two years (summer means from PTR-MS measurements = 310±6, 452±2, and 355±2 pptv for 2004, 2005, and 2006, respectively. Means from all three years significantly different, independent means t-test: p<0.001, SPSS v.15.0.1.1). Lower biogenic flux rates could reflect the cool, cloudy conditions that generally prevailed in the summer of 2004 as both the mean July–August temperature and the J_{NO2} photolysis rate measured at Thompson Farm were significantly lower than in 2005 and 2006 (mean temperature = 19.4±0.1, 20.8±0.1, and 20.7±0.1 °C and mean J_{NO2}=0.00229±0.00005, 0.00272±0.00006, and 0.00287±0.00006 s^{-1} for summer 2004, 2005, and 2006, respectively, independent means t-test: p<0.001, SPSS v.15.0.1.1). In contrast, the high monoterpene levels observed during summer 2005 suggest that tree and/or crop toluene emissions may have been underestimated for that year. If this were the case, it could explain why the combined biogenic and fuel evaporation estimates (33±9 pptv) were less than the observed summer toluene enhancement in 2005 (43±9 pptv).

Table 4. Biogenic and anthropogenic toluene emission estimates for New England.

Toluene Source	Emissions (Mg d^{-1})	
	New England (CT, RI, MA, NH, VT, ME)	Northern New England (NH, VT, ME only)
Crop Emissions	0.07	0.05
Forest Emissions	6	5
On-road Mobile Source	42	15
Non-road Mobile Source	32	18
Non-point Mobile Source	13	5
Point Source	5	0.5
Total Anthropogenic	92	39

The combined emission estimates for summer 2006 (fuel evaporation + biogenic = 50±10 pptv) were actually in good agreement with the observed summer toluene enhancement that year (50±10 pptv). However, the crop emissions estimates presented in Table 3 may have been underestimated for that summer as the alfalfa toluene fluxes during the height of the growing season are expected to be greater than the late September flux rates used in our calculations. It should be noted that the higher slope (Fig. 5c) used to calculate the fuel evaporation emission estimate in 2006 indicated the influence of an additional toluene source on the background toluene and i-pentane relationship and an underestimate in biogenic emissions may have been balanced by overestimate of fuel evaporative emissions.

These initial estimates indicate the need for a more comprehensive study of the seasonal cycle and environmental controls of toluene fluxes from crops and trees to fully explain the interannual variability in vegetative toluene emissions suggested here. However, our measurements indicate the significant impact this unexpected source might have on toluene variability in rural areas.

Comparison of Biogenic and Anthropogenic Toluene Emissions in New England

Extrapolating the toluene flux rates measured from alfalfa and loblolly pines (Table 2) to estimated forest and cultivated land areas for New England (Agricultural Statistics Board, 2006; US Department of Agriculture Forest Service, 2006) also allows an initial comparison of the contribution of potential biogenic and anthropogenic emissions (US Environmental Protection Agency, 2002) to the regional ambient levels of toluene. Three assumptions were made in this comparison. First, the toluene flux rates measured from North Carolina loblolly pine are comparable to toluene emissions from the mixed coniferous-hardwood forests throughout New England (Department of Agriculture Forest Service, 2006). Second, the late September toluene flux rates measured from alfalfa immediately after harvesting are representative of emissions throughout the growing season for all crop types. Third, the seasonality in anthropogenic emission rates (including on-road, non-road, point, and non-point sources) reported for each state is negligible.

The estimated daily emission rates of toluene from crops, forests, and anthropogenic sources are presented in Table 4 for all New England states (i.e., Connecticut, Rhode Island, Massachusetts, New Hampshire, Vermont and Maine) as well as the more rural states of northern New England alone (New Hampshire, Vermont, and Maine). It was estimated that biogenic emissions of toluene in New England (6 and $0.07 Mgd^{-1}$ for forest and crop emissions) represent as much as 7% of total anthropogenic emissions ($92 Mgd^{-1}$) with forest emissions on the same order of magnitude as industrial point source emissions ($5 Mgd^{-1}$). The northern states of New Hampshire, Vermont and Maine encompass the largest areas of forest and crop land in New England and therefore experience the most influence from biogenic sources. As anthropogenic emissions are also significantly less in these rural states, biogenic toluene emission rates are as much as 13% of total daily anthropogenic emissions in northern New England ($39 Mgd^{-1}$). These extrapolations further indicate the significant influence that biogenic emissions can have on toluene mixing ratios in rural New England and point to the need for more comprehensive studies of toluene emissions from vegetation.

Conclusions

The summertime enhancements in toluene mixing ratios evident in long-term daily measurements at Thompson Farm are driven by a combination of fuel evaporation emissions coupled with seasonal changes in RFG toluene content and biogenic emissions. Toluene releases from local industrial processes are also likely to impact ambient mixing ratios at the site but these emissions occur year-round and are unlikely to produce seasonal enhancements. Similar patterns of spring increases and summer correlations between i-pentane and toluene indicate that fuel evaporation emissions were the major influence on summer toluene enhancements in 2004. However, estimates of biogenic emissions from coniferous trees and crops were also significant and could not be fully dismissed, particularly in 2005 and 2006. The evidence of crop emission influences on seasonal toluene enhancements was strongest in 2006 when alfalfa was first planted in the Thompson Farm fields. Static chamber enclosure measurements revealed significantly increased toluene emissions from alfalfa after harvest. These flux measurements were made late in September and further studies are necessary to characterize the variability of toluene emissions from alfalfa and other vegetation more fully throughout the growing season. Extrapolations of the initial toluene flux measurements from pine and alfalfa to the forested and cultivated land areas of northern New England indicate that daily biogenic emission rates could be as much as 13% of total anthropogenic daily emission rates in rural areas further emphasizing the need for more comprehensive studies of toluene emissions from vegetation.

Acknowledgements

Financial support for this work was provided through the Office of Oceanic and Atmospheric Research at the National Oceanic and Atmospheric Administration under grants #NA05OAR4601080 and #NA06OAR4600189, and the US EPASTAR program through grant #RD-8314540. Research at the Duke FACE site was also supported by the Office of Science (BER), US Department of Energy, Grant No. DE-FG02-95ER62083. A special thanks to K. Allain, T. Finnigan-Allen, A. Csakai, P. Beckman, and S. Wadsworth at UNH.

References

1. Agricultural Statistics Board: Acreage, National Agricultural Statistics Service, US Department of Agriculture, Cr Pr 2–5 (6– 06), 2006.

2. Cheng, L., Fu, L., Angle, R. P., and Sandhu, H. S.: Seasonal variations of volatile organic compounds in Edmonton, Alberta, Atmos. Environ., 31, 239–246, 1997.

3. deGouw, J. A., Middlebrook, A. M., Warneke, C., Goldan, P. D., Kuster, W. C., Roberts, J. M., Fehsenfeld, F. C., Worsnop, D. R., Canagaratna, M. R., Pszenny, A. A. P., Keene, W. C., Marchewka, M., Bertman, S. B., and Bates, T. S.: Budget of organic carbon in a polluted atmosphere: results from the

New England Air Quality Study in 2002, J. Geophys. Res., 110, D16305, doi:10.1029/2004JD005623, 2005.

4. Dewulf, J. and Van Langenhove, H.: Analytical techniques for the determination and measurement data of 7 chlorinated C1- and C2-hydrocarbons and 6 monocyclic aromatic hydrocarbons in remote air masses: an overview, Atmos. Environ., 31, 3291–3307, 1997.

5. Draxler, R. R. and Rolph, G. D.: HYSPLIT (HYbrid Single-Particle Lagrangian Integrated Trajectory) Model access via NOAA ARL READY Website (http://www.arl.noaa.gov/ready/hysplit4.html), 15 August 2008, 2003.

6. FACTS-1 LAI-2000 Leaf Area Index Measurements: http://face. env.duke. edu/products.cfm, access date: 15 June 2007, 2006.

7. Gary, J. H. and Handwerk, G. E.: Petroleum Refining: Technology and Economics, Marcel Dekker, Inc., New York, 8–13, 2001.

8. Gelencser, A., Siszler, K., and Hlavay, J.: Toluene-benzene concentration ratio as a tool for characterizing the distance from vehicular emission sources, Environ. Sci. Technol., 31, 2869–2872, 1997.

9. Goldhaber, S., Wolf, S., and Laney, M.: Health Effects Notebook for Hazardous Air Pollutants, US Environmental Protection Agency, http://www.epa.gov/ ttn/atw/hlthef/hapindex.html, 1995.

10. Griffin, R. J., Johnson, C. A., Talbot, R. W., Mao, H., Russo, R. S., Zhou, Y., and Sive, B. C.: Quantification of ozone formation metrics at Thompson Farm during the New England Air Quality Study (NEAQS) 2002, J. Geophys. Res., 109, D24302, doi:10.1029/2004JD005344, 2004.

11. Heiden, A. C., Kobel, K., Komenda, M., Koppmann, R., Shao, M., and Wildt, J.: Toluene emissions from plants, Geophys. Res. Lett., 26, 1283–1286, 1999.

12. Lobert, J. M., Scharffe, D. H., Hao, W. M., and Crutzen, P. J.: Importance of biomass burning in the atmospheric budgets of nitrogen-containing gases, Nature, 346, 552–554, 1990.

13. Lough, G. C., Schauer, J. J., Lonneman, W. A., and Allen, M. K.: Summer and winter nonmethane hydrocarbon emissions from on-road motor vehicles in the midwestern United States, J. Air Waste Manage., 55, 629–646, 2005.

14. Mao, H. and Talbot, R.: O3 and CO in New England: Temporal variations and relationships, J. Geophys. Res., 109, D21304, doi:10.1029/2004JD004913, 2004a.

15. Mao, H., and Talbot, R.: Role of meteorological processes in two New England ozone episodes during summer 2001, J. Geophys. Res., 109, D20305, doi:10.1029/2004JD004850, 2004b.

16. Mao, H., Talbot, R., Nielsen, C., and Sive, B.: Controls on methanol and acetone in marine and continental atmospheres, Geophys. Res. Lett., 33, L02803, doi:10.1029/2005GL024810, 2006.

17. McClenny, W. A., Daughtrey, E. H., Adams, J. R., Oliver, K. D., and Kronmiller, K. G.: Volatile organic compound concentration patterns at the New Hendersonville monitoring site in the 1995 Southern Oxidants Study in the Nashville, Tennessee, area, J. Geophys. Res., 103, 22509–22518, 1998.

18. Na, K. and Kim, Y. P.: Seasonal characteristics of ambient volatile organic compounds in Seoul, Korea, Atmos. Environ., 35, 2603– 2614, 2001.

19. Ng, N. L., Kroll, J. H., Chan, A. W. H., Chhabra, P. S., Flagan, R. C., and Seinfeld, J. H.: Secondary organic aerosol formation from m-xylene, toluene, and benzene, Atmos. Chem. Phys., 7, 3909–3922, 2007, http://www.atmos-chem-phys.net/7/3909/2007/.

20. Romanow, S.: Fuel Trends Report: Gasoline 1995–2005, Compliance and Innovative Strategies Division, Office of Transportation and Air Quality, US Environmental Protection Agency, EPA420- R-08-002, 2008.

21. Russo, R. S. , Zhou, Y., White, M. L., Talbot, R., and Sive, B. C., et al.: Long Term Measurements of Nonmethane Hydrocarbons and Halocarbons in New Hampshire (2004–2008): Seasonal Variations and Regional Sources, Atmos. Chem. Phys. Discuss., in preparation, 2008.

22. Schauer, J. J., Fraser, M. P., Cass, G. R., and Simoneit, B. R. T.: Source reconciliation of atmospheric gas phase and particle phase pollutants during a severe photochemical smog episode, Environ. Sci. Tech., 36, 3806–3814, 2002.

23. Singh, H. B., Salas, L. J., Cantrell, B. K., and Redmond, R. M.: Distribution of aromatic hydrocarbons in the ambient air, Atmos. Environ, 19, 1911–1919, 1985.

24. Singh, H. B. and Zimmerman, P. B.: Atmospheric distribution and sources of nonmethane hydrocarbons, in: Gaseous Pollutants: Characterization and Cycling, edited by: Nriagu, J. O., John Wiley & Sons, Inc., New York, 177–235, 1992.

25. Sive, B. C., Zhou, Y., Troop, D.,Wang, Y., Little,W. C.,Wingenter, O. W., Russo, R. S., Varner, R. K., and Talbot, R.: Development of a cryogen-free concentration system for measurements of volatile organic compounds, Anal. Chem., 77, 6989–6998, 2005.

26. Sive, B. C., Varner, R. K., Mao, H., Blake, D. R., Wingenter, O. W., and Talbot, R.: A large terrestrial source of methyl iodide, Geophys. Res. Lett, 34, L17808, doi:10.1029/2007GL030528, 2007.

27. Smyth, S., Bradshaw, J., Sandholm, S., Liu, S. C., McKeen, S. A., Gregory, G. L., Anderson, B. E., Talbot, R., Blake, D., Rowland, F. S., Browell, E., Fenn, M., Merrill, J. T., Bachmeier, S., Sachse, G., Collins, J., Thornton, D. C., Davis, D. D., and Singh, H. B.: Comparison of free tropospheric western Pacific air mass classification schemes for the PEM-West A experiment, J. Geophys. Res., 101, 1743–1762, 1996.

28. Talbot, R., Mao, H., and Sive, B.: Diurnal characteristics of surface level O3 and other important trace gases in New England, J. Geophys. Res., 110, D09307, doi:10:1029/2004JD005449, 2005.

29. Taylor, J. R.: An introduction to error analysis, in: the study of uncertainties in physical measurements, Books in Physics, edited by: Commins, E. D., University Science Books, Sausalito, CA, 270 pp., 1982.

30. US Department of Agriculture Forest Service, Forest Inventory Data Online: http://199.128.173.26/fido/mastf/index.html, 15 August 2008, 2006.

31. US Environmental Protection Agency, National Emissions Inventory: http://www.epa.gov/ttn/chief/net/2002inventory.html, 15 August 2008, 2002.

32. US Environmental Protection Agency, Toxic Release Inventory (TRI) Explorer Online Database: http://www.epa.gov/ triexplorer/,15 June 2007, 2005.

33. Varner, R. K., Crill, P. M., and Talbot, R. W.: Wetlands: a potentially significant source of atmospheric methyl bromide and methyl chloride, Geophys. Res. Lett., 26, 2433–2436, 1999.

34. Wang, Y., Jacob, D. J., and Logan, J. A.: Global simulation of tropospheric O3-NOx-hydrocarbon chemistry: 3. Origin of tropospheric ozone and effects of nonmethane hydrocarbons, J. Geophys. Res., 103, 10 757–10767, 1998.

35. Warneke, C., De Gouw, J. A., Stohl, A., Cooper, O. R., Goldan, P., Kuster,W., Holloway, J.,Williams, E. J., Lerner, B. M., Mckeen, S. A., Trainer, M., Fehsenfeld, F. C., Atlas, E. L., Donnelly, S. G., Stroud, V., Lueb, A., and Kato, S.: Biomass burning and anthropogenic sources of CO over New England in the summer 2004, J. Geophys. Res., 111, D23S15, doi:10.1029/2005JD006878, 2006.

36. Warneke, C., McKeen, S. A., Gouw, J. A. D., Goldan, P. D., Kuster, W. C., Holloway, J. S., Williams, E. J., Lerner, B. M., Parrish, D. D., Trainer, M., Fehsenfeld, F. C., Kato, S., Atlas, E. L., Baker, A., and Blake, D. R.: Determination of urban volatile organic compound emission ratios and comparison with an emissions database, J. Geophys. Res., 112, D104S107, doi:110.1029/2006JD007930, 2007.

37. White, M. L., Russo, R. S., Zhou, Y., Mao, H., Varner, R. K., Ambrose, J., Veres, P., Wingenter, O. W., Haase, K., Stutz, J., Talbot, R., and Sive, B. C.:

Volatile Organic Compounds in Northern New England Marine and Continental Environments during the ICARTT 2004 Campaign, J. Geophys. Res., 113, D08S90, doi:10.1029/2007JD009161, 2008.

38. Zhou, Y., Varner, R. K., Russo, R. S., Wingenter, O. W., Haase, K. B., Talbot, R., and Sive, B. C.: Coastal water source of short-lived halocarbons in New England, J. Geophys. Res., 110, D21302, doi:10.1029/2004JD005603, 2005.

39. Zhou, Y., Mao, H., Russo, R. S., Blake, D. R., Wingenter, O. W., Haase, K., Ambrose, J. L., Varner, R. K., Talbot, R., and Sive, B.: Bromoform and dibromomethane measurements in the seacoast region of New Hampshire, 2002–2004, J. Geophys. Res., 113, D08305, doi:10.1029/2007JD009103, 2008.

Mathematical Modeling of Perfect Decoupled Control System and its Application: A Reverse Osmosis Desalination Industrial-Scale Unit

C. Riverol and V. Pilipovik

ABSTRACT

This short paper outlines the computer simulation using real data of a decoupled control system for a desalination unit. The control strategy incorporated a perfect decoupled controller for the control of the fresh water flow and conductivity. The model was estimated using real data and empirical tools instead of mass balances. The success is demonstrated in the reduction of wide fluctuations in the variables of the process and decreasing of the sensibility to the changes of pressure and/or pH and allows predicting problems of quality of water and waste of energy in the future.

Introduction

When we try to separate pure water and a salt solution through a semipermeable membrane, the pure water diffuses through the membrane and dilutes the salt solution. The membrane rejects most of the dissolved salts, while allowing the water to permeate.

This phenomenon is known as natural osmosis. As water passes through the membrane, the pressure on the dilute side drops, and the pressure of the concentrated solution rises. The osmotic flux continues until an equilibrium is reached, where the net water flux through the membrane becomes zero at equilibrium the liquid level in the saline water will be higher than that on the waterside. The amount of water passing in either direction will be equal. The hydrostatic pressure difference achieved is equal to the effective driving force causing the flow, called osmotic pressure. This pressure is a strong function of the solute concentration and the temperature, and depends on the type of ionic species present. Applying a pressure in excess of the osmotic pressure to the saline water section slows down the osmotic flow, and forces the water to flow from the salt solution into the waterside. Therefore, the direction of flow is reversed, and that is why this separation process is called reverse osmosis (Figure1).

Continuous progress in desalination technology makes it a prime, if not the only, candidate for alleviating severe water shortages across the globe[1]. Desalination costs have been continuously decreasing over the years as a result of advances in system design and operating experience, and the associated reductions in specific unit size and specific power consumption.

The most widely used desalination processes are membrane separation via reverse osmosis (RO), and three types of thermal separation: multistage flash desalination (MSF), multiple-effect evaporation, with thermal vapor compression (MEE-TVC) and without (MEE), and mechanical vapor compression (MVC). The MSF and RO processes dominate the market for both brackish water and seawater desalination, with a total share of more than 90%. All three types of thermal desalination systems are equipped with condenser tube bundles. The MEE and MVC systems are divided into evaporating effects, while MSF systems are divided into flashing stages. All of the systems employ a number of large pumping units, including pumps for seawater intake, distillate product, brine blowdown, and chemical dosing. The MSF and MEE systems have additional pumps for the cooling seawater. In addition, MSF has pumps for brine recycle. In MSF and MEE, steam extracted from low-and medium-pressure turbine lines provides the heat necessary for flashing or evaporation. In MSF, the heating steam is routed to the brine heater; in MEE, the heating steam is routed to the first evaporating effect. The MSF process operates with a top brine temperature in the range of 90–110°C. The MEE and MVC processes are operated with lower top brine temperatures in the range of 64–70°C. MVC is distinguished from the other processes by the presence of a mechanical vapor compressor, which compresses the vapor formed within the evaporator to the desired pressure and temperature. The system also includes plate heat exchangers for preheating the feed using heat recovered from the brine blow down stream and the distillate product.

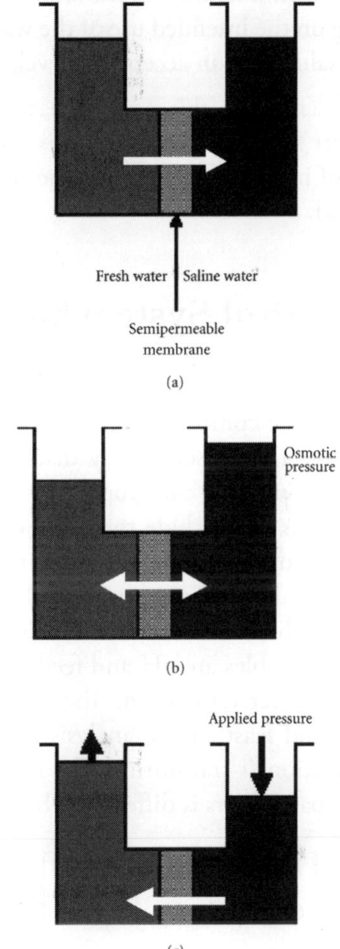

Figure 1. Principles of reverse osmosis:(a) natural osmosis,(b) osmotic equilibrium,(c) reverse osmosis.

All thermal processes produce a high-purity distillate product, with a salinity of less than10ppm.This is achieved by a wire-mesh mist eliminator, which removes entrained brine droplets formed in the distillate stream. The average conventional sizesare3000m³/d forMVC,33000m³/d for MSF, and12000m³/d for MEE. The RO process, see Figures 1 and 2, which employs membranes, has a simple layout, and is compact and modular. Existing units can be expanded to handle larger capacities. However, RO membranes are more sensitive to the conditions of the feed seawater, scaling, fouling, and pH than thermal processes. Furthermore, unlike thermal processes, RO membranes do not provide high purity water. On average, the permeate salinity varies over a range of 30–150ppm. The actual value depends on the process recovery, which is defined as the amount of product per unit mass of feed water. Depending on the intended use of the water, a second RO pass may be needed to reduce the salinity to an acceptable level.

The action control described in this paper is based on decoupled controllers applied to parallel transfer function processes. This method shows sensitivity to disturbance and tuning of inner loops, however the conditions were studied and the sensibility was reduced.

Design of the Control System based on Empirical Model

Four system parameters can be controlled in an RO unit: feed temperature, pH, conductivity, and pressure. The present study discusses two parameters which should be monitored and controlled for proper RO: conductivity (C) and inlet freshwater flow (F), and does not include the effect of the temperature over the behavior of the system. At industrial scale, the pressure affects the behavior of the system more significantly than the temperture. Additional parameters were not considered, however several literature can be consulted [2,3] with respect other models. The manipulated variables are pH and feed pressure (P). The model was got empirically using parameter estimation. The zero/poles discrete model was obtained using real data and least square analysis and later it was transformed incontinuous transfer function (s-transform). The result is similar to that in [2] although the value of the parameters is different. The system is

$$\begin{bmatrix} F \\ C \end{bmatrix} = \begin{bmatrix} G_{11} & G_{12} \\ G_{21} & G_{22} \end{bmatrix} \times \begin{bmatrix} P \\ pH \end{bmatrix}, \tag{1}$$

where

$$G_{11} = \frac{0.0045(0.104s+1)}{0.012s^2+s+1}, \quad G_{12} = zero,$$

$$G_{21} = \frac{(-0.12s+0.22)}{0.1s^2+0.3s+1}, \quad G_{22} = \frac{10(-3s+1)}{s^2+5s+1}. \tag{2}$$

The system has to work well in a range of P = 800–1000 kPa, F= 33000–54000 m³/d and pH = 6–7.2.

As in the desalination plant, a manipulated input affects more than one controlled output. One approach to handling this problem is known as decoupling[4]. The idea is to develop "synthetic" manipulated inputs that affect only one process output each. This approach is illustrated in Figure 3 where a multi variable process is controlled with a perfect decoupler and single-loop controllers with r_1 and r_2 the set points, respectively.

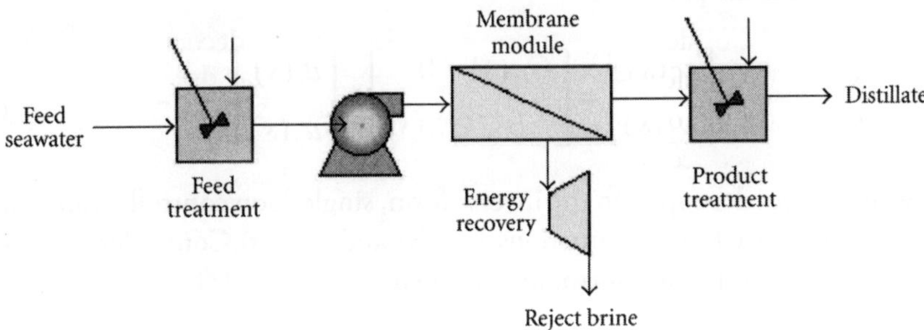

Figure 2. Desalination plant.

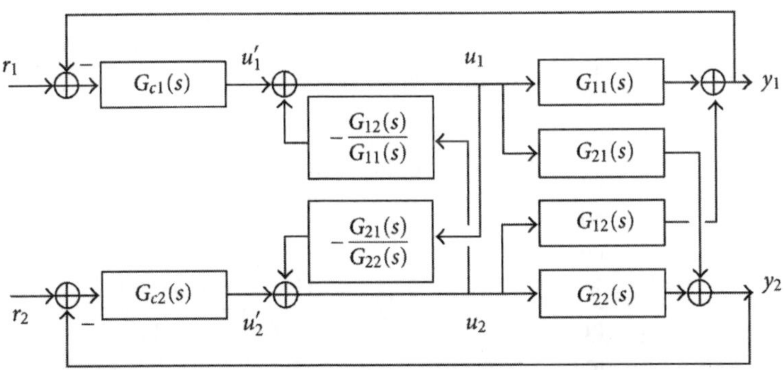

Figure 3. Control system.

The first step is to define the process transfer function matrix:

$$\begin{bmatrix} y_1(s) \\ y_2(s) \end{bmatrix} = \begin{bmatrix} G_{11}(s) & G_{12}(s) \\ G_{21}(s) & G_{22}(s) \end{bmatrix} \times \begin{bmatrix} u_1(s) \\ u_2(s) \end{bmatrix}$$

$$\Rightarrow \begin{bmatrix} F \\ C \end{bmatrix} = \begin{bmatrix} G_{11} & G_{12} \\ G_{21} & G_{22} \end{bmatrix} \times \begin{bmatrix} P \\ pH \end{bmatrix}.$$

(3)

The perfect decoupler is built in the inverse form where each branch of the decoupler is fed before the other branch pickup point. Its transfer function is

$$\begin{bmatrix} u_1(s) \\ u_2(s) \end{bmatrix} = \frac{\begin{bmatrix} G_{11}(s)G_{22}(s) & -G_{12}(s)G_{22}(s) \\ -G_{11}(s)G_{21}(s) & G_{11}(s)G_{22}(s) \end{bmatrix}}{G_{11}(s)G_{22}(s) - G_{12}(s)G_{21}(s)} \cdot \begin{bmatrix} u'_1(s) \\ u'_2(s) \end{bmatrix}. \tag{4}$$

This decoupler is realizable only if the decoupler transfer function matrix is open-loop stable. It can be shown that the transfer function matrix of the decoupler in series with the process is

$$\begin{bmatrix} y_1(s) \\ y_2(s) \end{bmatrix} = \begin{bmatrix} G_{11}(s) & 0 \\ 0 & G_{22}(s) \end{bmatrix} \times \begin{bmatrix} u'_1(s) \\ u'_2(s) \end{bmatrix}. \tag{5}$$

For the perfect decoupler in the inverse form, single-loop controllers are tuned on the single direct transfer functions G11(s) and G22(s).Controllers are then properly tuned even if one control loop is open.

If the set points r_1 and r_2 are constrained variables, a controller structure can be used to modify the control strategy in order to cancel the y_1 and y_2 feedback and to cancel the decoupler when the constraints are not active. Inversion of the controllers at the inputs permits to reconstruct the manipulated variables u_1 and u_2 which can then be used by two other controllers to achieve control objectives with less priority than the constraints while respecting the allowable r_1 and r_2 set point values.

Pole-zero cancellation method led to the following controllers for the parallel control method:

$$G_{c1} = \frac{(0.012s^2 + s + 1)}{(0.1s + 1)}, \qquad G_{c2} = \frac{10(s^2 + 4.3s + 1)}{(-2.7s + 1)}. \tag{6}$$

The primary objective of the control system was to keep the pressure at 800 psi and the conductivity at 450 µS/cm in the face of disturbances by the pressure or/and pH. The decoupled controller has the advantage that it does not need tuning as the classical PID; only if the model is changed the controller has to be modified. With respect to the behavior of the conductivity using real data, its value is practically constant along the time using different disturbances in the pressure. A resume of the strategy is shown in Figure 4 with conductivity records collection at 5-minute intervals (22000 values) to a 30% step.

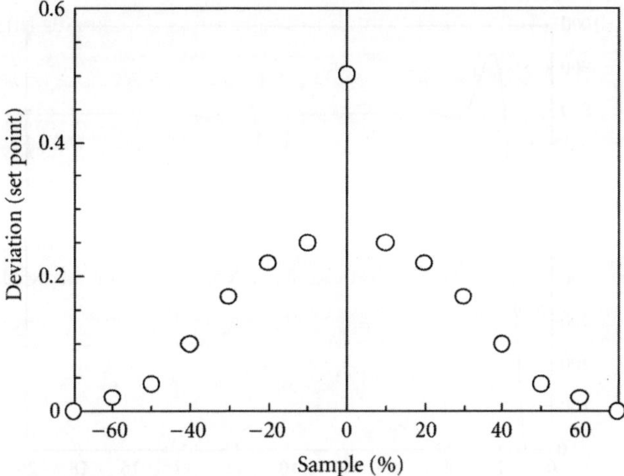

Figure 4. Behavior of the conductivity.

Table 1. Resume of the results using a long-term control (ISE 1 stands for integral square error).

Step over acid inlet flow (%)	10	20	30	40
Flow rate (ISE)	0.001	0.002	0.002	0.002
Conductivity (ISE)	0.001	0.003	0.007	0.007
Trans-memb pressure (ISE)	0.002	0.002	0.002	0.002

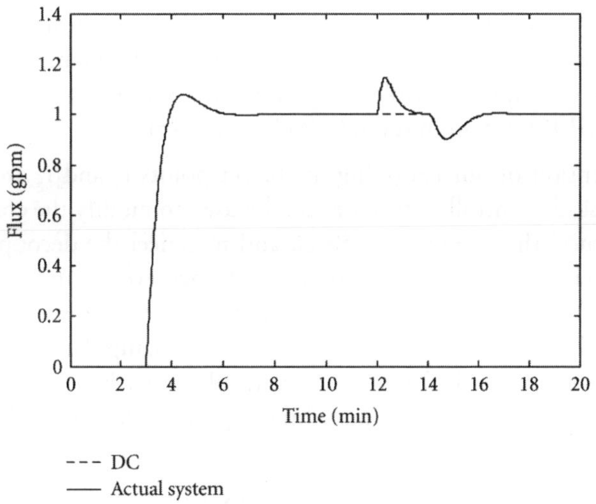

Figure 5. Comparison of the behavior of the waterflow using both systems.

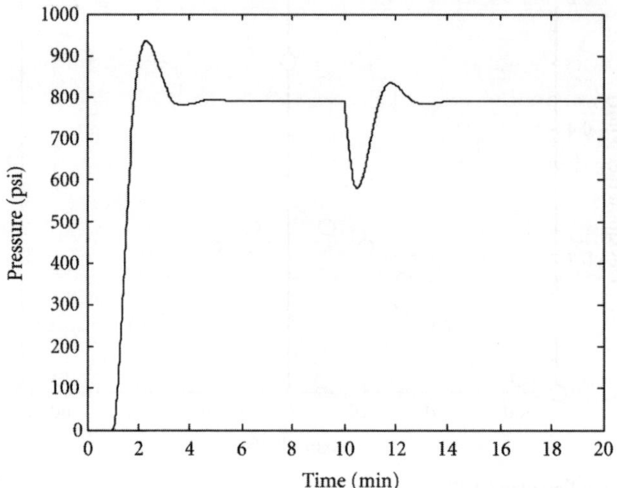

Figure 6. Behavior of the manipulated variable in real conditions.

The PID control originally gave very good results with system response. This needed to be improved to remain competitive in a modern market, and has been achieved by the use of digital control. The decoupled control system offers an easy solution to the new demand of the market and it is fast implementation. It is important to note that the process has a delay time of 2.88 minutes as can be observed in Figure 5, compared to Figure 6 that shows the behavior of one of the manipulated variables (pressure). The response is similar; however, in the classical system the response tried to become unstable or took a long time period to settle. The classical system is very sensible to any change in the manipulated variables; the use of the decoupled variables can reduce the sensibility because it tries to become the MIMO system in several SISO systems[5].

As an extension of our reasoning, if the set points r_1 and r_2 are constrained variables, a cascade controller structure can be used to modify the control strategy in order to cancel the y_1 and y_2 feedback and to cancel the decoupler when the constraints are not active. The decoupling can be achieved at the controller input with a suitable modification to the decoupler transfer functions. Inversion of the controllers at the inputs permits to reconstruct the manipulated variables u_1 and u_2 which can then be used by two other controllers to achieve control objectives with less priority than the contraints while respecting the allowable r_1 and r_2 set point values.

When saturation in not active, the variables $\hat{u}_i$ differ from u_i only by additive contants equal to the value of the controller integrators. Therefore, the variable $\hat{u}_i$

can be used as manipulated variables by another controller to regulate additional variables.

Finally, a brief demonstration of that the decoupled control system was a success is depicted in Table 1 where the behavior of the controlled variables was evaluated using different disturbances in the pH (acid flow) in the treatment area. It is easy to observe that as the conductivity is affected directly with the changes in the pH, however the water flow is not affected for the pH, such that the conductivity may be controlled with the pH and the fresh water flow with the pressure.

Conclusion

A decoupled control system was developed for an RO desalination unit. Testing showed that the control system could reasonably be used for evaluating the performance of this unit. The performance of the control system was demonstrated through evaluation of some short-and long-term control strategies. Some of the interesting results obtained from this evaluation are the following: (i) there is a delay time around 2 minutes in the unit that must be considered in the design of any model or control system; (ii) in future Journal of Automated Methods & Management in Chemistry developments for the evaluation of the long-term control strategies, benchmarks need to be able to assess settlers' performance.

References

1. H. H. Ettouney, H. T. El-Dessouky, R. S. Faibish, and P. J. Gowin, "Evaluating the economics of desalination," Chemical Engineering Progress, vol. 98, no. 12, pp. 32–40, December 2002.

2. I. Alatiqi, A. ghabris, and S. Ebrahim, "Measurement and control in reverse osmosis desalination," Desalination, vol. 75, pp. 119–140, 1989.

3. A. Mindlerand A. Epstein, "System identification and control of reverse osmosis desalination," Desalination, vol. 59, pp. 343– 379, 1986.

4. W. Ray, "Some recent applications of distributed parameter system theory—A survey," Automatica, vol. 14, pp. 281–299, 1978.

5. J. A. Mandler, "Modeling for control analysis and design in complex industrial separation and liquefaction processes," Journal of Process Control, vol. 10, no.2, pp. 167–175, 2000.

Tula Industrial Complex (Mexico) Emissions of SO$_2$ and NO$_2$ During the MCMA 2006 Field Campaign Using a Mobile Mini-DOAS System

C. Rivera, G. Sosa, H. Wöhrnschimmel, B. de Foy, M. Johansson and B. Galle

ABSTRACT

The Mexico City Metropolitan Area (MCMA) has presented severe pollution problems for many years. There are several point and mobile emission sources inside and outside the MCMA which are known to affect air quality in the area. In particular, speculation has risen as to whether the Tula industrial complex, located 60 km northwest of the MCMA has any influence on high SO2 levels occurring on the northern part of the city, in the winter season *mainly. As part of the MILAGRO Field Campaign, from 24 March to 17 April 2006, the differential vertical columns of sulfur dioxide (SO_2) and nitrogen dioxide (NO_2) were measured during plume transects in the neighborhood of the Tula industrial complex using mobile mini-DOAS instruments. Vertical profiles of wind speed and direction obtained from pilot balloons and radiosondes were used to calculate SO_2 and NO_2 emissions. According to our measurements, calculated average emissions of SO_2 and NO_2 during the field campaign were 384±103 and 24±7 tons day^{-1}, respectively. The standard deviation of these estimations is due to actual variations in the observed emissions from the refinery and power plant, as well as to the uncertainty in the wind fields at the exact time of the measurements. Reported values in recent inventories were found to be in good agreement with calculated emissions during the field campaign. Our measurements were also found to be in good agreement with simulated plumes.*

Introduction

The Tula industrial complex is located northwest of the Mexico City Metropolitan Area (MCMA), in the State of Hidalgo, Mexico. It is close to a number of other industries in the Tula-Vito-Apasco industrial corridor (Fig. 1). According to the latest information from the federal environmental authority 323 ktons per year (ktpy) of SO_2 and 44 ktpy of NO_x are released in this region. The main emitters are the Miguel Hidalgo Refinery (MHR) and the Francisco Pérez Ríos Power Plant (FPRPP) (SEMARNAT, 2002). Other industries such as cement plants, open-sky mines and agricultural activities are also responsible of important particle matter emissions into the atmosphere and for soil degradation of the area. The MHR processes 296 thousand barrels per day (TBD) of crude oil, representing 20% of the total refining capacity in the country. Final products are mainly gasoline, diesel, turbosine, kerosene, and other subproducts used to improve fuel specifications. To satisfy internal energy demand, the refinery consumes gas and liquid residuals of the refining processes, often of poor quality (3.8% weight of sulfur content). The FPRPP has an installed capacity of 2000 megawatt (MW), distributed in 9 units combining vapor (five) and combined cycle (four) technologies. These two industries contribute almost 90% of SO_2 and 80% of NO_x of the total emission in Hidalgo State (IMP, 2006).

NO_2 is of special interest due to its potential for undergoing photochemical reactions and producing, together with volatile organic compounds; ozone, peroxyacetyl nitrate, nitric acid, formaldehyde and formic acid, among others (Finlayson-Pitts and Pitts, 2000). Long term exposure of humans to NO_2 has negative health effects such as lung function decrease and higher risk of respiratory symptoms.

SO_2 was shown to lead to reductions in FEV1 (Forced Expiratory Volume in 1 s) and other indices of ventilatory capacity, as well as to increased mortality and hospital emergency admissions for total respiratory cases and chronic obstructive pulmonary disease at lower levels of exposure (WHO, 2000). In addition to their negative effects on human health, SO_2 and NO_2 tend to form sulfuric and nitric acids respectively, which through dry and wet deposition contribute to damage of plants and buildings.

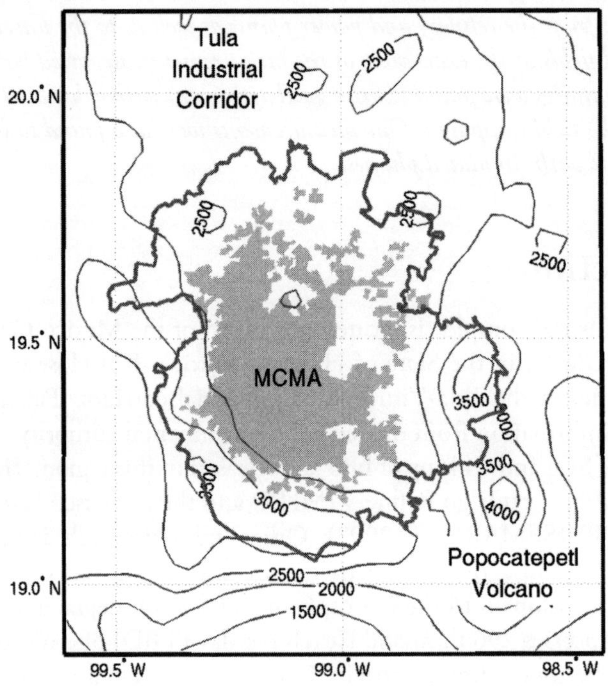

Figure 1. Location of the Tula industrial corridor, urban area of the MCMA in beige. Terrain contours every 500 m.

Because of its large emissions, the Tula industrial complex is thought to affect air quality in the MCMA. Since the early 2000's, the atmospheric monitoring network of the MCMA, has been reporting unusual high SO2 concentrations during night time at the northern part of the MCMA. According to a 2003 report of the "Program to Improve Air Quality in Mexico City Metropolitan Area 2002–2010" (CAM, 2003) in some occasions SO_2 concentrations in the north part of the city have exceeded the Mexican Air Quality Standard (0.13 ppm 24 h-average). It has not been possible to attribute this to irregular operations of industries located in the surrounding area. For this reason, it has been questioned whether the Tula industrial zone is responsible for worsening air quality in the MCMA. Mexican Petroleum (PEMEX) and Federal Commission of Electricity (CFE) companies located at the Tula industrial complex, however, claim to comply with emission regulations, and therefore not to affect air quality.

De Foy et al. (2007) used Concentration Field Analysis using backward trajectories during the MCMA-2003 field campaign to identify possible source regions of SO_2. These were found to be to the northwest of the MCMA, in the direction of the Tula industrial zone. Forward modeling using measured emissions from the complex showed that these could account for the large SO_2 peaks observed in the MCMA, but that they only contribute 20% of the long term average concentration. Local sources are the dominant cause of baseline SO_2 levels. The Popocatépetl volcano is another large point source that affects the MCMA and leads to increased sulfate aerosol production in the city (Raga et al., 1999) even though impacts during MCMA-2003 were shown to be possible but could not be differentiated from the local levels (de Foy et al., 2007).

To address the issue of emissions from the refinery and the power plant, the total fluxes of SO_2 and NO_2 were determined by measurements of their respective integrated vertical columns in the neighborhood of the Tula industrial zone. The importance of this study relies on the possibility to verify published emission inventories and provide detailed emission information for modeling studies.

Methods

Mobile Mini-DOAS

SO_2 and NO_2 emissions have been determined using Differential Optical Absorption Spectroscopy (DOAS). DOAS is a spectroscopic technique based on the absorption of electromagnetic radiation by matter, allowing the remote detection of trace gases. DOAS instruments have been developed in a wide variety of designs and measurements can be performed using several experimental setups (Platt and Stutz, 2008).

In this field campaign, scattered UV sunlight in the zenith position was collected by mobile mini-DOAS instruments installed on a car, while traversing underneath the plumes of the industrial sites. SO_2 and NO_2 emissions were determined by integrating the total number of molecules in a vertical cross-section of the gas plume, and multiplying them by the wind speed at plume height. The instrument is referred to as mini-DOAS due to its small size and low weight. The mobile mini-DOAS collects scattered ultraviolet sunlight using a telescope equipped with a quartz lens; the light enters a spectrometer through an optical fiber (Galle et al., 2002). The spectrometers used in this study have a spectral resolution of about 0.6 nanometers (nm) and spectral range of 280– 420 nm for SO_2 and 336–480 nm for NO_2. Similar equipments have been successfully used to quantify SO_2 emissions from volcanoes (Bobrowski et al., 2003; Edmonds et al., 2003; Galle et al., 2002; McGonigle et al., 2002; Mori et al., 2006; O'Dwyer et al., 2003), industries (McGonigle et al., 2004; Rivera et al., 2009) and urban areas (Johansson et al., 2008, 2009).

A general spectral evaluation procedure was applied to every spectrum collected during a traverse, starting with dark current correction of every recorded spectrum, division with a "clean-air" reference spectrum, the application of a high pass filter to separate broad and narrow band spectral structures and finally a logarithm of the spectrum. For each traverse, the "clean-air" reference spectrum has been selected as the first spectrum in the measurement, as this reduces the time difference between the collection of the "clean-air" reference and each of the subsequently measured spectra. As already pointed out by Sinreich et al. (2005), this procedure can help to eliminate the contribution of stratospheric NO_2 to the signal and give less influence of instrumental instabilities. Afterwards a non-linear fitting of several absorption cross sections is made to the measured spectra, obtaining vertical columns of the gases of interest which are differential with respect to the vertical columns in the "clean-air" reference spectrum. The fitting intervals used correspond to 307– 317 nm and 415–455 nm for SO_2 (Bogumil et al., 2003) and NO_2 (Vandaele et al., 1998) cross sections, respectively. The O_3 cross section (Voigt et al., 2001) was included in the fitting procedure for both SO_2 and NO_2 as well.

The mobile mini-DOAS instrument was complemented with a Global Positioning System (GPS), both of them connected to a laptop computer and controlled by custom-built software -MobileDOAS-(Johansson and Zhang, 2004) which collects and evaluates acquired spectra in real time. The mobile mini-DOAS instrument was mounted on a car and spectra were recorded, at different distances from the source, both encircling the source and traversing downwind the pollutants plume. For each collected spectrum, GPS data was recorded providing time and position before and after the spectrum was collected.

Fluxes from traverses were calculated multiplying differential vertical columns by the distance traversed perpendicular to the wind direction, and by the wind speed at plume height. The corresponding methods for obtaining meteorological information are discussed in the following section.

Meteorological Measurements

For accurately quantifying emissions, the mobile mini-DOAS technique requires wind speed and wind direction information at the height of the plume. Therefore, these parameters were measured at surface and aloft, using different methods. The meteorological equipment was deployed at a Mexican Petroleum Institute (IMP) site (Longitude 99°16' 24.4" W and Latitude 20°2' 48.6" N), located inside the MHR facilities. A surface meteorological station (MAUS-210 from Vaisala) registered continuously those variables 10 m above the ground, while vertical measurements were performed using pilot balloons and radiosondes.

Wind data from pilot balloons were used to calculate SO_2 and NO_2 emissions for measurements performed between 24 and 26 March 2006 because they were more frequently launched than radiosondes. For measurements performed after 27 March 2006, results from radiosondes were used instead, being our only source of wind data available at plume height.

Pilot Balloons

Pilot balloons were launched during 24–26 March 2006 with a frequency of 1–2 h during daytime, in order to obtain information on the vertical distribution of wind speed and direction at plume height. The pilot balloons were filled with a pre-determined amount of commercial helium, resulting in a standardized uplift force. The ascent rate was estimated by intercomparisons with radiosonde data (see Fig. 2). Once the balloons were launched, they were tracked by theodolites (Tamaya model TD3), and both azimuth and elevation were registered in 10 s intervals. The position of the balloon in space was calculated from these angles from which wind speed and direction were derived.

Radiosondes

A Digicora II radiosonde system from Vaisala (Mod. SPS220) was used for upper air sounding measurements of pressure, temperature, relative humidity and the horizontal wind vector. The radiosonde system consists of a ground-base station that receives and stores the incoming signal from the radiosonde transmitter; a radiosonde that supports all meteorological sensors, and a meteorological balloon that raises the sonde from the ground to the upper atmosphere. The radiosondes were launched four times a day from 16 March to 22 April 2006 at 08:00, 12:00, 15:00 and 18:00 h (local time) by using 300 g latex-helium-filled balloons.

Modeling

Forward plume simulations were carried out using Lagrangian particle trajectories. The mesoscale meteorology was simulated with the Weather Research and Forecast (WRF) model version 3.0.1 (Skamarock et al., 2005). Three nested grids were used with resolutions of 27, 9 and 3 km, and wind fields were saved every hour. The simulation options and model evaluations are described in de Foy et al. (2009). Stochastic particle trajectories were calculated with WRF-FLEXPART (Doran et al., 2008; Stohl et al., 2005). 1800 particles per hour were released from a single stack representing FPRPP and 720 particles per hour from a stack representing MHR. Plume rise was calculated using the algorithm in CAMx (EN-VIRON, 2009) on an hourly basis. This yielded particle release heights varying from 175 to 1400 m AGL for FPRPP and from 75 to 1000 m AGL for MHR, depending on the time of day and on stability conditions.

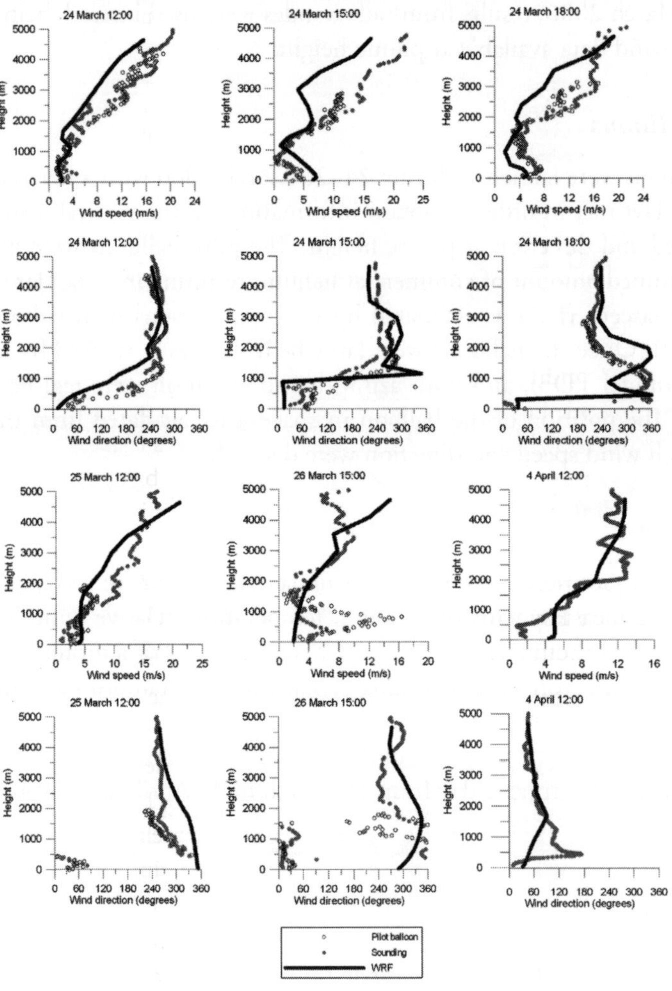

Figure 2. Wind speed and wind direction comparison between pilot balloon, sounding and WRF for 24, 25 and 26 March 2006. Comparisons are presented only when coincident information of the three methods is available. Additionally wind speed and wind direction results from a sounding launched on 4 April 2006 at 12:00 local time are presented together with WRF results.

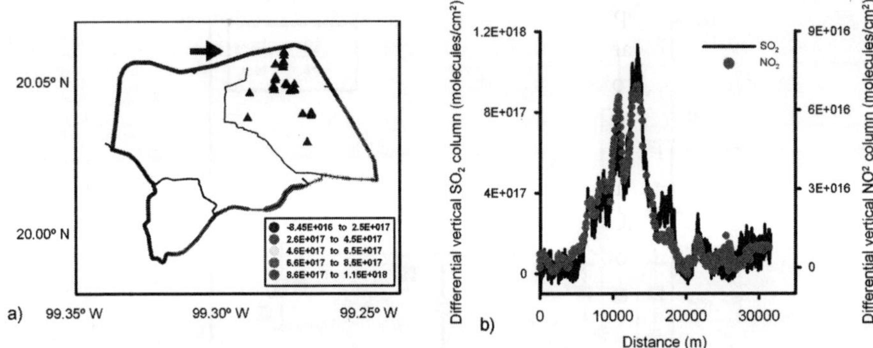

Figure 3. Typical measurement at Tula industrial complex performed on 26 March 2006 between 14:18 and 15:03 local time. Panel (a) presents the spatial distribution of differential SO$_2$ columns (color coded) quantified during the measurement, whereas panel (b) shows variation in differential vertical SO$_2$ (black) and NO$_2$ (grey) columns for the measurement as function of travelled distance. The arrow in panel (a) shows the starting point of the measurement and points towards the direction of travel.

Results and Discussion

Meteorological Conditions

Meteorological conditions in the Tula industrial complex were variable during the monitoring period. From 23 March 2006 a cold surge over the Gulf of Mexico produced strong northerly winds persisting for two days (Fast et al., 2007). Cold surges are a characteristic feature in Central Mexico, bringing cold humid air south into the MCMA. This leads to reduced vertical mixing and increased rain and cloudiness. After the cold surge, conditions in the basin remained humid with weak southward transport at night and afternoon convection events until the end of March (de Foy et al., 2008). During April, most of the days were clear or partly cloudy without rain due to a persistent high pressure system.

Wind speeds were higher during the first days of the field campaign and less intense at the end. Minimum wind speeds were recorded during night time and early morning hours while maximum values were recorded between 19:00–20:00 local time. Maximum temperatures occurred between 16:00– 17:00 and minimum temperatures between 07:00–08:00 local time. Relative humidity followed an expected behavior with a maximum during night time and a minimum coinciding with high temperatures. Wind direction before 09:00 was from south-southeast, a transition period was systematically observed between 09:00–11:00 when northerly winds appeared. After midday, wind direction turned from north-northwest to north to north-northeast and this condition was maintained until 22:00 where winds turned again from south-southeast until 9:00 hours of the next day. This cycle was continually observed over the entire measurement period.

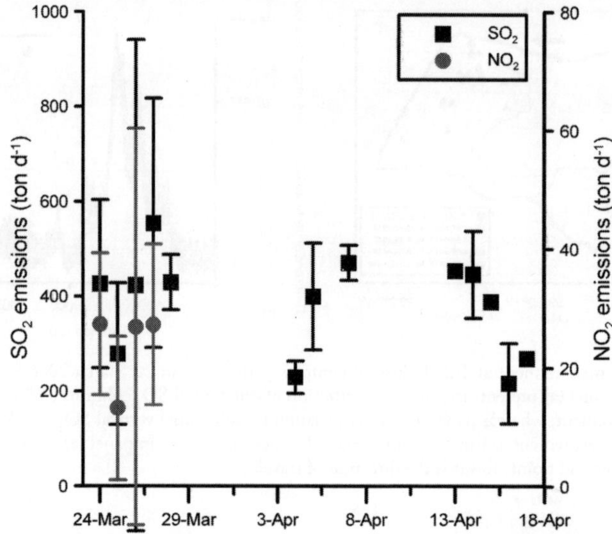

Figure 4. Daily averages of SO_2 and NO_2 emissions from Tula industrial complex. Vertical lines represent standard deviation, referring to the variability within single measurements conducted during the same day.

Table 1. Summary of measurements at Tula industrial complex.

Date	NO_2 transects	NO_2 emission (tones day^{-1})[a]	SO_2 transects	SO_2 emission (tones day^{-1})[a]
24 March 2006	2	27±12	3	427±177
25 March 2006	4	13±12	5	279±149
26 March 2006	11	27±33	28	423±518
27 March 2006	10	27±14	26	554±263
28 March 2006			4	430±58
4 April 2006			2	230±34
5 April 2006			5	400±112
7 April 2006			5	470±37
13 April 2006			1	454
14 April 2006			5	446±92
15 April 2006			1	389
16 April 2006			10	217±85
17 April 2006			1	269
TOTAL	27	...	96	...

[a] The standard deviation refers to the variability of single measurements conducted during the same day and not to the uncertainty of the measurements per se.

Observed Emission Factors

A typical measurement around the Tula industrial complex for 26 March 2006 is depicted in Fig. 3. Figure 3a shows the spatial distribution of differential SO_2

columns (color coded) quantified during the measurement. The measurement was conducted between 14:18 and 15:03 local time. The known emission sources located inside the measurement circuit are shown with black triangles. Figure 3b shows the differential SO_2 and NO_2 columns quantified during the measurement, which follow each other during the entire measurement. The traverse started at the north-western part of the Tula industrial complex (black arrow in Fig. 3a) and continued towards the east. A main peak was found during the traverse at the southeast from the known sources. This measurement yielded an emission of 44 906 kg/h of SO_2 and 2210 kg/h of NO_2. Figure 4 shows a daily average time series of NO_2 and SO_2 emissions calculated from the conducted measurements during the field campaign. In addition Table 1 shows a summary of all the measurements performed during the field campaign. NO_2 emissions were measured between 24– 27 March, and SO_2 emissions were quantified between 24 March–17 April 2006.

A total number of 96 transects were performed in order to determine the emissions of SO_2 and NO_2 in the region. Actual emission variations due to changes in the processes, as well as atmospheric perturbations along the measurements in a particular transect are mainly responsible for the observed variation in the estimated fluxes. From the statistical point of view, the larger the number of transects measurements, the better the flux determination.

During MCMA 2006, SO_2 emissions from Tula's industrial complex yielded 384 ± 103 tons day^{-1}; and NO_2 emissions accounted for 24 ± 7 tons day^{-1}. Both SO_2 and NO_2 quantified emissions present large standard deviations, and SO_2 emissions are an order of magnitude larger than the quantified NO_2 fluxes. The high variability on the flux determinations is associated to the actual emission variation on the SO_2 sources in the region, as well as to the uncertainty associated to the wind field at the specific time of the measurements.

In order to give insight as to whether emissions from the Tula industrial complex during the period of this study had any influence on the air quality of the MCMA, data from the Ambient Air Monitoring Network (Red Automática de Monitoreo Atmosférico, RAMA) was accessed online (http://www.sma.df.gob.mx/simat/). During the period of this study, no extraordinary levels of SO_2 and NO_2 (1-h average data), that would lead to exceedance of the Mexican Norms, were registered by the network. However, Concentration Field Analyses using backward trajectories calculated from some of the RAMA sites during the MCMA 2006 field campaign point towards the Tula industrial complex as a potential source area of SO_2 for the MCMA (de Foy et al., 2009).

Table 2. SO_2 and NO_x emission inventories and NO_2 measurements of Tula industrial complex.

Year	SO_2 (tpy)	NO_x (tpy) [a]	Point Sources	Reference
1999	356 966	37 834	Tula-Vito-Apasco industrial complex	SEMARNAT-INE (2006) [b]
2002	158 330	15 040	Power plant only	Miller and Van-Atten (2004); Vijay et al. (2004)
2003	145 000	–	Tula-Vito-Apasco industrial complex, 2003	de Foy et al. (2007)
2005	112 934 [c]	24 259	Stack emissions from Pemex refinery and Power Plant	SEMARNAT (2008)
2006	135 232	5697	Total emission from Pemex refinery	PEMEX (2006)
2006	140 046 ±37 533	8647 ±2549	Tula-Vito-Apasco industrial complex	This study 2006

[a] Emission inventories give values of NO_x whereas measurements conducted during the field campaign give NO_2 values.

[b] NO_x and SO_x reported emissions for Hidalgo State from point sources where power generation facilities and refineries are the main contributors.

[c] SO_x emissions.

Comparison with Emission Inventories

Table 2 shows a comparison of our results with published emission inventories. In order to develop the first National Inventory of Emissions in Mexico (base year 1999), a multidisciplinary effort between national and international institutions was made. As part of this inventory, PEMEX provided most of the information regarding combustion and process emissions. As for emissions from power generation plants, data was provided by Mexico's Energy Secretariat (SENER). Almost 70% of emissions from combustion were calculated using emission factors from U.S. EPA, 1995, section 1 (AP42), and the rest were based on measurements reported by the power stations themselves (SEMARNAT-INE, 2006).

Vijay et al. (2004) and Miller and Van-Atten (2004) estimated emissions from power generation plants based on fuel consumption and energy generation data provided by SENER as well as emission factors for specific power generation plants. The methodology used followed the recommendations of the Emissions Inventory Improvement Program of the US EPA; the emission factors used were obtained from the EPA's AP-42 (1998).

During MCMA 2003 field experiment, SO_2 emissions from the Tula-Vito-Apasco industrial complex were quantified using zenith sky UV spectroscopy, the same technique applied in this study.

The Mexican Environment and Natural Resources Secretariat (SEMARNAT) provided emission data for 2005, based on DATGEN (General Data). DATGEN is a database containing emissions inventories information (principally from combustion processes) from fixed sources of federal and state jurisdiction, located in areas where air quality management plans have been developed.

Surprisingly, regardless the difference in emission determination approaches applied, both SO_2 and NO_2 emissions quantified by the DOAS technique are comparable with emission inventories reported, particularly those after the year 2002. In the Sustainable Development Annual Report (2006), the Mexican Petroleum Oil Company (PEMEX) informs an annual average reduction of 6.3% of emissions into the atmosphere during the period 2001–2006 in the company (PEMEX, 2006). Assuming that this annual average rate applies also for the MHR, current emissions reported in 2006 are consistent with those reported in 1999 (approximately 40% in reduction in six years).

Because almost 100% of the sulfur content in fuels is emitted as SO_x, uncertainty on the reported emissions by the Mexican industries applying the AP-42 emission factors is considered to be low. This may be the reason for the accordance of SO_2 reported emissions and measurements.

It is important to note that emission inventories are given in NO_x while our instruments quantify NO_2. The knowledge of the NO_2/NO_x ratio is then important in order to derive NO_x emissions from our measurements. Because of the well known reactions in which NO is oxidized to NO_2 in plumes exiting the stacks of industrial facilities (Finlayson-Pitts and Pitts, 2000), we therefore may assume that the quantified NO_2 represents only part of the total released NO_x and the discrepancies between NO_x inventories and quantified NO_2 may be partially explained by the amount of NO that has not yet been oxidized.

Comparison of Measured and Modeled Plumes

Comparisons between measured and simulated emissions from the Tula industrial complex were made for 26 March 2006 (Fig. 5) and 4 April 2006 (Fig. 6). On 26 March, according to the DOAS measurements, the plume from the industrial complex was continuously shifting: early in the morning the plume dispersed towards the east, moving towards the south-southeast by noon and turning back towards the southeast by late afternoon. This is correctly simulated by the model, with the plumes initially moving to east, and then turning to the south and back towards the east again.

There are slight differences in the timing and strength of the shifts. The model is not able to represent the split plume at 11:55 with one part going east and the other south, although this is to be expected as these features are smaller than resolution of the wind simulations (3 km grid cells, output every 1 hour). At 14:18, the simulated plume is much narrower than the measurements, suggesting that there is insufficient dispersion in the model. As the spatial and temporal scales of this are below the resolution of the model, this suggests that turbulent mixing should be increased in the trajectory simulations. On 4 April, a wide plume was observed at noon, gradually narrowing towards the afternoon. In general, the plume was moving straight to the southwest throughout the day. This was correctly represented in the model, including the narrowing of the plume as the winds became stronger and the transport faster. In summary, there is good agreement between the measured and simulated plumes suggesting that the model is capable of representing the plume transport, and that the measurements correctly captured the entire plume.

Conclusions

Calculated emissions from Tula's industrial complex during the MCMA 2006 field campaign yielded an average of 384±103 tons day^{-1} for SO_2 and 24±7 tons day–1 for NO_2. These emissions are comparable to the more recent emission inventories, but lower than the inventories from 1999 and earlier. According to the inventories, SO_2 emissions show average reduction over the years (approximately 40% in six years) and compare well with 2006 emissions reported by PEMEX for the MHR (PEMEX, 2006). On the other hand NO_2 calculated emissions show lower values than previous NOx emission inventories. This discrepancy may be explained by an incomplete oxidation of NO to NO_2. An important outcome of this study is the fact that our measurements are in good agreement with values reported in recent inventories, implying that the use of emission factors for determination of emissions from power plants and refineries, is an adequate procedure when reliable information of energy consumption and power generation is available.

Although, during the period of this study no extraordinary levels of SO_2 and NO_2 (1-h average data), were registered on the northern part of the MCMA by the Ambient Air Monitoring Network, Concentration Field Analyses using backward trajectories calculated from some of the RAMA sites during the MCMA 2006 field campaign point towards the Tula industrial complex as a potential source area of SO_2 for the MCMA (de Foy et al., 2009).

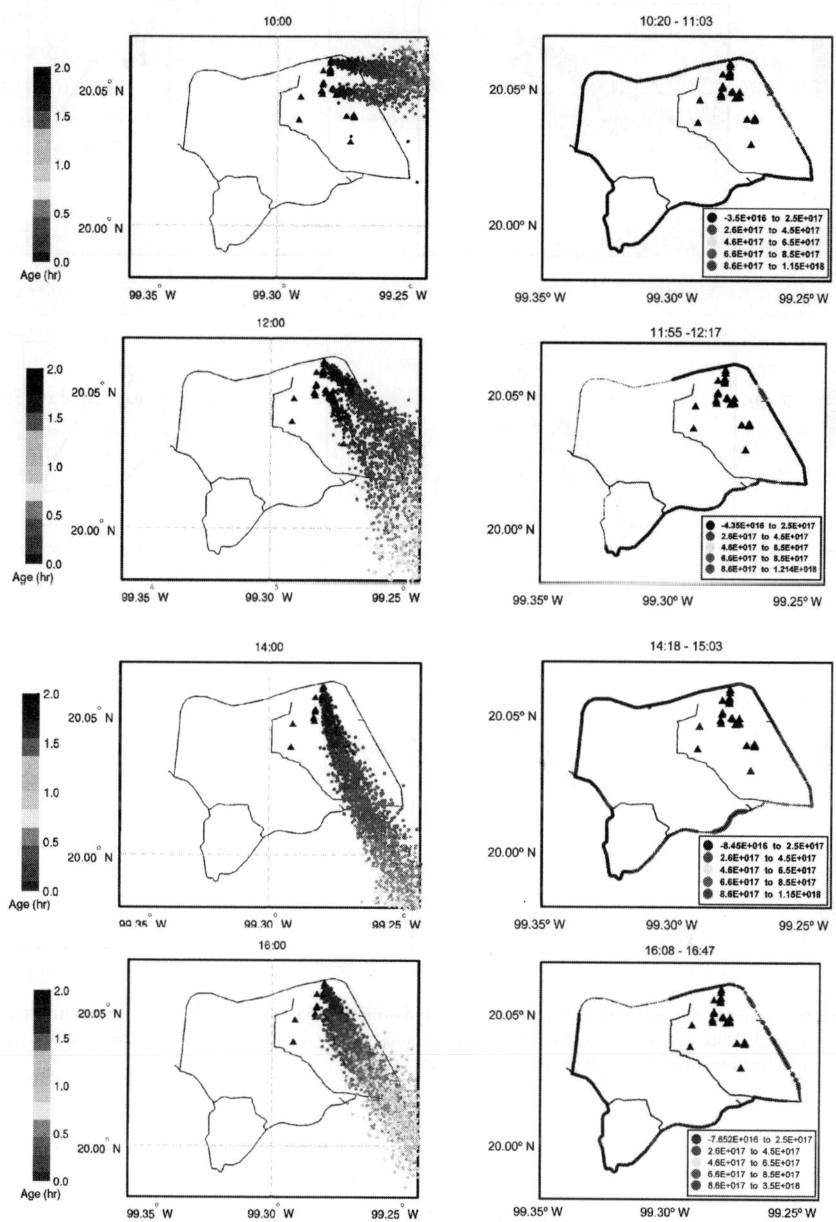

Figure 5. Comparison between modeled plumes (left) and observed spatial distribution of SO_2 columns (right) during 26 March 2006. Known sources of the power plant and refinery are shown with black triangles. The units of the differential vertical SO_2 columns are molecules/cm2.

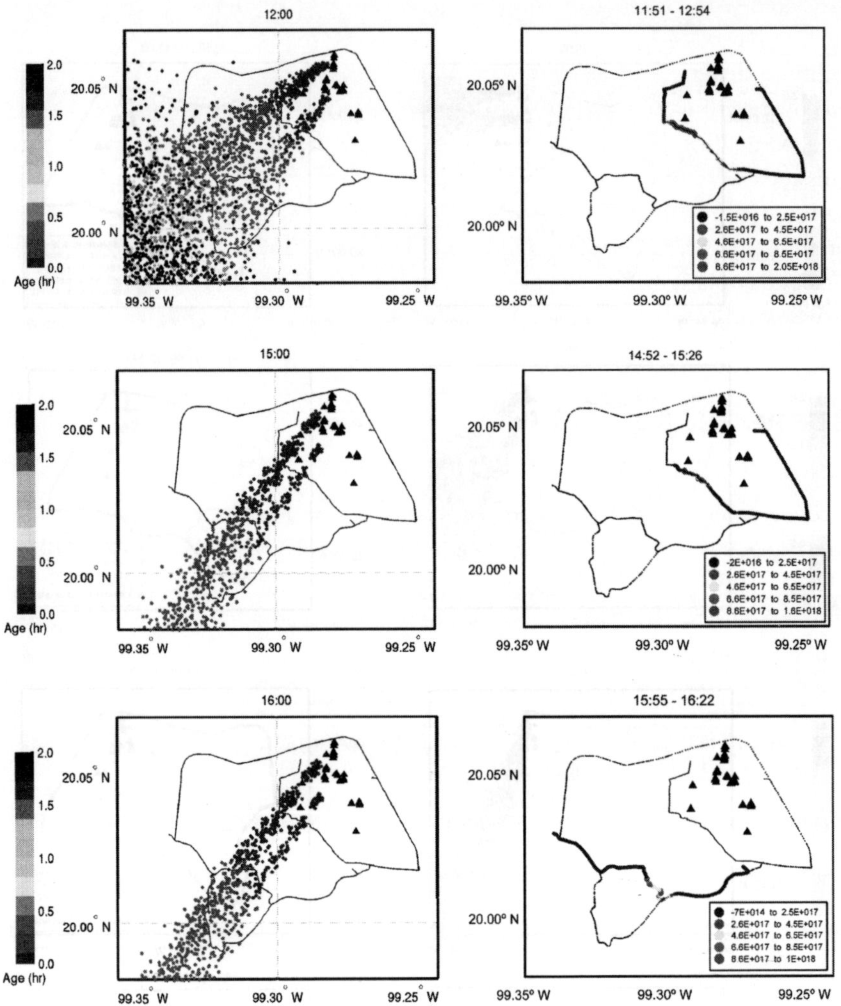

Figure 6. Comparison between modeled plumes (left) and observed spatial distribution of SO_2 columns (right) during 4 April 2006. Known sources of the power plant and refinery are shown with black triangles. The units of the differential vertical SO_2 columns are molecules/cm2.

In addition, the industrial plumes were simulated with forward particle trajectories using the measured emission rates. The good agreement between the simulated plume transport and the column measurements suggests that the model is capable of reproducing dispersion from the Tula industrial zone and brings supporting evidence that the column measurements correctly captured the plume. A remaining question is the large standard deviation of the measurements performed during the field campaign. It is thought that the reasons for them are associated with changes in real emissions from the refinery and power plant, as well as unavailability of wind fields at the exact time of every measurement. Variability in fluxes could be caused by plume meandering as well. Both simulations and observations during March attest for plumes shifting over short periods of time, coinciding with measurement days where the largest standard deviations in emissions were found. Quantified emissions during April show less standard deviations, coinciding with more defined plumes. Detailed information about production and performance of the Miguel Hidalgo refinery and the Francisco Pérez Ríos power plant during the field campaign would yield an improved comparison between our measurements and reported values.

Acknowledgements

This work was partly funded by the Comisión Ambiental Metropolitana, the Alliance for Global Sustainability, the Swedish International Development Cooperation Agency, the US National Science Foundation (awards ATM-0511803 and ATM-0810950), the Molina Center for Strategic Studies in Energy and the Environment, and the Mexican Oil Company, PEMEX. The participation of technical staff and students is highly acknowledged: Felipe Angeles, Olivia Rodríguez and Ana Lilia Alonso. We acknowledge SEMARNAT for providing validated emissions data. Edited by S. Madronich

References

1. Bobrowski, N., H¨onninger, G., Galle, B., and Platt, U.: Detection of bromine monoxide in a volcanic plume, Nature, 423, 273–276, 2003.

2. Bogumil, K., Orphal, J., Homann, T., Voigt, S., Spietz, P., Fleischmann, O. C., Vogel, A., Hartmann, M., Kromminga, H., Bovensmann, H., Frerick, J., and Burrows, J. P.: Measurements of molecular absorption spectra with the SCIAMACHY preflight model: instrument characterization and reference data for atmospheric remote-sensing in the 230–2380 nm region, J. Photoch. Photobio. A, 157, 167–184, 2003.

3. CAM: Programa para Mejorar la Calidad del Aire de la Zona Metropolitana del Valle de M´exico 2002–2010, Comisi´on Ambiental Metropolitana, M´exico, D.F., 2003.

4. de Foy, B., Fast, J. D., Paech, S. J., Phillips, D., Walters, J. T., Coulter, R. L., Martin, T. J., Pekour, M. S., Shaw, W. J., Kastendeuch, P. P., Marley, N. A., Retama, A., and Molina, L. T.: Basin-scale wind transport during the MILA-

GRO field campaign and comparison to climatology using cluster analysis, Atmos. Chem. Phys., 8, 1209–1224, 2008, http://www.atmos-chem-phys.net/8/1209/2008/.

5. de Foy, B., Lei, W., Zavala, M., Volkamer, R., Samuelsson, J., Mellqvist, J., Galle, B., Mart´ınez, A.-P., Grutter, M., Retama, A., and Molina, L. T.: Modelling constraints on the emission inventory and on vertical dispersion of CO and SO2 in the Mexico City Metropolitan Area using Solar FTIR and zenith sky UV spectroscopy, Atmos. Chem. Phys., 7, 781–801, 2007, http://www.atmos-chem-phys.net/7/781/2007/.

6. de Foy, B., Zavala, M., Bei, N., and Molina, L. T.: Evaluation of WRF mesoscale simulations and particle trajectory analysis for the MILAGRO field campaign, Atmos. Chem. Phys., 9, 4419– 4438, 2009, http://www.atmos-chem-phys.net/9/4419/2009/.

7. Doran, J. C., Fast, J. D., Barnard, J. C., Laskin, A., Desyaterik, Y., and Gilles, M. K.: Applications of lagrangian dispersion modeling to the analysis of changes in the specific absorption of elemental carbon, Atmos. Chem. Phys., 8, 1377–1389, 2008, http://www.atmos-chem-phys.net/8/1377/2008/.

8. Edmonds, M., Herd, R. A., Galle, B., and Oppenheimer, C. M.: Automated high-time resolution measurements of SO2 flux at Soufri`ere Hills Volcano, Montserrat, B. Volcanol., 65, 578–586, 2003.

9. ENVIRON: CAMx, Comprehensive Air Quality Model with extensions, User's Guide, Tech. Rep. Version 5.40, ENVIRON International Corporation, 2009.

10. Fast, J. D., de Foy, B., Acevedo Rosas, F., Caetano, E., Carmichael, G., Emmons, L., McKenna, D., Mena, M., Skamarock, W., Tie, X., Coulter, R. L., Barnard, J. C., Wiedinmyer, C., and Madronich, S.: A meteorological overview of the MILAGRO field campaigns, Atmos. Chem. Phys., 7, 2233–2257, 2007, http://www.atmos-chem-phys.net/7/2233/2007/.

11. Finlayson-Pitts, B. J. and Pitts, J. N.: Chemistry of the upper and lower atmosphere: Theory, experiments and applications, Academic Press, San Diego, California, 2000.

12. Galle, B., Oppenheimer, C., Geyer, A., McGonigle, A. J. S., Edmonds, M., and Horrocks, L.: A miniaturised ultraviolet spectrometer for remote sensing of SO2 fluxes: a new tool for volcano surveillance, J. Volcanol. Geoth. Res., 119, 241–254, 2002.

13. IMP: Estudio de las emisiones de la zona industrial de Tula y su impacto en la calidad del aire regional, IMP, PS-MA-IF-F213931, 2006.

14. Johansson, M., Galle, B., Yu, T., Tang, L., Chen, D., Li, H., Li, J. X., and Zhang, Y.: Quantification of total emission of air pollutants from Beijing using mobile mini-DOAS, Atmos. Environ., 42, 6926–6933, 2008.

15. Johansson, M., Rivera, C., de Foy, B., Lei, W., Song, J., Zhang, Y., Galle, B., and Molina, L.: Mobile mini-DOAS measurement of the outflow of NO2 and HCHO from Mexico City, Atmos. Chem. Phys., 9, 5647–5653, 2009, http://www.atmos-chem-phys.net/9/5647/2009/.

16. Johansson, M. and Zhang, Y.: Mobile DOAS, Version 4.1, Optical Remote Sensing Group, Chalmers University of Technology, 2004.

17. McGonigle, A. J. S., Oppenheimer, C., Galle, B., Mather, T. A., and Pyle, D. M.: Walking traverse and scanning DOAS measurements of volcanic gas emission rates, Geophys. Res. Lett., 29, 1985, doi:10.1029/2002GL015827, 2002.

18. McGonigle, A. J. S., Thomson, C. L., Tsanev, V. I., and Oppenheimer, C.: A simple technique for measuring power station SO2 and NO2 emissions, Atmos. Environ., 38, 21–25, 2004.

19. Miller, P. J. and Van-Atten, C.: North American Power Plant Air Emissions, Commission for Environmental Cooperation of North America, Montreal (Quebec) Canada, 2004.

20. Mori, T., Mori, T., Kazahaya, K., Ohwada, M., Hirabayashi, J., and Yoshikawa, S.: Effect of UV scattering on SO2 emission rate measurements, Geophys. Res. Lett., 33, L17315, doi:10.1029/2006GL02685, 2006.

21. O'Dwyer, M., Padgett, M. J., McGonigle, A. J. S., Oppenheimer, C., and Inguaggiato, S.: Real-time measurement of volcanic H2S and SO2 concentrations by UV spectroscopy, Geophys. Res. Lett., 30, 1625, doi:10.1029/2003GL017246, 2003.

22. PEMEX: Petr´oleos Mexicanos Sustainable Development Annual Report, Petr´oleos Mexicanos, 2006. Platt, U. and Stutz, J.: Differential Optical Absorption Spectroscopy: Principles and Applications, Springer, 2008.

23. Raga, G. B., Kok, G. L., Baumgardner, D., Baez, A., and Rosas, I.: Evidence for volcanic influence on Mexico City aerosols, Geophys. Res. Lett., 26, 1149–1152, 1999.

24. Rivera, C., Garcia, J. A., Galle, B., Alonso, L., Zhang, Y., Johansson, M., Matabuena, M., and Gangoiti, G.: Validation of optical remote sensing measurement strategies applied to industrial gas emissions, Int. J. Remote Sens., 30(12), 3191–3204, 2009.

25. SEMARNAT: Cedulas de Operaci´on Anual (COAs) for federal legislation point sources at Hidalgo State, Mexico, SEMARNAT, 2002.

26. SEMARNAT: DATGEN: Base de datos del sector industrial en 2005, SEMARNAT, 2008.

27. SEMARNAT-INE: Inventario Nacional de Emisiones de M´exico, 1999, Secretar´ıa de Medio Ambiente y Recursos Naturales, Instituto Nacional de Ecolog´ıa, 2006.

28. Sinreich, R., Friess, U., Wagner, T., and Platt, U.: Multi axis differential optical absorption spectroscopy (MAX-DOAS) of gas and aerosol distributions, Faraday Discuss., 130, 153–164, 2005.

29. Skamarock, W. C., Klemp, J. B., Dudhia, J., Gill, D. O., Barker, D. M., Wang, W., and Powers, J. G.: A description of the advanced research WRF version 2, NCAR Technical Note, NCAR/TN468+STR, 8 pp., 2005.

30. Stohl, A., Forster, C., Frank, A., Siebert, P., and Wotawa, G.: Technical Note: The Langrangian particle dispersion model FLEXPART version 6.2, Atmos. Chem. Phys., 5, 2461–2474, 2005, http://www.atmos-chem-phys.net/5/2461/2005/.

31. Vandaele, A. C., Hermans, C., Simon, P. C., Carleer, M., Colin, R., Fally, S., M´erienne, M. F., Jenouvrier, A., and Coquart, B.: Measurements of the NO2 absorption cross-section from 42 000 cm-1 to 10 000 cm-1 (238–1000 nm) at 220 K and 294 K, J. Quant. Spectrosc. Ra., 59, 171–184, 1998.

32. Vijay, S., Molina, L. T., and Molina, M. J.: Estimating Air Pollution Emissions from Fossil Fuel Use in the Electricity Sector in Mexico, Integrated Program on Urban, Regional, and Global Air Pollution at the Massachusetts Institute of Technology. Prepared for: North American Commission for Environmental Cooperation, 2004.

33. Voigt, S., Orphal, J., Bogumil, K., and Burrows, J. P.: The temperature dependence (203–293 K) of the absorption cross sections of O3 in the 230–850 nm region measured by Fourier-transform spectroscopy, J. Photochem. Photobio. A, 143, 1–9, 2001.

34. WHO: Air quality guidelines for Europe, second edition, World Health Organization, Copenhagen, 2000.

Hit from Both Sides: Tracking Industrial and Volcanic Plumes in Mexico City with Surface Measurements and OMI SO_2 Retrievals During the MILAGRO Field Campaign

B. de Foy, N. A. Krotkov, N. Bei, S. C. Herndon, L. G. Huey,
A.-P. Martínez, L. G. Ruiz-Suarez, E. C. Wood,
M. Zavala and L. T. Molina

ABSTRACT

Large sulfur dioxide plumes were measured in the Mexico City Metropolitan Area (MCMA) during the MILAGRO field campaign. This paper seeks to *identify the sources of these plumes and the meteorological processes that affect their dispersion in a complex mountain basin. Surface measurements of SO_2 and winds are analysed in combination with radar wind profiler data to identify transport directions. Satellite retrievals of vertical SO_2 columns from the Ozone Monitoring Instrument (OMI) reveal the dispersion from both the Tula industrial complex and the Popocatepetl volcano. Oversampling the OMI swath data to a fine grid (3 by 3 km) and averaging over the field campaign yielded a high resolution image of the average plume transport. Numerical simulations are used to identify possible transport scenarios. The analysis suggests that both Tula and Popocatepetl contribute to SO_2 levels in the MCMA, sometimes on the same day due to strong vertical wind shear. During the field campaign, model estimates suggest that the volcano accounts for about one tenth of the SO_2 in the MCMA, with a roughly equal split for the rest between urban sources and the Tula industrial complex. The evaluation of simulations with known sources and pollutants suggests that the combination of observations and meteorological models will be useful in identifying sources and transport processes of other plumes observed during MILAGRO.*

Introduction

Sulfur dioxide (SO_2) might well be thought to be the least of Mexico City's air quality problems. And yet, two large point sources on either side of the urban area provide a natural experiment in basin dispersion and a valuable tracer for wind transport in the region. Tracking the movement of SO_2 in the basin reveals meteorological features that are difficult to observe directly, and it identifies transport episodes for use in interpreting measurements made during the MILAGRO field campaign.

SO_2 Emissions and Detection

The Mexico City Metropolitan Area (MCMA) lies in an elevated basin surrounded by mountains with an opening to the Mexican Plateau to the north, and is home to over 20 million people. The MILAGRO field campaign took place in March 2006 to characterise the atmospheric pollution in the basin and the export and transformation of pollutants to the surrounding regions as a way of evaluating possible megacity impacts on the global atmosphere and climate.

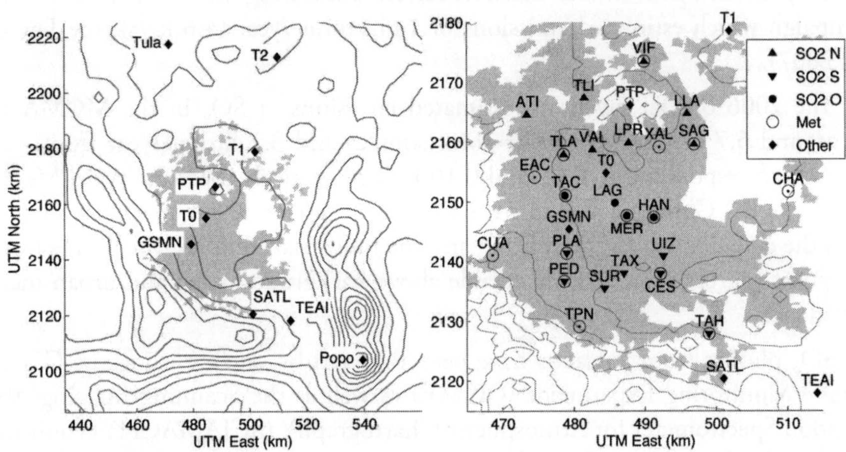

Figure 1. Map of the basin and the MCMA showing sites used in this study. RAMA SO_2 and meteorology sites shown on the right, classified by groups used for plotting (N=North, S=South, O=Central). Urban area of the MCMA shown in beige, terrain contours every 250 m.

The Popocatepetl volcano is a passively degassing eruptive volcano rising to 5426 m m.s.l. which forms part of the southeastern rim of the Mexico City basin and is approximately 70 km southeast of the MCMA centre. It emits SO_2 continuously in the absence of any visible eruptions (Delgado-Granados et al., 2001). During the MILAGRO field campaign, Grutter et al. (2008) estimated SO_2 emission rates from the volcano using a scanning DOAS instrument located on the northern flank of the volcano. These emission rates were compared with estimates from a COSPEC instrument and from transects of an airborne DOAS aboard an ultra-light aircraft. Daily average values were in the range of 0.6 to 4.4 Gg/day, corresponding to 7 to 50 kg/s. These values are similar to measurements made in April 2003 during the MCMA-2003 field campaign, where two transects yielded estimates of around 0.8 Gg/day (10 kg/s) (de Foy et al., 2007) and to COSPEC estimates of 2 to 3 Gg/day (20 to 25 kg/s) during a pre-eruptive period leading up to August 1995, and 9 to 13 Gg/day (100 to 150 kg/s) during an effusive-explosive period from March 1996 to January 1998 (Delgado-Granados et al., 2001).

The Tula industrial complex is the home of a number of industries including a power plant and a refinery, and is located about 70 km northwest of the MCMA centre—diametrically opposite to the Popocatepetl volcano, see Fig. 1. The total official inventory for the area estimates SO2 emissions of 323 ktonne/year, corresponding to 10 kg/s (Rivera et al., 2009). Mini-DOAS transects from 24 March to 17 April 2006 estimated average SO_2 fluxes of 155±120 ktonne/year (4.9±3.8 kg/s) for the refinery and the power plant together (Rivera et al., 2009). These values are in agreement with similar transects carried out during the MCMA-2003 field campaign which estimated emissions of 145 ktonne/year (4.6 kg/s), (de Foy et al., 2007).

The 2006 official inventory estimated emissions of SO_2 in the MCMA to be around 5.7 ktonne/year from point sources and 3.2 ktonne/year from area sources, corresponding to 0.18 and 0.10 kg/s respectively (Secretar´ıa del Medio Ambiente del Gobierno del Distrito Federal, 2008). As these are much smaller than the emissions from the volcano and the industrial complex, the plume from the point sources should be detectable above background rural and urban measurements.

SO_2 plumes from volcanos have been detected by satellite using the Global Ozone Monitoring Experiment (GOME) as well as the Scanning Imaging Absorption Spectrometer for Atmospheric ChartographY (SCIAMACHY) confirming that the SO_2 plumes from the Popocatepetl are some of the largest on earth (Khokhar et al., 2005), (Loyola et al., 2008). The Ozone Monitoring Instrument (OMI) on NASA's Aura satellite provides higher spatial and spectral resolution combined with daily coverage providing retrievals of SO_2 column amounts (Krotkov et al., 2006). In addition to detecting volcano plumes (Yang et al., 2007), it has also been able to detect SO_2 plumes from copper smelters (Carn et al., 2007). Evaluation of the retrievals over Northeast China found that OMI could distinguish between background conditions and heavy pollution on a daily basis, with noise in the data of around 1.5 DU (Dobson Units), which can be reduced to 0.3 DU with spatial and temporal averaging (Krotkov et al., 2008). The algorithm has been further refined to improve retrievals of very large loadings from volcanic plume, and detected over 1000 DU from the Sierra Negra eruption in Ecuador in October 2005 (Yang et al., 2009). Given the emissions of the Tula industrial complex and the Popocatepetl, it should be possible to detect these under routine monitoring conditions.

Basin-Scale Wind Transport

The MCMA is located in the subtropics where there is weak synoptic forcing and at high elevation surrounded by mountains leading to weak winds and complex flow patterns. Jauregui (1988) describe the drainage flow into the basin that is decoupled from the westerlies aloft and accentuated by the urban heat island. At a time when SO_2 emissions were much larger in the city itself, this led to the highest SO_2 concentrations located at the centre of the heat island. Williams et al. (1995) simulated SO_2 dispersion in the MCMA and identified complex mixing suggesting that an elevated plume was entrapped in the drainage flow down Pico de Tres Padres and transported to the basin floor at night. Starting in 1992, the

SO_2 content of fuels was reduced in the MCMA leading to a dramatic reduction of average concentrations from around 60 ppb to below 10 ppb currently (see Sistema de Monitoreo Atmosf'erico, http://www.sma.df.gob.mx/simat2/ informaciontecnica). In terms of SO_2, this has shifted the concern from urban sources to regional point sources.

Particle trajectories were used by Bossert (1997) to show how an undercutting plain-to-plateau density current could transport pollutants into the basin with minimal mixing even though the urban plume was being vented aloft, moving above the surface current in the opposite direction. Fast and Zhong (1998) describe the recirculation patterns in the basin where the plume is transported along the surface, up the mountain slopes, and back over the urban area where it could mix back down to the surface, but was usually efficiently vented. A conceptual model of wind transport for the MCMA-2003 field campaign found stable drainage flows on most nights, accompanied by weak, stable winds from the north. These met with a gap flow from the southeast to cause a convergence line and rapid venting of the urban plume (de Foy et al., 2006c). The location and movement of these convergence lines determined the location of high pollution events in the basin (Jazcilevich et al., 2005), (de Foy et al., 2006a). These studies suggest that both the Tula plume below the basin and the Popocatepetl plume above could have significant impacts in the MCMA.

Sulfur Transport

Episodes of high SO_2 concentrations and sulfate aerosol loadings were measured in the south of the MCMA in November 1997 and were attributed to emissions from Popocatepetl based on estimates of emission rates and dilution due to vertical mixing (Raga et al., 1999). These findings were corroborated by measurements during 2001 which identified high sulfate formation at southwestern measurement sites in the basin during moist periods from April to June when the volcano was active (Moya et al., 2003). In contrast, aerosol measurements during the IMADA field campaign were compared at boundary and urban sites, suggesting that transport from north to south accounted for about two-thirds of the sulfate in the MCMA (Chow et al., 2002), and that these might be from the Tula industrial complex.

During the MCMA-2003 field campaign (Molina et al., 2007), aerosol measurements found high particulate sulfate loadings associated with transport from the north (Salcedo et al., 2006). Concentration field analysis of SO_2 time series data suggested that the Tula industrial complex accounted for the high SO_2 episodes during the campaign (de Foy et al., 2007). Forward Eulerian modelling of

Popocatepetl emissions suggested that there could be urban impacts, but that these could not be differentiated from local emissions during April 2003.

With prevailing winds during the dry season from the west, the Popocatepetl plume would be more likely to be transported to the east past Puebla. It was detected there during a field campaign in April and May 1999 (Jimenez et al., 2004). Measurements of ozone and carbon monoxide were used to distinguish between urban and volcanic air masses, showing increases in sulfate aerosols due to the volcano. Juarez et al. (2005) found air quality impacts in the city of Puebla itself during an intense volcanic activity between December 2000 and January 2001. Measurements at the end of February 2001 in Pico de Orizaba National Park, over 200 km to the east, were carried out to determine the air quality impacts of neighbouring cities (Marquez et al., 2005). Pyle and Mather (2005) point out that in addition to urban impacts, the measurements indicated impacts of both SO_2 and sulfate aerosols from Popocatepetl. While these studies are focused on longer range transport, they do show that the plume can have surface impacts through downmixing, and that consequently with winds aloft to the west, Popocatepetl should significantly influence MCMA's air quality. Furthermore, the large variability in emissions opens up the possibility of very large MCMA impacts during episodes with particularly large emissions.

Outline

The synoptic meteorological conditions and meteorological measurements available during MILAGRO are described in Fast et al. (2007). The basin scale conditions were shown to be climatologically representative of the warm dry seasons of the last 10 years (de Foy et al., 2008). Cluster analysis was used to identify both surface wind features and vertical stratification of wind layers, leading to a conceptual model of the basin transport with six main categories (de Foy et al., 2008) which were similar to those of MCMA-2003 (de Foy et al., 2005). Overall, the analysis shows that there were days with venting both to the south and to the north, with complex mixing and stratification in the vertical.

So as to identify the sources of individual plumes in the basin, we carry out detailed analysis of ground measurements of SO2 concentrations in combination with hourly maps of surface winds and daily evolution of vertical pro-filer winds. Column measurements from satellite remote sensing provide a spatial view of the plume dispersion. Model comparisons are then used to integrate the different measurements available and to evaluate basin dynamics and plume impacts. At the same time, the measurements provide constraints on model performance and identify both model weaknesses and sources of uncertainty. This paper will describe specific episodes, but the entire set of surface wind vectors and radar wind profiler data is shown in the supplementary material http://www.atmos-chemphys.net/9/9599/ 2009/acp-9-9599-2009-supplement.pdf for readers who desire extra supporting evidence or who are interested in other episodes.

Measurements

Figure 1 shows the location of the measurement sites used in this study. The Ambient Air Monitoring Network (Red Autom´atica de Monitoreo Atmosf´erico, RAMA) operates a network of surface stations measuring meteorological parameters and criteria pollutants throughout the city. Quality-assured data at 1-hour intervals was used for the statistical comparisons and the hourly plots. This was available for the full month of March from 19 stations. Data at one-minute intervals was used for the plume time series plots to show the detailed transport in the basin. Wind vectors were available from 14 stations during the campaign.

SO_2 measurements were made using pulsed UV fluorescence (Teledyne API models 100 and 100A). UV radiation of 214 nm is passed through the sample chamber. UV photons are absorbed by SO_2 molecules which return to their ground state by emitting a lower energy photon with a wavelength of 330 nm. When the temperature is known, the amount of fluorescent light is directly related to the SO_2 concentration in the sample chamber. The measurements were digitised with 1 ppb increments, and had a stated instrument accuracy of 1% but likely overall measurement accuracy within 10%.

Two mobile laboratories were deployed with similar equipment, one at Santa Ana Tlacotenco (SATL), a small village on the southeastern edge of the basin overlooking the MCMA, and one at Tenango del Aire (TEAI), in the mountain pass to the southeast below the Popocatepetl volcano. The Aerodyne mobile laboratory (Kolb et al., 2004) was located at the summit of Pico de Tres Padres (PTP) from 8 to 19 March. SO_2 data were available starting on 13 March. This site is approximately 750 m above the basin floor in the north of the MCMA. It therefore serves as a background site at night, observes the mixing of the morning emissions during the day and the outflow of the urban area on afternoons with strong gap flows from the southeast.

At PTP and at T1, SO_2 was measured with a Thermo 43C pulsed fluorescence instrument which was periodically calibrated by standard addition. The background signal of the instrument was periodically measured with a Na_2CO_3 impregnated filter. The detection limit was of the order of 50 pptv for a one minute average. Owing to less frequent SO_2 calibration while at PTP, the overall PTP measurement accuracy is likely below 20%.

A detailed description of the meteorological data collected during the campaign can be found in Fast et al. (2007) and in de Foy et al. (2008). In addition to the RAMA, SATL and TEAI wind vectors, this study uses winds from the five surface stations of the Mexican National Weather Service (SMN) located in the basin, as well as meteorological measurements from temporary stations at the T0, T1 and T2 sites.

Radar wind profilers were installed at T0, T1 and T2. These were 915 MHz models manufactured by Vaisala. They were operated in a 5-beam mode with nominal 192-m range gates. As described in Doran et al. (2007), the NCAR Improved Moment Algorithm was used to obtain 30-min average consensus winds. Plots of horizontal winds aloft also show the radiosonde observations from the SMN headquarters (GSMN) launched every 6 h. The timezone in the MCMA was Central Standard Time (CST = UTC–6) during the entire campaign, all times reported in this study will be in CST.

The Ozone Monitoring Instrument (OMI) provides SO_2 retrievals with a nadir resolution of 13 by 24 km and daily overpasses of the MCMA between 12:00 and 14:00. This study uses the level 2, version 3 swath data available online from NASA's Goddard Earth Sciences Data and Information Services Center. The total planetary boundary layer SO2 column product was used (Krotkov et al., 2006), as we were interested in the urban impacts in the MCMA.

Modelling

Mesoscale meteorological simulations were carried out with the Weather Research and Forecast model version 3.0.1 (WRF, Skamarock et al., 2005) using the Global Forecast System (GFS) as initial and boundary conditions. There were three domains in the simulation with grid resolutions of 27, 9 and 3 km, 41 vertical levels and one-way nesting. Diffusion in coordinate space was used for domains 1 and 2, and in physical space for domain 3 (Z¨angl et al., 2004). The following options were used: the YSU boundary layer scheme (Hong et al., 2006), the Kain-Fritsch convective parameterisation (Kain, 2004), the WSM6 microphysics scheme, the Dudhia shortwave scheme and the RRTM longwave scheme. High resolution satellite remote sensing was used to improve the land surface representation in the NOAH land surface model for domains 2 and 3 by using landuse, surface albedo, vegetation fraction and land surface temperature from the Moderate Resolution Imaging Spectroradiometer (MODIS) as described in de Foy et al. (2006b).

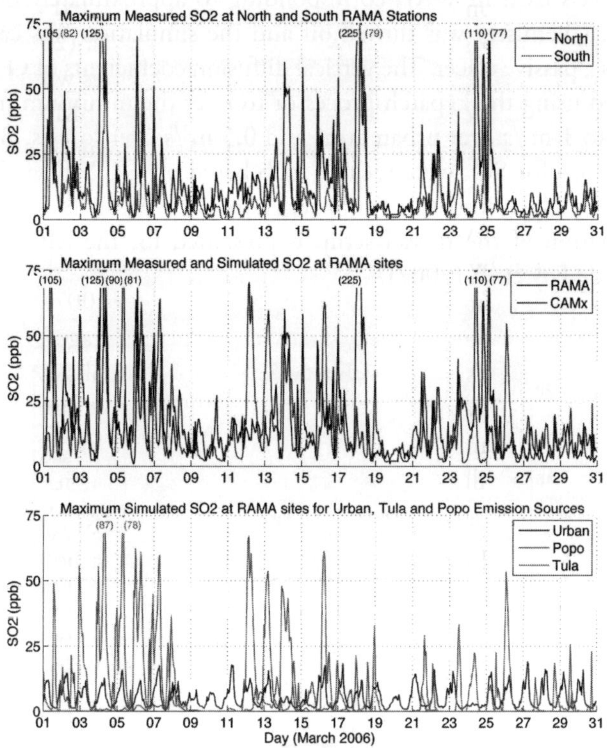

Figure 2. Measured and simulated SO2 in the MCMA. Maximum 1-hour concentrations for the North and South station groups defined in Fig. 1 (Top). Comparison of maximum domain-wide RAMA measurements and CAMx simulations (Middle). Domain-wide maximum of CAMx simulations for all RAMA stations measuring SO_2, for separate source groups (Bottom). Maximum of values off the chart shown in brackets.

The full details of the simulations and their evaluation are presented in de Foy et al. (2009). This compared the results of the simulations used in this study ("WRFb") with an alternative set-up of WRF ("WRFa") and with results from MM5. By analysing wind roses segregated by clusters, it was shown that the drainage flows in the basin were under-represented in the model. It was further shown that the model had too much vertical stratification of winds. Nevertheless, by evaluating the model against transport of carbon monoxide and SO_2, it was shown that the simulations were a representative approximation of actual transport in the basin. On the basis of this, it was suggested that following Oreskes (1998), the simulations met the criteria for "Aristotelian Accuracy" by being of sufficient quality for the purposes at hand (de Foy et al., 2009).

Eulerian pollutant transport was calculated using the Comprehensive Air-quality Model with eXtensions (CAMx, ENVIRON (2008)), version 4.51. This was run on the finest WRF domain at 3 km resolution with the first 18 of the

41 vertical levels used in WRF, corresponding to approximately 6000 m above ground level. Chemistry was turned off and the simulation was carried out for SO_2 acting as a passive tracer. The vertical diffusion coefficients of O'Brien (1970) were modified using the kvpatch processor to reset the minimum in the bottom 500 m layer to 1 m^2/s over urban areas and 0.5 m^2/s over forests. Boundary and initial conditions for SO_2 were set to 1 ppb based on GOME satellite retrievals available at the Belgian Institute for Space Aeronomy (IASB-BIRA). More details and an evaluation of the model setup is presented for the MCMA-2003 field campaign in de Foy et al. (2007).

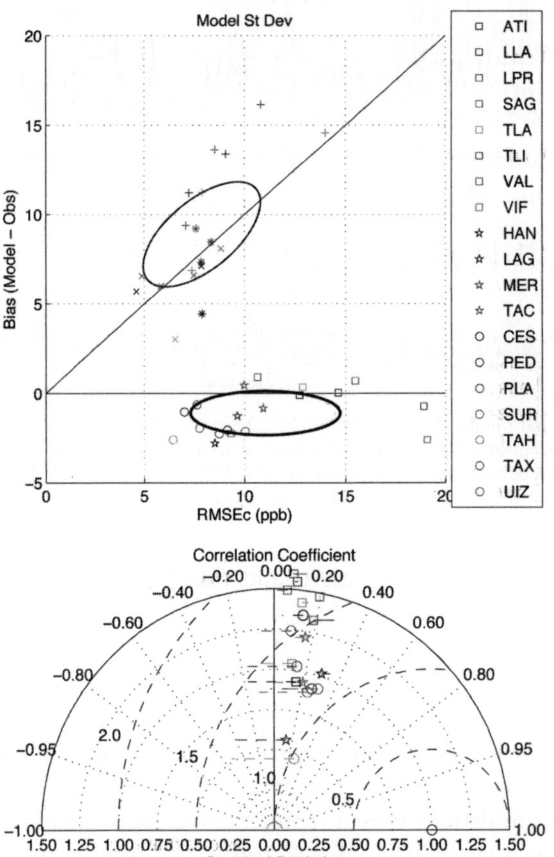

Figure 3. Statistics diagrams of simulated versus measured hourly SO2 by station in the MCMA. Top: RMSEc-bias diagram, open symbols show the model bias versus the centred root mean square error (RMSEc). Closed symbols show the model standard deviation versus that of the observations (squares match '+', circles match 'x'). Ellipses are centred on the mean with semi-axes given by the standard deviations of the metrics. Bottom: Taylor diagram showing the correlation coefficient and the standard deviation, solid horizontal line indicates positive bias, dashed line for negative bias.

The emissions for the Tula industrial complex were taken from Rivera et al. (2009) and those for the Popocatepetl were interpolated on an hourly basis from the daily values reported in Grutter et al. (2008). For the Popocatepetl volcano, the maximum terrain height in the model is 4438 m m.s.l.. The emissions were therefore released at a height of 1027 m above ground, to correspond to the actual summit of the mountain at 5465 m m.sl. MCMA urban emissions were much lower than these point sources and were based on the 2006 official emissions inventory for the MCMA (Comisi´on Ambiental Metropolitana, 2008).

Results

Figure 2a shows the maximum hourly SO_2 concentrations measured by RAMA stations in the north and south of the MCMA during the whole MILAGRO campaign. The stations used for each group are shown in Fig. 1. We show the maximum concentrations by domain because we are interested in looking at the sources of large plumes. Baseline levels are low, with a clear diurnal cycle starting at 5 ppb at night and rising to 20 ppb during the day. The main feature in the time series are the short spikes in concentrations at night rising up to a campaign maximum of 225 ppb. Concentrations are clearly higher in the north of the MCMA. The baseline levels in the south vary from 0 to 5 ppb, and the spikes rarely reach the same levels as those of the northern domain.

The comparison between measured and simulated maximum domainwide concentrations is shown in Fig. 2b. Qualitatively, this is in agreement with the measurements in terms of both the base line levels and the presence of high concentration episodes. Three time periods exhibit relatively high numbers of SO_2 spikes: the South-Venting flow of the early campaign (1–8 March), the days following the Cold Surge events on 14 March and again after 21 March. Figure 3 shows the statistical metrics for the hourly time series by station using the Taylor diagram (Taylor, 2001) and RMSEc-bias diagram (de Foy et al., 2006b). Overall, the simulated maximum SO_2 levels are too low by 1.8 ppb, the centred Root Mean Square Error is 24 ppb, Pearson's correlation coefficient is 0.14 and the Index of Agreement (Willmott, 1982) is 0.39. For the domainwide mean, these values are 1.1 ppb too high, 8.4 ppb, 0.24 and 0.48 respectively. Figure 2b shows that these low performance indices are mainly due to false positives and false negatives. Case-by-case analysis below will show that this is because the point sources are 70 km from the urban centre, and that small differences in the wind fields can make the difference between the plumes missing or hitting the measurement sites, but that the model nevertheless represents the dominant flow features and transport directions in the basin.

Finally, Fig. 2c shows the maximum 1-h CAMx simulated concentration levels for all RAMA stations for three different sources: the urban sources, the Popocatepetl volcano and the Tula industrial complex. This suggests that urban sources are responsible for a small, regular, diurnal variation in SO_2, that the volcano causes occasional peaks in the model and that most of the high SO_2 peaks are due to transport from the Tula industrial complex.

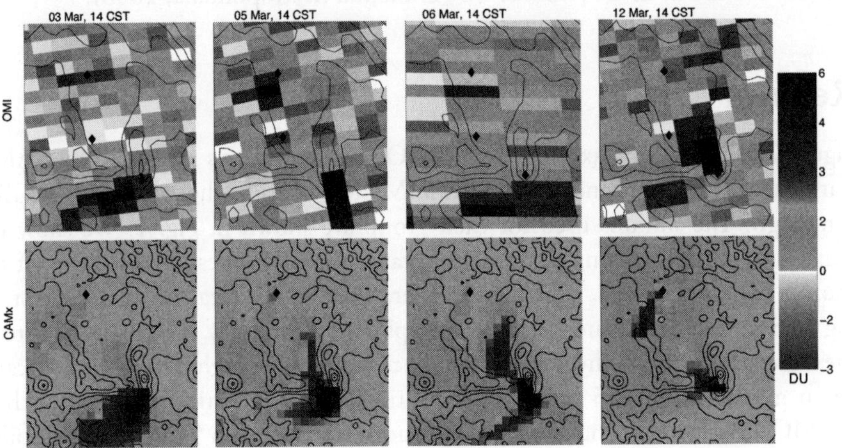

Figure 4. SO_2 total columns from OMI swath data and CAMx regional simulations with 9 by 9 km grid cells with all sources (urban, Tula and Popocatepetl). Black diamonds shows the location of Tula and Popo. Terrain contours every 500 m.

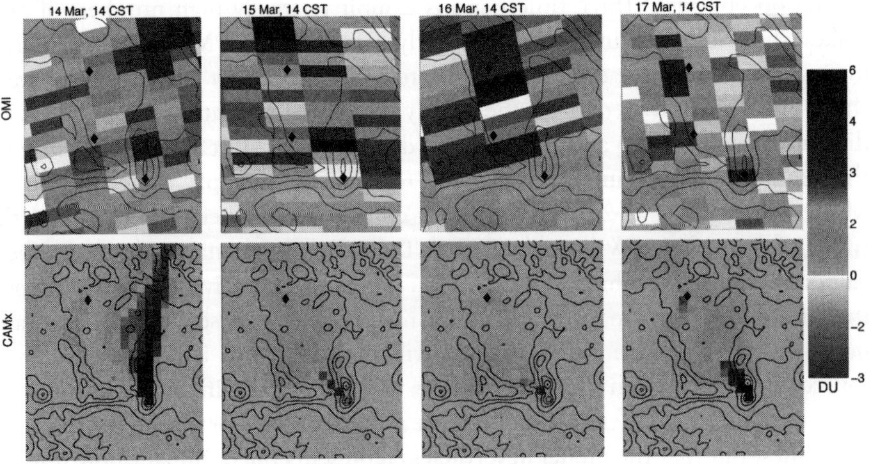

Figure 5. SO_2 total columns from OMI swath data and CAMx regional simulations with 9 by 9 km grid cells with all sources (urban, Tula and Popocatepetl). Black diamonds shows the location of Tula and Popo. Terrain contours every 500 m.

OMI Evaluation

Figures 4 and 5 show the total planetary boundary layer columns of SO2 measured by the OMI sensor as well as columns simulated by CAMx for all sources (urban, Tula and Popocatepetl) for eight days during the campaign. These cases clearly show high SO2 columns over the Popocatepetl volcano and over the Tula industrial complex. Both the direction and the intensity of each plume varies from day to day. The industrial plume rapidly dilutes to below detection level of the OMI sensor, but the volcano plume can be tracked for longer distances.

Figure 6 shows the monthly composite of all the swath data available mapped on the CAMx grid, with the corresponding average model result. Based on the OMI user's guide (OMI Team, 2009), only swath pixels between cross track positions 10 and 50 were used for which the radiative cloud fraction was less than or equal to 0.2. These swath pixels were projected onto a 3 by 3 km grid using nearest neighbor interpolation. A monthly average was created from teh 29 available daily grids.

By oversampling the SO_2 data at a fine resolution, this image fusion method is able to provide an image of the average plume at a higher resolution that the original data. This is similar to the goal of super-resolution (Capel, 2004), which has been successfully applied to land cover mapping (Li et al., 2009), (Boucher et al., 2008). Super-resolution works because the underlying map can be assumed to be constant. In the present case however the SO_2 plumes are always changing and consequently the more sophisticated methods cannot be directly applied.

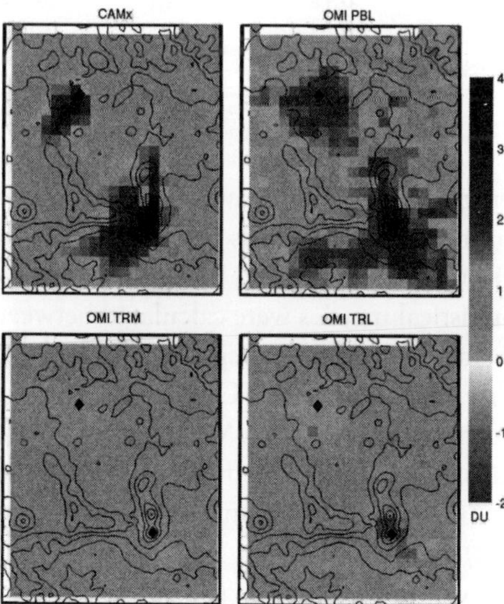

Figure 6. Average SO_2 total columns for 3 to 31 March 2009 from OMI swath data and CAMx regional simulations with 9 by 9 km grid cells and all sources (urban, Tula and Popocatepetl). Black diamonds shows the location of Tula and Popo. Terrain contours every 500 m.

The background SO_2 concentration of 1 ppb in the simulations leads to a vertical column of 0.4 DU. The OMI retrievals use a sliding median residual correction method to remove the along-and cross-track biases (Yang et al., 2007). Because of the high values in the MCMA region, this leads to a background of –0.5 DU. In order to account for this, we have added an offset of 0.9 DU to the satellite retrievals.

The values of the OMI retrieved PBL columns depend on the estimation of the air mass factor (AMF) defined as the ratio of the satellite measured slant column amount to the vertical column amount (Krotkov et al., 2008). Using basin averaged pressure and elevated vertical profiles of the Tula plume and a surface albedo value of 0.02 at 315 nm (Corr et al., 2009) increases the AMF by 10% compared to the operational AMF, which would reduce the PBL columns accordingly. Accounting for aerosols in the boundary layer using AERONET measurements and a single scattering albedo of 0.8 at 315 nm (Corr et al., 2009) decreases the AMF by 10% bringing it back to the same value as that used by default in the derivation of the PBL product (AMF = 0.36). No correction was therefore needed for the operational values due the local conditions in the MCMA.

Table 1. Basin averaged bias (OMI minus model), centred Root Mean Square Error and Pearson correlation coefficient comparing the OMI PBL, TRM and TRL retrievals with the CAMx simulation.

Metric	OMI PBL	OMI TRM	OMI TRL
Bias	0.18	−0.18	−0.16
RMSEc	0.63	0.78	0.74
Pearson r	0.66	0.57	0.58

The monthly averages were spatially averaged to the same 9 by 9 km grid as the CAMx simulation for visual comparison. Model monthly averages were created by using simulation data at each grid point for which a corresponding OMI pixel was available. Statistical metrics were calculated between the simulated grids and the satellite retrievals by further aggregating the grid points to 18 by 18 km to account for the fact that the highest OMI pixel resolution is 13 by 24 km. Table 1 shows the basin averaged bias (OMI minus model), centred Root Mean Square Error and Pearson correlation coefficient between the OMI retrievals and the CAMx simulations (negative bias means the satellite retrievals were lower than the simulated columns).

This shows that OMI clearly detected both the Popocatepetl and the Tula industrial complex plumes. The main difference is the higher resolution of the plume afforded by the simulations leading to higher column values in the vicinity of the sources. The transport directions of the plumes are in qualitative agreement, with the Tula plume transported to the southwest and the Popocatepetl plume transported either to the north or to the southwest. While the plumes are detected in all three satellite products, it is clear that the PBL retrieval is the one that is most sensitive to the plumes, and closest to the simulations.

Detailed results will be presented for the 2, 5 and 6 March which are part of the first group of days with strong winds to the south, and for the period from 12 until 18 March which cover the first Cold Surge episode. Areas of agreement and discrepancy between the measurements and the simulations will be used to evaluate the transport processes and model performance on an individual basis.

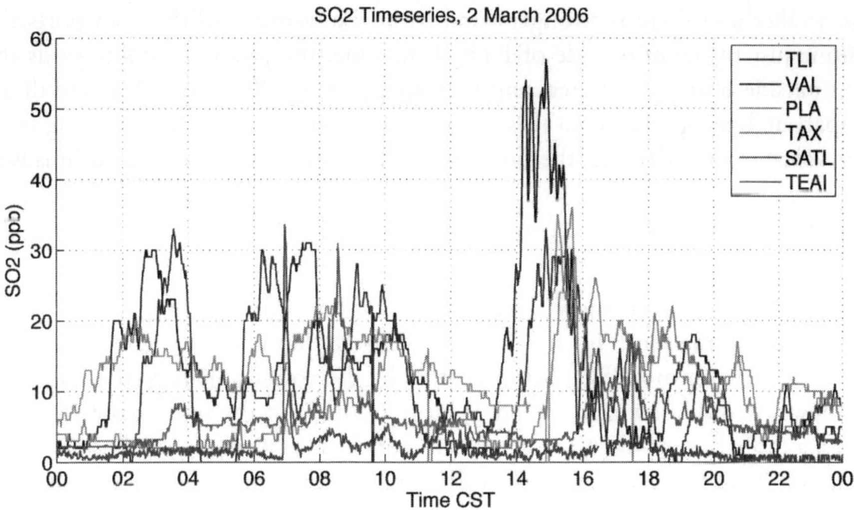

Figure 7. Time series of measured SO_2 concentrations for 2 March.

Industrial Impacts

A straightforward case of plume transport took place on 2 March, which was a day with strong winds from the north—a "South-Venting" day. Figure 7 shows variable SO_2 levels during the first part of the day followed by a uniform increase starting at 14:00 in the north at TLI, impacting urban sites at very similar times and SATL a little later. Figure 8 shows the transport to the south with a narrow plume extending through the centre of the MCMA. The combination of the surface wind vectors, the surface contours of SO_2 and the simulated plume provide strong evidence that the plume originated in Tula. Note that the plume extends from north to south, but moves from east to west in time as the wind direction changes very slightly. This illustrates how sensitive impacts are to small changes in the winds. Furthermore, in this case, the timing of the impact reflects the lateral movement of the plume rather than the speed of transport across the basin.

Figure 9 shows the time series of SO_2 concentrations for 5 and 6 March. Meteorologically, these are South-Venting days similar to 2 March. On the 5 March, levels remain low, but there is a well defined plume over the urban area from 08:00 to 12:00 and a second shorter one from 15:00 to 17:00. Surface winds and simulated contour plots, shown in Fig. 8, suggest that these are transport events from Tula.

On 6 March, the two plumes are much more clearly defined with levels reaching 70 ppb. Measured impacts are to the west of Pico de Tres Padres from 00:00 to 04:00. In the simulations, the drainage flows from the southwest basin rim are weaker and there is stronger wind from the northwest. This transports the plume just to the other side of PTP. With time, the plume moves towards the east outside of the urban area, and then returns at 08:00, with a clear and direct impact at Tenango del Aire. This is too early in the day to be downmixing from the volcano. Furthermore, the progression along the east side of the basin is well captured in the time series data. Note that the simulations capture both the westward transport at the surface, and the southward flow through the mountain gap in the southeast.

The OMI columns show clear transport of both the Tula and the Popocatepetl plume to the south for the 3, 5 and 6 March. Simulations are in agreement, with clearly a lot more SO_2 being emitted by the volcano than by the industrial complex. It would seem that there is insufficient SO_2 in the simulations for these days, although it is difficult to draw hard conclusions given the resolution of the features. Note however that part of the simulated volcano plume is entrained in the gap flow that forms northwards in the early afternoon. There is no evidence of this in the data, and setting a higher plume release height eliminates this feature entirely.

The episode from the 21 to 27 March shows similar transport of the Tula plume into the MCMA, albeit with more complex flows due to the weaker, moister winds causing afternoon convection. The simulated volcano spike on 21 March is most likely a false positive due to entrainment in an overly developed gap flow.

Volcano Impacts

SO_2 concentrations were low on 12 March, which had weak drainage flows into the basin followed by northerly surface flows and then a strong gap flow from the south in the late afternoon ("O3-South"). Figure 10 shows low levels of SO_2 throughout the day, but a distinct plume signature at Santa Ana (SATL) starting at 04:00 in the morning. There are short impacts at T1 around 06:00 followed by a uniform increase at sunrise at both Tenango del Aire (TEAI) and T1 which are at either end of the basin.

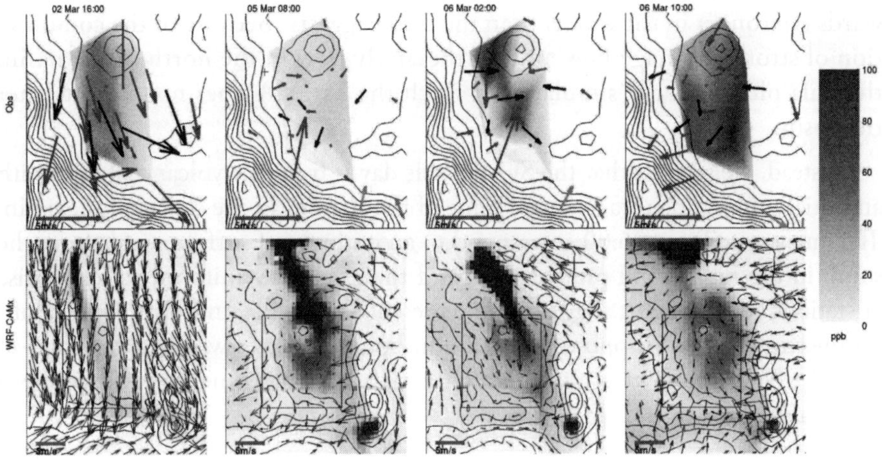

Figure 8. Measured winds and hourly surface concentration of SO2 in the MCMA and simulated winds (WRF) and SO2 (CAMx, all sources) in the basin. Observed winds are coloured according to network: green -RAMA, blue -SMN, magenta -T1, tan -St. Ana, orange -Tenango del Aire. Terrain contours every 100 m (top) and 500 m (bottom), measurement domain shown as a box in model domain, red diamonds show T0, T1, T2 and Tula.

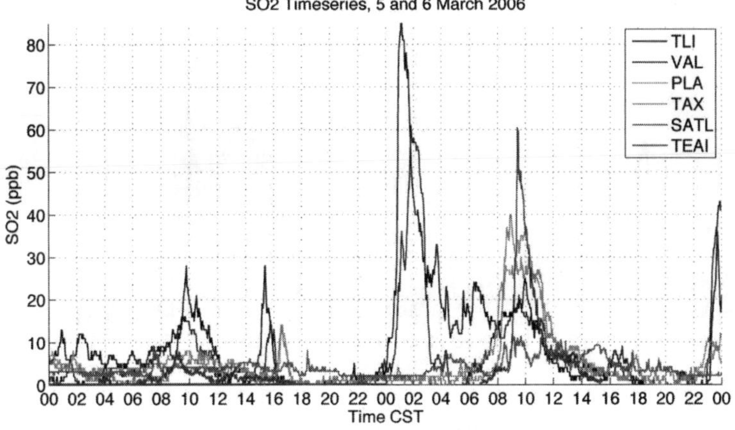

Figure 9. Time series of measured SO_2 concentrations for 5 and 6 March.

The OMI columns in Fig. 4, show very clearly a split in the volcano plume. Part of it is transported towards the southwest, while the bulk of it is transported northwards where it covers the eastern side of the MCMA. This is represented by the model, albeit with lower total columns of SO_2. The Tula plume is simulated to move southwest, but was not detected by OMI.

Wind vectors show the strong drainage flow in the basin, see Fig. 11. The simulations do not represent this feature, and the gap flow moving northwards is stronger in the model than the data suggests. Because of the combination of strong drainage flow and northeasterly flow in the north of the basin, the Tula plume that is simulated to reach the MCMA does not in fact enter the basin.

Instead, we suggest that the SO_2 on this day is from the volcano plume, with an initial impact at Santa Ana and some of the stations in the centre of the basin. The impacts at T1 at 06:00 correspond to an outburst of surface winds from the south in the radar wind profiler data, and the levels rise uniformly after sunrise at stations very far apart suggesting the presence of a wide, uniform plume aloft. A concentration of 10 ppb over a 4000 m boundary layer would correspond to a total SO_2 column of around 2 DU, which is in agreement with the OMI retrievals.

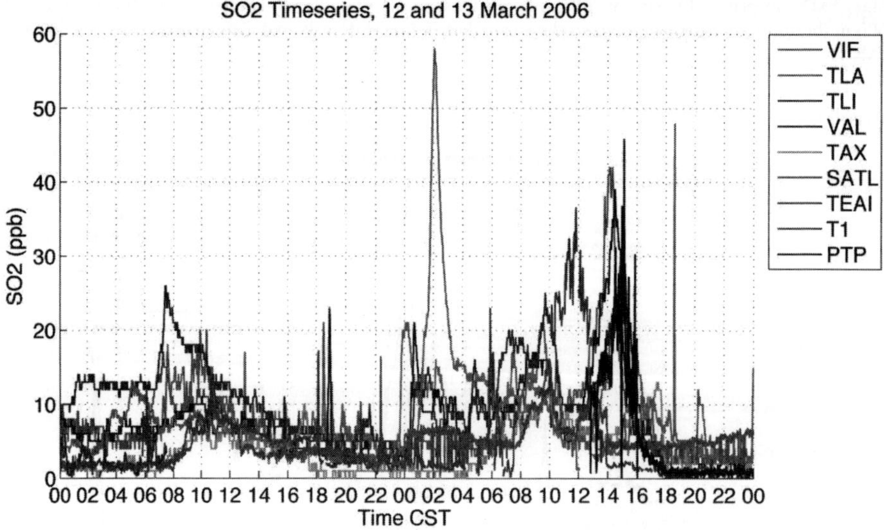

Figure 10. Time series of measured SO_2 concentrations for 12 and 13 March.

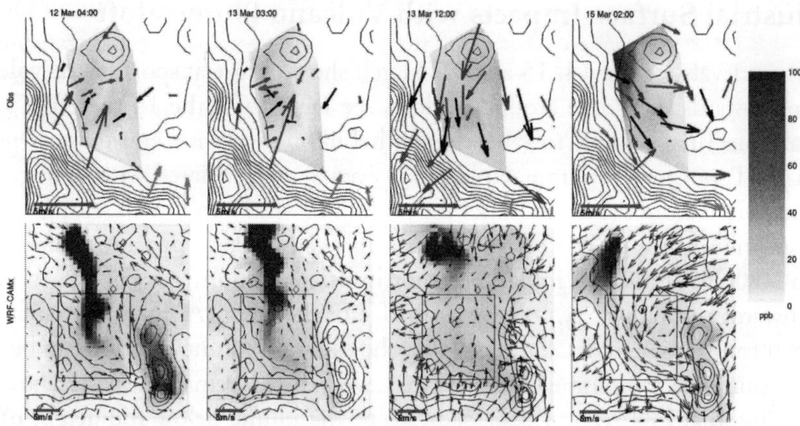

Figure 11. Measured winds and hourly surface concentration of SO_2 in the MCMA and simulated winds (WRF) and SO_2 (CAMx, all sources) in the basin. Observed winds are coloured according to network: green -RAMA, blue -SMN, magenta -T1, tan -St. Ana, orange -Tenango del Aire. Terrain contours every 100 m (top) and 500 m (bottom), measurement domain shown as a box in model domain, red diamonds show

On March 13, there is a sharp plume at Santa Ana (SATL) at 02:00 in the morning. This occurs during strong drainage flows from the south at SATL. Levels of SO_2 rise across the basin suggesting a volcanic impact, although the excess SO_2 above 20 ppb could be due to a local source. Vertical stratification of the plume probably prevented impacts at the stations on the basin floor as well as at Tenango del Aire. Later in the day, there are strong winds from the north providing clear evidence of a Tula impact. This is the first day with SO_2 data at PTP. Concentrations similar to those at the basin floor show that the plume is well mixed within the boundary layer. SO_2 concentrations everywhere drop after sunset as the winds aloft start coming from the east and then from the south.

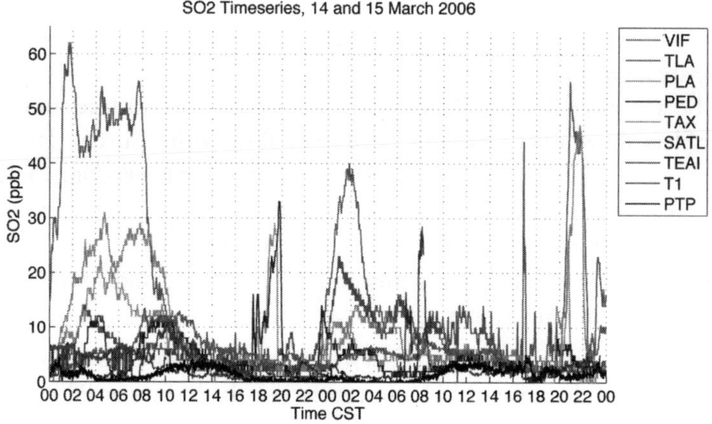

Figure 12. Time series of measured SO_2 concentrations for 14 and 15 March.

Industrial Surface Impacts with Volcano Plume aloft

OMI retrievals for the 14, 15 and 16 March show clear transport of the volcano plume towards the north around midday, see Fig. 5. On the 14 there is a weak signal from Tula, possibly towards the south, but on the 15 it seems to move north with the Popocatepetl plume. Cold Surge events are associated with strong, cold surface winds from the north under the prevailing winds aloft, with strong vertical stratification of the flow. On 14 March, this is clearly reflected in a Tula plume from 00:00 to 10:00 that impacts the northwestern stations, see Fig. 12. It moves southwards through the whole western edge of the MCMA, reaching Santa Ana at 08:00 and Tenango del Aire at 09:00. The extent of plume dilution can be seen before sunrise as the plume moves south. Levels at PTP remain near zero however indicating that low vertical dispersion keeps the plume below the height of the stations (750 m above the basin floor). After sunrise, vertical dispersion rapidly dilutes the plume everywhere. From 18:00 to 20:00, there is a short impact as a plume skims the western edge of the basin. Wind vectors indicate that this is due to direct transport from Tula, similar to the situation on 2 March. Overall therefore, the Popocatepetl plume was not detected at the surface—although it might have impacted the eastern side of the basin.

On 15 March, transport is slightly more to the northwest suggesting possible volcano impacts over the city. The time series show a clear Tula plume starting at midnight, moving south into the basin without impacting PTP aloft, and then diluting during the day, see Figs. 11 and 12. There is a second Tula impact at 20:00 that is similar in structure. In between, there is no surface evidence of a Popocatepetl impact that would be significantly above the background levels. In particular, PTP does not register any SO_2 that cannot be readily attributed to local transport from vertical mixing.

On 16 March, SO_2 plumes impact mainly the north of the MCMA, see Fig. 13. The OMI retrievals show the volcano plume moving northwest over the basin and there are high columns both to the north and to the south of Tula. This would seem to be a perfect day for Popocatepetl impacts at the surface, but the radar wind profilers show a very strongly decoupled flow with a surface layer moving south and a layer above the boundary layer moving north, see Fig. 14. The high SO_2 is therefore clearly from Tula, moving around during the day with variable impacts. The pattern in the OMI retrievals around Tula is most likely due to the superposition of the two plumes, with the Tula plume moving south and the Popocatepetl plume moving north.

There is a convergence line in the north of the city which is accompanied by rapid increase and decrease in SO_2 concentrations including at PTP from 16:00 to 18:00. Later in the evening, at 21:00, there is a short lived spike at PTP as the plume moves over the north of the MCMA. This seems like a meteorological curiosity, where the edge of the plume was transported briefly over the mountain as it moved west but the bulk of the plume behind has gone around because of the nighttime stability, leaving low SO_2 conditions at PTP. The same feature was observed at 23:00 and then again at 05:00 the following day.

Conditions on 17 March are very similar to the previous two days, with the radar wind profilers showing winds from the north at the surface and from the south aloft, see Fig. 16. From 19:00 to 20:00, there is uniform flow from the south, and this coincides with the lowest levels of SO_2 of the day, further indicating that Tula, rather than Popocatepetl, is the source of the SO_2.

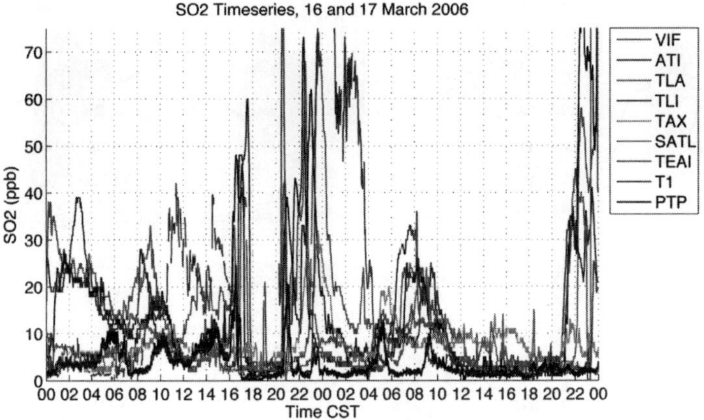

Figure 13. Time series of measured SO_2 concentrations for 16 and 17 March.

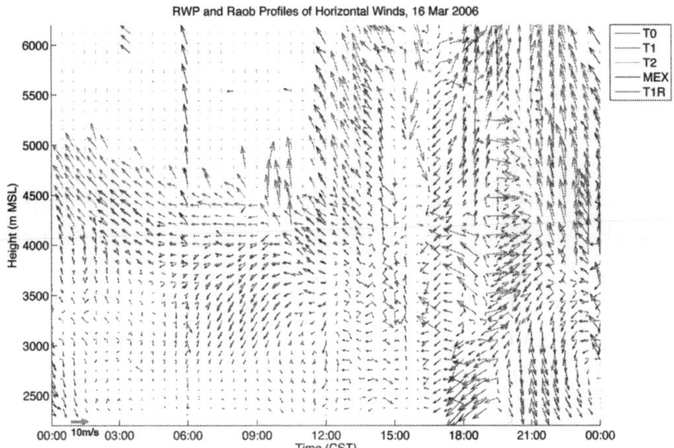

Figure 14. Radar wind profiler and Radiosonde wind vectors at GSMN, T0, T1 and T2 on 16 March 2006. This shows the evolution of horizontal winds with height, an arrow pointing up means winds moving north. MEX are the radiosondes launched at GSMN, T1R are the radiosondes launched at T1.

Double Impacts

Following a couple of hours of clean southerly air on 17 March, there is a north-erly surface layer that brings with it the Tula plume and the highest SO_2 levels of the campaign on 18 March. Levels drop to around 50 ppb during the morning and by 14:00, strong winds from the south have cleaned the basin. Combined with the impact of a holiday week-end, this now sets the stage for the cleanest day of the campaign on 19 March. Unfortunately, the plumes cannot be seen in the OMI retrievals for the 18 March because the MCMA is on the edge of the swath.

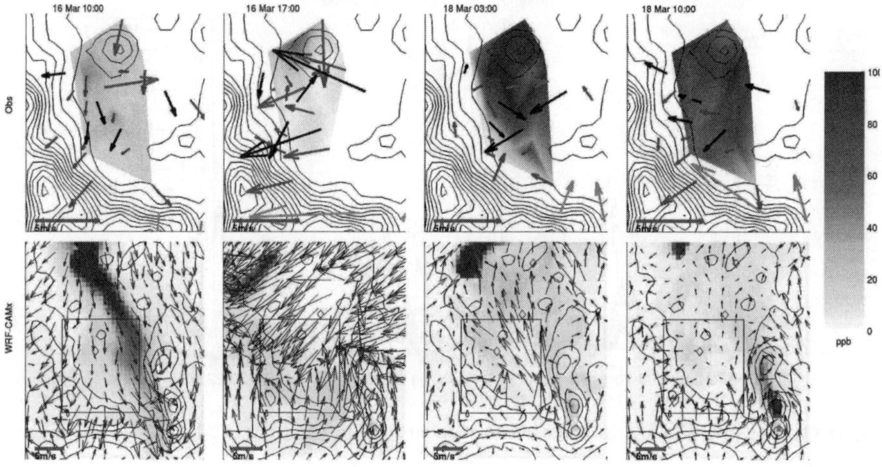

Figure 15. Measured winds and hourly surface concentration of SO2 in the MCMA and simulated winds (WRF) and SO_2 (CAMx, all sources) in the basin. Observed winds are coloured according to network: green -RAMA, blue -SMN, magenta -T1, tan -St. Ana, orange -Tenango del Aire. Terrain contours every 100 m (top) and 500 m (bottom), measurement domain shown as a box in model domain, red diamonds show T0, T1, T2 and Tula.

Figure 17 shows the time series at selected stations. One minute concentrations reach above 200 ppb at a number of stations before sunrise. At PTP, levels rise to 10 ppb at 01:30 and remain at this level until they increase to 70 ppb from 05:30 to 07:00. By this time, the levels have dropped at the stations below, and it is only after sunrise that the surface concentrations rise again to values between 30 and 50 ppb. There is a sudden spike at T1 from 09:00 to 10:00 after which the concentrations return to similar values as other northern sires, and then shortly before 12:00 the concentrations drop to under 10 ppb.

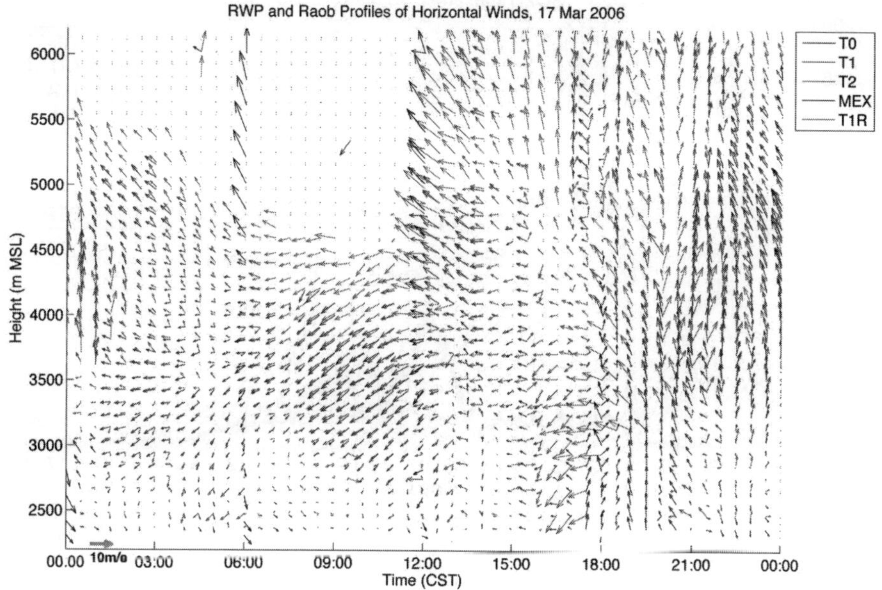

Figure 16. Radar wind profiler and Radiosonde wind vectors at GSMN, T0, T1 and T2 on 17 March 2006. This shows the evolution of horizontal winds with height, an arrow pointing up means winds moving north. MEX are the radiosondes launched at GSMN, T1R are the radiosondes launched at T1.

The surface wind vectors provide a clear picture of transport from Tula around PTP, see Fig. 15. There is a strong gap flow indicated by southerly winds at the eastern stations. The radar wind profiler data clearly show a layer 500 m thick or less that is from the north west, with strong southerly flow aloft, see Fig. 18. By 04:00, the southerly flow at T0 has caused SO_2 concentrations at TAX in the south. At PTP, the concentrations rose at 01:30 when the winds started to come from the southsoutheast at 3000 m m.s.l. The increase around 06:00 coincides with winds turning slightly to be more from the direction of the volcano. The concentrations are now higher aloft than anywhere at the surface, and it is only after sunrise, with the start of vertical mixing, that concentrations rise to levels comparable to those at PTP. At T1, the shallow northerly layer lasts longer and includes a brief period of westerly winds that brings high concentrations from 09:00 to 10:00 before the concentrations subsequently return to levels comparable to the other northern stations. Fig. 15 shows strong surface wind vectors from the east at 10:00 with a strong gap flow from Tenango del Aire.

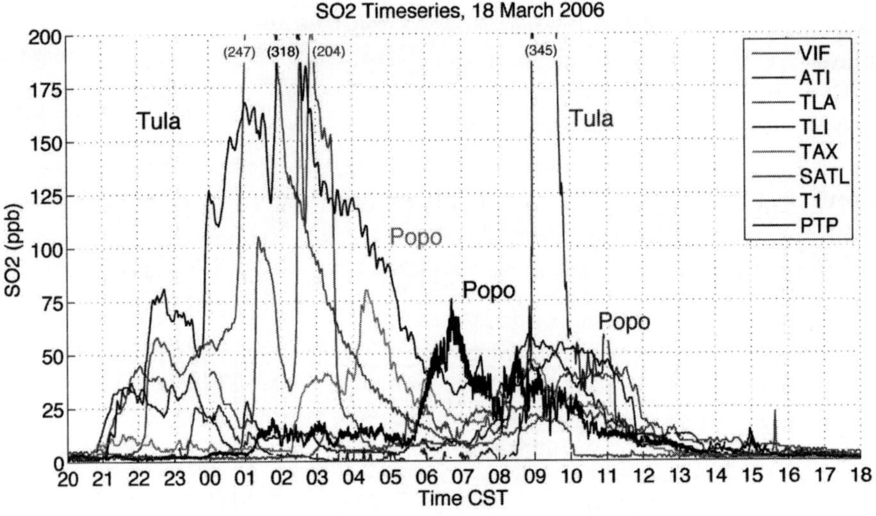

Figure 17. Time series of measured SO$_2$ concentrations for 18 March. Maximum of values off the chart shown in brackets. Tula and Popo plumes labelled with the colour of the corresponding site.

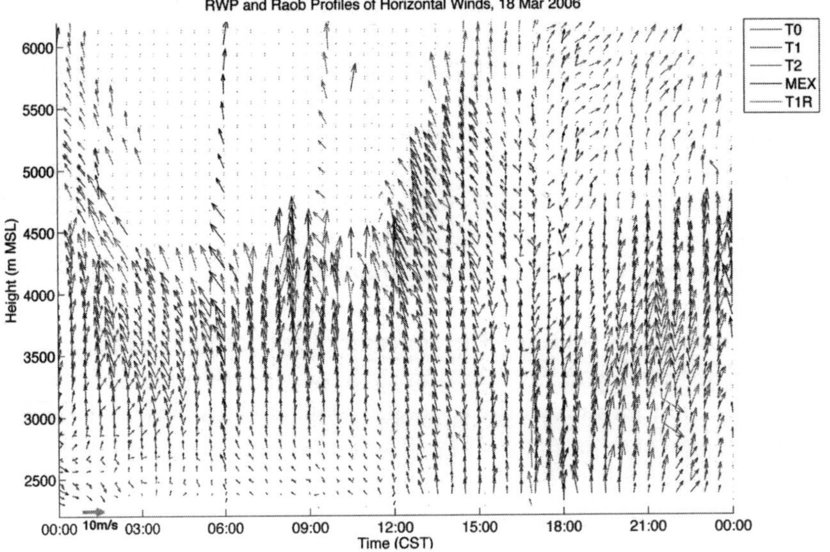

Figure 18. Radar wind profiler and Radiosonde wind vectors at GSMN, T0, T1 and T2 on 18 March 2006. This shows the evolution of horizontal winds with height, an arrow pointing up means winds moving north. MEX are the radiosondes launched at GSMN, T1R are the radiosondes launched at T1.

Combining the evidence, one can see that the day starts with a Tula plume at the surface. Concentrations are higher than usual because the surface layer is shallower than normal, suppressing vertical dispersion. Meanwhile, the volcano plume is transported over the basin, causing impacts at PTP starting at 01:30, at TAX after 02:00 and at SATL after 03:00. The impacts subsequently increase first at TAX, then at PTP. After sunrise vertical mixing starts and the plume is mixed down to the surface at the northern stations. By 12:00, the winds now blow due north and the volcano plume no longer impacts the basin. The 09:00 spike at T1 must be of Tula origin, when one considers the surface layer flow in the radar wind profiler data, and the concentrations that are higher than any other station, especially PTP above it.

In terms of simulations, the early morning gap flow is over-represented, preventing the formation of a shallow northerly layer, sweeping the basin clean and causing the Tula plume to vent to the north. The easterly basin flows in the morning are under-represented, limiting the simulated Popocatepetl impacts to the eastern edge of the basin. This case provides a clear example of how the model tries to represent features of the basin flow, but discrepancies in the detailed representation lead to large differences in the industrial and the volcanic plume.

Conclusions

Detailed case by case analysis of the SO2 plumes in the MCMA using both meteorological observations and numerical simulations suggest that most of the large peaks observed during MILAGRO originate in the Tula industrial complex. In comparison, the Popocatepetl volcano had smaller impacts on fewer occasions. Numerical models of plume dispersion were able to simulate impacts from both plumes. The analysis confirms past meteorological studies and illustrates the night-time flow into the basin from the north under stable conditions. Vertical mixing during the day was observed at PTP occurring from both the ground up, in the case of the industrial plume, and from the layer above down to the ground in the case of the volcano plume. 18 March in particular illustrates how complex wind patterns can be in the MCMA with the highest impacts of the campaign from both sources occurring in immediate succession. There were both false positives and negatives of simulated Tula impacts in the basin which were shown to result from small variations in wind direction. Subtle changes in the strength of the down-valley flow from Pachuca to the northeast, and the up-valley flow to Tula also from the northeast could totally change the resulting plume transport at Tula between going south towards the MCMA or being vented northwards. While this provides a cautionary tale in the evaluation of model output, it also shows that by interpreting the results using both data and models a reasonable degree of confidence can be reached.

Table 2. Percentage of mean SO_2 concentrations in the MCMA due to each source group during March 2006. "WRFa" and "MM5" refer to simulations discussed in de Foy et al. (2009), Popocatepetl releases at 4438 m and 6438 m correspond to 0 and 2000 m release heights above the model surface, Tula no plume rise has releases set at stack height, which made a negligible difference.

Model Run	Urban	Popocatepetl	Tula Ind. Complex
WRF	37%	10%	53%
"WRFa"	46%	14%	40%
"MM5"	39%	5%	57%
Popo 4438 m release	34%	18%	48%
Popo 6438 m release	37%	3%	57%
Tula no plume rise	37%	10%	53%

Table 2 shows the percentage of the mean simulated SO_2 impacts in the MCMA due to the three different source groups in CAMx. This was calculated by summing SO_2 concentrations from CAMx simulations with individual source groups for every hour of the campaign. In this instance we are interested in mean values as way of identifying the relative importance of the emission inventories, but using the maxima (see Fig. 2c) yields the same results. For comparison, impact fractions are presented using the different simulations discussed in de Foy et al. (2009), as well as for different release heights for the Popocatepetl and the Tula industrial complex. As expected, the lower volcano release height leads to higher estimated impacts and conversely the higher release lowers the impacts. The Tula plume rise however makes no significant difference to the results, suggesting that the findings are robust with respect to estimates of plume parameters. Overall, about one tenth of the SO_2 in the MCMA during MILAGRO could be due to the volcano with the rest split between the Tula industrial complex and local urban sources. This finding is not significantly affected by model setup. This represents an increase over the 20% of impacts thought to be caused by Tula during MCMA2003. Such variation can easily be caused by changes in wind patterns, with the higher numbers during MILAGRO due to the week of strong northerly flows at the beginning of March 2006. Furthermore, both of these episodes are during the warm dry season when the westerlies are the prevailing winds aloft. The numbers could change significantly during the wet season when the weak, easterly trade winds dominate aloft. Further analysis and modelling work would be needed to explore the impacts of more transport to the east and increased stratification in the basin due to weaker winds.

Satellite retrievals of total SO_2 columns from OMI were shown to detect both the industrial and the volcanic plumes, and were in broad agreement with simulations. The OMI data was of sufficiently high resolution to identify some of the features of the plume transport. Furthermore, oversampling the swath data to a fine grid (3 by 3 km) and averaging over one month yielded a higher resolution image of the average plume transport than could be seen in the original swath data. During one episode, the volcano plume was clearly identified aloft but no impacts were detected at the surface, and on another day, the plume was split by the southeastern gap flow with part of it mixing down into the basin. Expanding the current analysis to a longer time period would show variations in the emission strengths and expanding it to a larger domain would identify regional transport and transformation of the plume.

The formation of sulfate aerosols is a clear application of this work, although it is not addressed here. Dunn et al. (2004) identified nucleation events during MCMA-2003 at both Santa Ana, a rural site, and CENICA, an urban site. These accompanied clean air events with high SO_2 concentrations and were particularly pronounced during periods of high relative humidity, as had already been described by Raga et al. (1999) and Baumgardner et al. (2000). The events took place during periods of northerly flow, where it is likely that the SO_2 transport was from the Tula complex.

During MILAGRO, Kleinman et al. (2008) identified sulfate formation rates within the polluted air mass that were consistent with Salcedo et al. (2006). On a more regional basis, three different SO_2 plumes with sulfate formation showed different ages and origins (DeCarlo et al., 2008), in accord with the present description of surface plumes from Tula mixing with the urban air mass and elevated plumes from the volcano. Future work could use the known plumes identified in this paper to explore aerosol formation measured during MILAGRO at both T0 and PTP.

SO_2 emissions from biomass burning could further impact the MCMA and especially sulfate aerosol formation Baumgardner et al. (2009). Estimates of the maximum total biomass emissions of SO_2 for the Yucatan during March 2006 are equivalent to emissions on the order of 5 kg/s based on Wiedinmyer et al. (2006) and would increase to 30 kg/s using the emission factors in Yokelson et al. (2009) making this source comparable to the volcano. Dilution would however be much larger for the Yucatan sources given their greater distance from the MCMA.

We have shown that SO_2 serves as a useful tracer for plume transport in the MCMA. It is hoped that this will facilitate interpretation of other measurements by comparing the behaviour of different species during the events, and additionally, by making use of similar analyses for different plumes.

Acknowledgements

We are indebted to the large number of people involved in the MILAGRO field campaign as well as those involved in monitoring in the Mexico City basin without which this study would not exist. We would like to thank A. Retama, C. Ortuno, M. Jaimes, G. Granados and the operators and analyst personnel of the "Red Automatica de Monitoreo Atmosferico del Gobierno del Distrito Federal" for administering and gathering the surface air quality and meteorological data. M. Rosengaus, J. L. Razo, J. Olalde and P. Garcia of the Mexican National Meteorological Service kindly provided the EMA and radiosonde data. We are very grateful for the radar wind profiler data provided by S. J. Paech, D. Phillips and J. T. Walters of the University of Alabama in Huntsville for T0 and by R. L. Coulter and T. J. Martin of Argonne National Laboratory for T1. We thank B. Lamb of Washington State University for the loan of the SO2 monitor used aboard the Aerodyne mobile laboratory at PTP. We acknowledge the Goddard Earth Sciences Data and Information Services Center for providing the OMI data and J. E. Fritts for helpful discussions.

The MILAGRO field campaign was supported by the Comisi′on Ambiental Metropolitana of Mexico, NSF, DOE, NASA and USDA Forest Service among others. The financial support of the US National Science Foundation (awards ATM-0810931 and ATM-0810950) and the Molina Center for Strategic Studies in Energy and the Environment is gratefully acknowledged for this work. Edited by H. Singh

References

1. Baumgardner, D., Raga, G. B., Kok, G., Ogren, J., Rosas, I., Baez, A., and Novakov, T.: On the evolution of aerosol properties at a mountain site above Mexico City, J. Geophys. Res., 105, 22243– 22253, 2000.

2. Baumgardner, D., Grutter, M., Allan, J., Ochoa, C., Rappenglueck, B., Russell, L. M., and Arnott, P.: Physical and chemical properties of the regional mixed layer of Mexico's Megapolis, Atmos. Chem. Phys., 9, 5711–5727, 2009, http://www.atmos-chem-phys.net/9/5711/2009/.

3. Bossert, J. E.: An investigation of flow regimes affecting the Mexico City region, J. Appl. Meteorol., 36, 119–140, 1997.

4. Boucher, A., Kyriakidis, P. C., and Cronkite-Ratcliff, C.: Geostatistical solutions for super-resolution land cover mapping, IEEE T. Geosci. Remote Sens., 46, 272–283, doi:{10.1109/TGRS.2007. 907102}, 2008.

5. Capel, D.: Image Mosaicing and Super-resolution, Springer, 12–16, 2004. Carn, S. A., Krueger, A. J., Krotkov, N. A., Yang, K., and Levelt, P. F.: Sulfur dioxide emissions from Peruvian copper smelters detected by the Ozone Monitoring Instrument, Geophys. Res. Lett., 34, L09801, doi:{10.1029/2006GL029020}, 2007.

6. Chow, J. C., Watson, J. G., Edgerton, S. A., Vega, E., and Ortiz, E.: Spatial differences in outdoor PM10 mass and aerosol composition in Mexico City, J. Air Waste Manage. Assoc., 52, 423–434, 2002.

7. Comision Ambiental Metropolitana: Inventario de Emisiones de la Zona Metropolitana del Valle de Mexico, Tech. Rep. (Web), Secretar´ıa del Medio Ambiente, Gobierno de Mexico, 2008.

8. Corr, C. A., Krotkov, N., Madronich, S., Slusser, J. R., Holben, B., Gao, W., Flynn, J., Lefer, B., and Kreidenweis, S. M.: Retrieval of aerosol single scattering albedo at ultraviolet wavelengths at the T1 site during MILAGRO, Atmos. Chem. Phys., 9, 5813– 5827, 2009, http://www.atmos-chem-phys.net/9/5813/2009/.

9. de Foy, B., Caetano, E., Maga´na, V., Zit´acuaro, A., C´ardenas, B., Retama, A., Ramos, R., Molina, L. T., and Molina, M. J.: Mexico City basin wind circulation during the MCMA-2003 field campaign, Atmos. Chem. Phys., 5, 2267–2288, 2005, http://www.atmos-chem-phys.net/5/2267/2005/.

10. de Foy, B., Clappier, A., Molina, L. T., and Molina, M. J.: Distinct wind convergence patterns in the Mexico City basin due to the interaction of the gap winds with the synoptic flow, Atmos. Chem. Phys., 6, 1249–1265, 2006a.

11. de Foy, B., Molina, L. T., and Molina, M. J.: Satellite-derived land surface parameters for mesoscale modelling of the Mexico City basin, Atmos. Chem. Phys., 6, 1315–1330, 2006b.

12. de Foy, B., Varela, J. R., Molina, L. T., and Molina, M. J.: Rapid ventilation of the Mexico City basin and regional fate of the urban plume, Atmos. Chem. Phys., 6, 2321–2335, 2006c.

13. de Foy, B., Lei, W., Zavala, M., Volkamer, R., Samuelsson, J., Mellqvist, J., Galle, B., Martinez, A. P., Grutter, M., Retama, A., and Molina, L. T.: Modelling constraints on the emission inventory and on vertical dispersion for CO and SO2 in the Mexico City Metropolitan Area using Solar FTIR and zenith sky UV spectroscopy, Atmos. Chem. Phys., 7, 781–801, 2007, http://www.atmos-chem-phys.net/7/781/2007/.

14. de Foy, B., Fast, J. D., Paech, S. J., Phillips, D., Walters, J. T., Coulter, R. L., Martin, T. J., Pekour, M. S., Shaw, W. J., Kastendeuch, P. P., Marley, N. A., Retama, A., and Molina, L. T.: Basin-scale wind transport during the MILAGRO field campaign and comparison to climatology using cluster analysis, Atmos. Chem. Phys., 8, 1209–1224, 2008, http://www.atmos-chem-phys.net/8/1209/2008/.

15. de Foy, B., Zavala, M., Bei, N., and Molina, L. T.: Evaluation of WRF mesoscale simulations and particle trajectory analysis for the MILAGRO field

campaign, Atmos. Chem. Phys., 9, 4419– 4438, 2009, http://www.atmos-chem-phys.net/9/4419/2009/.

16. DeCarlo, P. F., Dunlea, E. J., Kimmel, J. R., Aiken, A. C., Sueper, D., Crounse, J., Wennberg, P. O., Emmons, L., Shinozuka, Y., Clarke, A., Zhou, J., Tomlinson, J., Collins, D. R., Knapp, D., Weinheimer, A. J., Montzka, D. D., Campos, T., and Jimenez,

17. J. L.: Fast airborne aerosol size and chemistry measurements above Mexico City and Central Mexico during the MILAGRO campaign, Atmos. Chem. Phys., 8, 4027–4048, 2008.

18. Delgado-Granados, H., Gonzalez, L. C., and Sanchez, N. P.: Sulfur dioxide emissions from Popocatepetl volcano (Mexico): case study of a high-emission rate, passively degassing erupting volcano, J. Volcanol. Geotherm. Res., 108, 107–120, 2001.

19. Doran, J. C., Barnard, J. C., Arnott, W. P., Cary, R., Coulter, R., Fast, J. D., Kassianov, E. I., Kleinman, L., Laulainen, N. S., Martin, T., Paredes-Miranda, G., Pekour, M. S., Shaw, W. J., Smith,

20. D. F., Springston, S. R., and Yu, X. Y.: The T1-T2 study: evolution of aerosol properties downwind of Mexico City, Atmos. Chem. Phys., 7, 1585–1598, 2007, http://www.atmos-chem-phys.net/7/1585/2007/.

21. Dunn, M., Jimenez, J., Baumgardner, D., Castro, T., McMurry, P., and Smith, J.: Measurements of Mexico City nanoparticle size distributions: Observations of new particle formation and growth, Geophys. Res. Lett., 31, L10102, doi:{10.1029/ 2004GL019483}, 2004.

22. ENVIRON: CAMx, comprehensive air quality model with extensions, User's Guide, Tech. Rep. Version 4.50, ENVIRON International Corporation, 2008.

23. Fast, J. D. and Zhong, S. Y.: Meteorological factors associated with inhomogeneous ozone concentrations within the Mexico City basin, J. Geophys. Res.-Atmos., 103, 18927–18946, 1998.

24. Fast, J. D., de Foy, B., Rosas, F. A., Caetano, E., Carmichael, G., Emmons, L., McKenna, D., Mena, M., Skamarock, W., Tie, X., Coulter, R. L., Barnard, J. C., Wiedinmyer, C., and Madronich, S.: A meteorological overview of the MILAGRO field campaigns, Atmos. Chem. Phys., 7, 2233–2257, 2007, http://www.atmos-chem-phys.net/7/2233/2007/.

25. Grutter, M., Basaldud, R., Rivera, C., Harig, R., Junkerman, W., Caetano, E., and Delgado-Granados, H.: SO2 emissions from Popocatepetl volcano: emission rates and plume imaging using optical remote sensing techniques,

Atmos. Chem. Phys., 8, 6655–6663, 2008, http://www.atmos-chem-phys. net/8/6655/2008/.

26. Hong, S. Y., Noh, Y., and Dudhia, J.: A new vertical diffusion package with an explicit treatment of entrainment processes, Mon. Weather Rev., 134, 2318–2341, 2006.

27. Jauregui, E.: Local wind and air pollution interaction in the Mexico basin, Atmosfera, 1, 131–140, 1988.

28. Jazcilevich, A. D., Garcia, A. R., and Caetano, E.: Locally induced surface air confluence by complex terrain and its effects on air pollution in the valley of Mexico, Atmos. Environ., 39, 5481– 5489, 2005.

29. Jimenez, J. C., Raga, G. B., Baumgardner, D., Castro, T., Rosas, I., Baez, A., and Morton, O.: On the composition of airborne particles influenced by emissions of the volcano Popocatepetl in Mexico, Nat. Hazards, 31, 21–37, 2004.

30. Juarez, A., Gay, C., and Flores, Y.: Impact of the Popocatepetl's volcanic activity on the air quality of Puebla City, Mexico, Atmosfera, 18, 57–69, 2005.

31. Kain, J. S.: The Kain-Fritsch convective parameterization: An update, J. Appl. Meteorol., 43, 170–181, 2004.

32. Khokhar, M., Frankenberg, C., Van Roozendael, M., Beirle, S., Kuhl, S., Richter, A., Platt, U., and Wagner, T.: Satellite observations of atmospheric SO_2 from volcanic eruptions during the time-period of 1996–2002, Adv. Space Res., 36, 879–887, doi:{10.1016/j.asr.2005.04.114}, 2005.

33. Kleinman, L. I., Springston, S. R., Daum, P. H., Lee, Y. N., Nunnermacker, L. J., Senum, G. I., Wang, J., Weinstein-Lloyd, J., Alexander, M. L., Hubbe, J., Ortega, J., Canagaratna, M. R., and Jayne, J.: The time evolution of aerosol composition over the Mexico City plateau, Atmos. Chem. Phys., 8, 1559–1575, 2008, http://www.atmos-chem-phys.net/8/1559/2008/.

34. Kolb, C. E., Herndon, S. C., McManus, B., Shorter, J. H., Zahniser, M. S., Nelson, D. D., Jayne, J. T., Canagaratna, M. R., and Worsnop, D. R.: Mobile laboratory with rapid response instruments for real-time measurements of urban and regional trace gas and particulate distributions and emission source characteristics, Environ. Sci. Technol., 38, 5694–5703, 2004.

35. Krotkov, N., Carn, S., Krueger, A., Bhartia, P., and Yang, K.: Band residual difference algorithm for retrieval of SO_2 from the aura Ozone Monitoring Instrument (OMI), IEEE T. Geosci. Remote Sens., 44, 1259–1266, doi:{10.1109/ TGRS.2005.861932}, 2006.

36. Krotkov, N. A., McClure, B., Dickerson, R. R., Carn, S. A., Li, C., Bhartia, P. K., Yang, K., Krueger, A. J., Li, Z., Levelt, P. F., Chen, H., Wang, P., and

Lu, D.: Validation of SO2 retrievals from the Ozone Monitoring Instrument over NE China, J. Geophys. Res.-Atmos., 113, doi:10.1029/2007JD008818, 2008.

37. Li, F., Jia, X., and Fraser, D.: Superresolution Reconstruction of Multispectral Data for Improved Image Classification, IEEE T. Geosci. Remote Sens., 6, 689–693, doi:{10.1109/LGRS.2009. 2023604}, 2009.

38. Loyola, D., van Geffen, J., Erbertseder, T., Roozendael, M. V., Thomas, W., Zimmer, W., and Wisskirchen, K.: Satellite-based detection of volcanic sulphur dioxide from recent eruptions in Central and South America, Adv. Geosci., 14, 35–40, 2008, http://www.adv-geosci.net/14/35/2008/.

39. Marquez, C., Castro, T., Muhlia, A., Moya, M., Martinez-Arroyo, A., and Baez, A.: Measurement of aerosol particles, gases and flux radiation in the Pico de Orizaba National Park, and its relationship to air pollution transport, Atmos. Environ., 39, 3877– 3890, 2005.

40. Molina, L. T., Kolb, C. E., de Foy, B., Lamb, B. K., Brune, W. H., Jimenez, J. L., Ramos-Villegas, R., Sarmiento, J., Paramo-Figueroa, V. H., Cardenas, B., Gutierrez-Avedoy, V., and Molina, M. J.: Air quality in North America's most populous city—overview of the MCMA-2003 campaign, Atmos. Chem. Phys., 7, 2447–2473, 2007, http://www.atmos-chem-phys.net/7/2447/2007/.

41. Moya, M., Castro, T., Zepeda, M., and Baez, A.: Characterization of size-differentiated inorganic composition of aerosols in Mexico City, Atmos. Environ., 37, 3581–3591, 2003.

42. O'Brien, J. J.: A note on the vertical structure of the eddy exchange coefficient in the planetary boundary layer, J. Atmos. Sci., 27, 1214–1215, 1970.

43. OMI Team: Ozone Monitoring Instrument (OMI) Data User's Guide, Tech. Rep. OMI-DUG-3.0, NASA, http://so2.umbc.edu/ omi/omi docs.html, 2009.

44. Oreskes, N.: Evaluation (not validation) of quantitative models, Environ. Health Perspect., 106, 1453–1460, 1998.

45. Pyle, D. M. and Mather, T. A.: The regional influence of volcanic emissions from Popocateptl, Mexico: Discussion of "Measurement of aerosol particles, gases and flux radiation in the Pico de Orizaba Nacional Park, and its relationship to air pollution transport", Marquez et al., 2005, Atmos. Environ., 39, 3877-3890, Atmos. Environ., 39, 6475–6478, 2005.

46. Raga, G. B., Kok, G. L., Baumgardner, D., Baez, A., and Rosas, I.: Evidence for volcanic influence on Mexico City aerosols, Geophys. Res. Lett., 26, 1149–1152, 1999.

47. Rivera, C., Sosa, G., W¨ohrnschimmel, H., de Foy, B., Johansson, M., and Galle, B.: Tula industrial complex (Mexico) emissions of SO2 and NO2 during the MCMA 2006 field campaign using a mobile mini-DOAS system, Atmos. Chem. Phys., 9, 6351–6361, 2009, http://www.atmos-chem-phys.net/9/6351/2009/.

48. Salcedo, D., Onasch, T., Dzepina, K., Canagaratna, M., Zhang, Q., Huffman, J., DeCarlo, P., Jayne, J., Mortimer, P., Worsnop, D., Kolb, C., Johnson, K., Zuberi, B., Marr, L., Volkamer, R., Molina, L., Molina, M., Cardenas, B., Bernabe, R., Marquez, C., Gaffney, J., Marley, N., Laskin, A., Shutthanandan, V., Xie, Y., Brune, W., Lesher, R., Shirley, T., and Jimenez, J.: Characterization of ambient aerosols in Mexico City during the MCMA2003 campaign with Aerosol Mass Spectrometry: results from the CENICA Supersite, Atmos. Chem. Phys., 6, 925–946, 2006, http://www.atmos-chem-phys.net/6/925/2006/.

49. Secretar´ıa del Medio Ambiente del Gobierno del Distrito Federal: Inventario de Emisiones de Contaminantes Criterio de la Zona Metropolitana del Valle de M´exico, Tech. Rep. (Web), Secretar´ıa del Medio Ambiente, Gobierno del Distrito Federal, Mexico, 2008.

50. Skamarock, W. C., Klemp, J. B., Dudhia, J., Gill, D. O., Barker, D. M., Wang, W., and Powers, J. G.: A Description of the Advanced Research WRF Version 2, Tech. Rep. NCAR/TN468+STR, NCAR, 2005.

51. Taylor, K. E.: Summarizing multiple aspects of model performance in a single diagram., J. Geophys. Res.-Atmos., 106, 7183–7192, 2001.

52. Wiedinmyer, C., Quayle, B., Geron, C., Belote, A., McKenzie, D., Zhang, X., O'Neill, S., and Wynne, K. K.: Estimating emissions from fires in North America for air quality modeling, Atmos. Environ., 40, 3419–3432, doi:{10.1016/j. atmosenv.2006.02.010}, 2006.

53. Williams, M. D., Brown, M. J., Cruz, X., Sosa, G., and Streit, G.: Development and testing of meteorology and air dispersion models for Mexico City, Atmos. Environ., 29(21), 2929–2960, 1995.

54. Willmott, C. J.: Some comments on the evaluation of model performance, B. Am. Meteor. Soc., 63, 1309–1313, 1982.

55. Yang, K., Krotkov, N. A., Krueger, A. J., Carn, S. A., Bhartia, P. K., and Levelt, P. F.: Retrieval of large volcanic SO2 columns from the Aura Ozone Monitoring Instrument: Comparison and limitations, J. Geophys. Res.-Atmos., 112, D24S43, doi: {10.1029/2007JD008825}, 2007.

56. Yang, K., Krotkov, N. A., Krueger, A. J., Carn, S. A., Bhartia, P. K., and Levelt, P. F.: Improving retrieval of volcanic sulfur dioxide from backscattered UV satellite

observations, Geophys. Res. Lett., 36, L03102, doi:{10.1029/2008GL036036}, 2009.

57. Yokelson, R. J., Crounse, J. D., DeCarlo, P. F., Karl, T., Urbanski, S., Atlas, E., Campos, T., Shinozuka, Y., Kapustin, V., Clarke, A. D., Weinheimer, A., Knapp, D. J., Montzka, D. D., Holloway, J., Weibring, P., Flocke, F., Zheng, W., Toohey, D., Wennberg, P. O., Wiedinmyer, C., Mauldin, L., Fried, A., Richter, D., Walega, J., Jimenez, J. L., Adachi, K., Buseck, P. R., Hall, S. R., and Shetter, R.: Emissions from biomass burning in the Yucatan, Atmos. Chem. Phys., 9, 5785–5812, 2009, http://www.atmos-chem-phys.net/9/5785/2009/.

58. Zangl, G., Chimani, B., and H¨aberli, C.: Numerical Simulations of the Foehn in the Rhine Valley on 24 October 1999 (MAP IOP 10), Mon. Weather Rev., 132, 368–389, 2004.

Characterization of a β-Glucanase Produced by Rhizopus microsporus var. microsporus, and Its Potential for Application in the Brewing Industry

Klecius R. Silveira Celestino, Ricardo B. Cunha
and Carlos R. Felix

ABSTRACT

Background

In the barley malting process, partial hydrolysis of β-glucans begins with seed germination. However, the endogenous 1,3-1,4-β-glucanases are heat in-activated, and the remaining high molecular weight β-glucans may cause severe problems such as increased brewer mash viscosity and turbidity. Increased viscosity impairs pumping and filtration, resulting in lower efficiency, reduced yields of extracts, and lower filtration rates, as well as the appearance of gelatinous precipitates in the finished beer. Therefore, the use of exogenous β-glucanases to reduce the β-glucans already present in the malt barley is highly desirable.

Results

The zygomycete microfungus Rhizopus microsporus var. microsporus secreted substantial amounts of β-glucanase in liquid culture medium containing 0.5% chitin. An active protein was isolated by gel filtration and ion exchange chromatographies of the β-glucanase activity-containing culture supernatant. This isolated protein hydrolyzed 1,3-1,4-β-glucan (barley β-glucan), but showed only residual activity against 1,3-β-glucan (laminarin), or no activity at all against 1,4-β-glucan (cellulose), indicating that the R. microsporus var. microsporus enzyme is a member of the EC 3.2.1.73 category. The purified protein had a molecular mass of 33.7 kDa, as determined by mass spectrometry. The optimal pH and temperature for hydrolysis of 1,3-1,4-β-glucan were in the ranges of 4–5, and 50–60°C, respectively. The Km and Vmax values for hydrolysis of β-glucan at pH 5.0 and 50°C were 22.39 mg.mL^{-1} and 16.46 mg.min^{-1}, respectively. The purified enzyme was highly sensitive to Cu^{+2}, but showed less or no sensitivity to other divalent ions, and was able to reduce both the viscosity and the filtration time of a sample of brewer mash. In comparison to the values determined for the mash treated with two commercial glucanases, the relative viscosity value for the mash treated with the 1,3-1,4-β-glucanase produced by R. microsporus var. microsporus. was determined to be consistently lower.

Conclusion

The zygomycete microfungus R. microsporus var. microsporus produced a 1,3-1,4-β-D-glucan 4-glucanhydrolase (EC 3.2.1.73) which is able to hydrolyze β-D-glucan that contains both the 1,3- and 1,4-bonds (barley β-glucans). Its molecular mass was 33.7 kDa. Maximum activity was detected at pH values in the range of 4–5, and temperatures in the range of 50–60°C. The enzyme was able to reduce both the viscosity of the brewer mash and the filtration time, indicating its potential value for the brewing industry.

Background

1,3-1,4-β-Glucans are polysaccharides, components of the cell walls of higher members of the Poaceae family. They areparticularly abundant in the endosperm cell walls of commercially valuable cereals such as barley, rye, sorghum, oats and wheat [1]. Structurally, these polysaccharides are linear glucans of up to 1,200 β-D-glucosyl residues linked through β-1,3 and β-1,4 glycosyl bonds. Variations in the proportions of β-1,3-(25–30%) and β-1,4-linkages, and in the length of the mixed-linked segments are currrently reported [2]. During malt production, partial hydrolysis of barley β-glucans begins with seed germination [13]. However, the endogenous 1,3-1,4-β-glucanases are heat inactivated, and the remaining high molecular weight β-glucans may cause severe problems such as increased brewer mash viscosity and turbidity[14]. Increased viscosity impairs pumping and filtration, causing lower efficiency, reduced yields of extracts, and lower filtration rates, as well as the appearance of gelatinous precipitates in the finished beer [2]. Thus, both the level of glucan-hydrolysing activities achieved during germination and the amounts of their substrates, mainly 1,3-1,4-β-glucan, are important factors in the production of high quality malts. Addition of exogenous 1,3-1,4-β-glucanases to the mash could therefore be an outstanding option for improving the brewing process. However, the β-glucanases currently marketed do not really meet the brewing industry's needs, mainly due to economic factors. Novel 1,3-1,4-β-glucanases with uncommon features would be highly desirable. Here we report on the production, purification and partial characterization of a 1,3-1,4-β-glucanase produced by R. microsporus var. microsporus, considering as well its potential for use in the brewing industry.

Results and Discussion

Enzyme Production

A 1,3-1,4-β-glucan-degrading filamentous fungus was isolated from a malt silo in a brewery. This zygomycete microfungus was identified as Rhizopus microsporus var. microsporus by rcDNA analysis. It grew strongly in liquid medium containing chitin as the sole carbon source, and produced substantial amounts of β-glucanase activity (Figure 1) which was able to fully hydrolyze barley β-glucan (1,3-1,4-β-glucan). The specificity of substrate hydrolysis by this enzyme (Table 2) fully supports the assumption that it belongs to the 3.2.1.73 category. The inducible nature of 1,3-1,4-β-glucanase production has already been reported for 1,3-β-glucanase from Trichoderma sp. [27]. Although cellulose and xylan were also inducers, the levels of enzyme secreted in the presence of these carbohydrates were considered smaller than the activity induced by chitin. In cultures grown under agitation (120 rpm) at 40°C, the enzyme activity increased from a minimum to a maximum level within 24h of growth. It has been reported that several other microrganisms, including Bacillus sp. [15]Trichoderma sp. [16], Talaromyces emersonii [17], and Rhizobium sp. [2], produce 1,3-1,4-β-glucanase enzymes which are currently used in the brewing industry. However, depending on the substrate (β-glucan) used as inducer, production of the glucanases for industrial application may be very costly, to the point of being considered economically prohibitive [28]. Chitin, on the other hand, is a relatively cheap and readily available carbon source in comparison to barley β-glucan and laminarin. The ability, therefore, of R. microsporus var. microsporus to produce 1,3-1,4-β-glucanase in the presence of chitin, favors its use on an industrial scale.

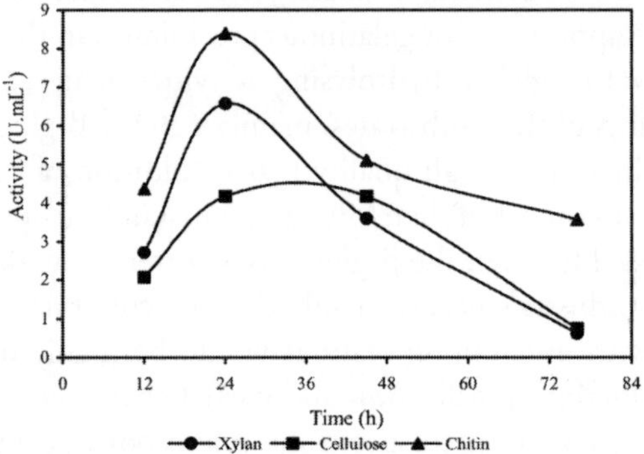

Figure 1. Time course of production of 1,3-1,4-β-glucanase by Rhizopus microsporus var. microsporus in the presence of 0.5% of either xylan, cellulose or chitin at a temperature of 40°C and at 120 rpm.

Table 1. Summary of the purification protocol of the 1,3-1,4-β-glucanase produced by Rhizopus microsporus var. microsporus.

Steps	Total Protein(mg)	Total Activity(U)	Specific activity(U.mg⁻¹)	Purification (-fold)	Yield(%)
Concentrated crude extract	1.574	0.228	0.197	1	100.000
Sephacryl S-100 eluate	0.235	0.227	0.966	4.915	99.561
SP – Sepharose eluate	0.021	0.262	12.596	55.529	114.912

Table 2. Hydrolysis of glucan substrates by the R. microsporus purified β-glucanase. $\Delta Abs_{550\ nm}$ represents the net absorbance of the reaction mixture after incubation for 0.5 h with the enzyme at 50°C.

Substrate	Δ**Abs** $_{550\ nm}$
Laminarin (1,3-β-glucan)	0.029
Chitin	0.001
CM-cellulose (1,4-β-glucan)	0.001
Xylan	0.026
Manan	0.000
Barley β-glucan (1,3-1,4-β-glucan)	0.826

Enzyme Purification

The culture supernatant of R. microsporus var. microsporus grown in liquid medium containing chitin was concentrated 10-fold by ultrafiltration, using a 10 kDa cut-off membrane. No 1,3-1,4-β-glucanase activity was found in the filtrate. Chromatography of the concentrate on a Sephacryl S-100 gel filtration column (not shown) followed by chromatography on an SP-Sepharose ion exchange column resulted in elution of two peaks of proteins (PGI and PGII) (Figure 2). While the PGI proteins were inactive, the PGII protein (fractions 22–34) showed substantial activity against 1,3-1,4-β-glucan. A summary of the purification steps of the 1,3-1,4-β-glucanase produced by the R. microsporus var. microsporus is shown in table 1. The enzyme was purified (Figure 3) 55.529-fold with a yield of 114.912% and a specific activity of 12.596 U.mg-1. The molecular mass of the PGII protein was 36.5 kDa, as indicated by SDS-PAGE analysis (Figure 3). This value is comparable to that (33.7 kDa) determined by mass spectrometry analysis for this enzyme (Figure 3). While 1,3-1,4-β-glucanases from Bacillus sp. have smaller molecular masses varying in the range of 25–30 kDa [2], the enzymes from Clostridium thermocellum (38 kDa) [34], Bacteroides succinogenes (37 kDa) [33] and Talaromyces emersonii (40.7 kDa) [2] showed comparable molecular mass values.

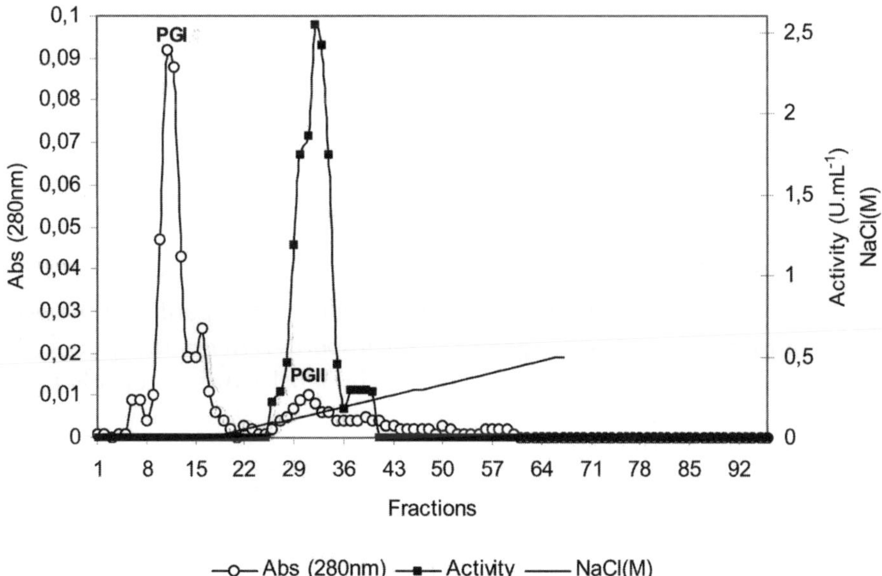

Figure 2. Ion exchange (SP-Sepharose column) chromatography of the concentrated culture filtrate of Rhizopus microsporus var. microsporus grown in liquid medium containing 0.5% chitin.

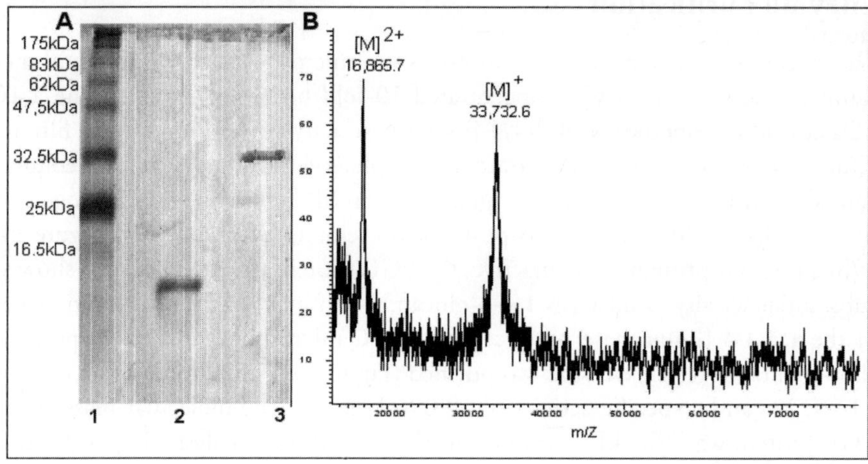

Figure 3. SDS-PAGE (A) and MALDI-TOF mass spectrometry (B) analysis of the purified 1,3-1,4-β-glucanase from Rhizopus microsporus var. microsporus. A: line 1, molecular weight markers; line 2, PGI protein fraction; line 3, PGII protein fraction.

Enzyme Specificity

The R. microsporus purified β-glucanase was tested for its ability to hydrolyze several other glucan substrates. As may be seen in table 2, only the barley β-glucan was efficiently hydrolyzed, as indicated by the much higher net absorbance. In comparison to the activity against the 1,3-1,4-β-glucan, very low or no activity at all was shown by the enzyme against the substrates laminarin (1,3-β-glucan) and CM-cellulose (soluble 1,4-β-glucan), indicating clearly that the enzyme may be taken as a member of the EC 3.2.1.73 enzyme category.

Effect of pH and Temperature Optima

The effect of pH and temperature on the activity of the purified 1,3-1,4-β-glucanase from Rhizopus microsporus var. microsporus is shown in figures 4 and 5, respectively. At 50°C, the enzyme showed substantial activity in the pH range of from 2 to 6. Maximal activity was recorded in the range of from 4 to 5. No enzyme activity was detected at pH higher than 6 (Figure 4). At pH 5.0, the purified enzyme was substantially active in the temperature range from 20°C to 65°C. Maximal activity was detected at 50°C and 60°C, indicating that the optimal temperature for glucan hydrolysis is 55°C (Figure 5). The optima pH and temperature values determined for the purified 1,3-1,4-β-glucanase from R. microsporus var. microsporus were similar to those determined for 1,3-1,4-β-glucanases from several other fungi and bacteria [2]. In addition, these values are comparable to those presented by enzymes currently being used in the brewing industry [11,2]. The purified 1,3-1,4-β-glucanase retained 100% and 87% of its activity after incubation for 2 h and 24 h, respectively, at 50°C. The half-lives of the enzyme at the temperatures of 60°C and 70°C were found to be 10 min and 1 min, respectively. At 50°C, the half-life was 72 h (data not shown). For hydrolysis of β-glucan by a novel 1,3-1,4-β-glucanase produced by Bacillus halodurans C-125, the pH optimum was between 6 and 8, and the temperature optimum was 60°C. After 2 h incubation at 50°C and 60°C, the residual activity remained 100% and 50%, respectively. The enzymatic activity was abolished after 3 min incubation at 70°C. The optimum temperature for hydrolysis of lichenan by a 1,3-1,4-β-glucanase from Bacteroides succinogenes at ph 6.0 was 50°C [33].

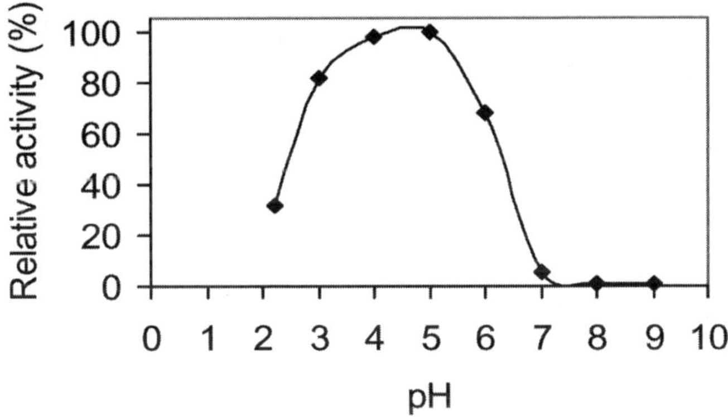

Figure 4. Effect of pH on the activity of the purified 1,3-1,4-β-glucanase from Rhizopus microsporus var. microsporus, at 50°C.

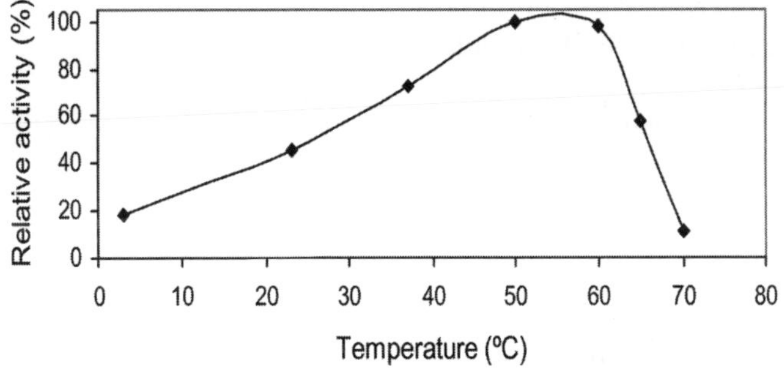

Figure 5. Effect of temperature on the activity of the purified 1,3-1,4-β-glucanase from Rhizopus microsporus var. microsporus, at pH 5.0.

Effect of Metal Ions

The effect of several ions on the activity of the purified 1,3-1,4-β-glucanase produced by R. microsporus var. microsporus is shown in Table 3. The enzyme was sensitive to copper and fairly sensitive to zinc and manganese, but insensitive to magnesium, calcium and aluminum (Table 3). Glucanases produced by Rhizopus oryzae [29], Bacillus clausii [30], Bacillus halodurans [32] and Trichoderma harzianum [31] show similar sensitivity to the divalent metal ion copper.

Table 3. Effect of metal ions on the activity of the purified 1,3-1,4-β-glucanase from Rhizopus microsporus var. microsporus.

Ion	Residual activity (%)
Control	100
Cu^{+2} (12 mM)	0.3
Mg^{+2} (12 mM)	95.2
Fe^{+3} (12 mM)	89.6
Zn^{+2} (12 mM)	65.0
Mn^{+2} (12 mM)	62.3
Ca^{+2} (12 mM)	105.9
Al^{+3} (12 mM)	109.8

Kinetic Parameters

The purified 1,3-1,4-β-glucanase produced by R. microsporus var. microsporus hydrolyzed 1,3—1,4-β-glucan in a Michaelis-Menten fashion (Figure 7). Kinetic parameters were calculated using a Michaelis-Menten plot with a non-linear regression data analysis program [10]. Values of 19.8 mg.mL^{-1}, 12.7s^{-1} and 16.5 U.mL^{-1} were determined for Km, Kcat and Vmax, respectively. Km values of 1.2—1.5 mg.mL^{-1} for hydrolysis of barley β-glucan and 0.8—2 mg.mL^{-1} for lichenan were reported for the 1,3-1,4-β-glucanase produced by Bacillus sp [2]. Values of 1,296 ± 51, 2.50 ± 0.09, and 518 were reported for Kcat (s^{-1}), Km (mg.mL^{-1}) and Kcat/Km (s^{-1}.M^{-1}) respectively, for hydrolysis of lichenan by a 1,3-1,4-β-glucanase produced by Bacteroides succinogenes. [33].

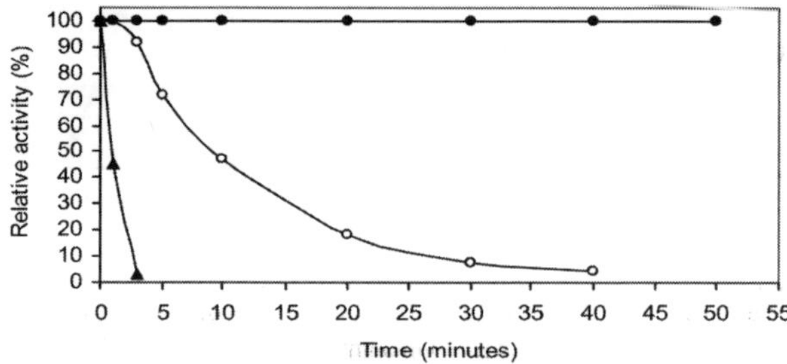

Figure 6. Thermostability of the purified 1,3-1,4-β-glucanase from Rhizopus microsporus var. microsporus, at temperature of 50ºC (●), 60ºC (○) and 70ºC (▲), at pH 5.0.

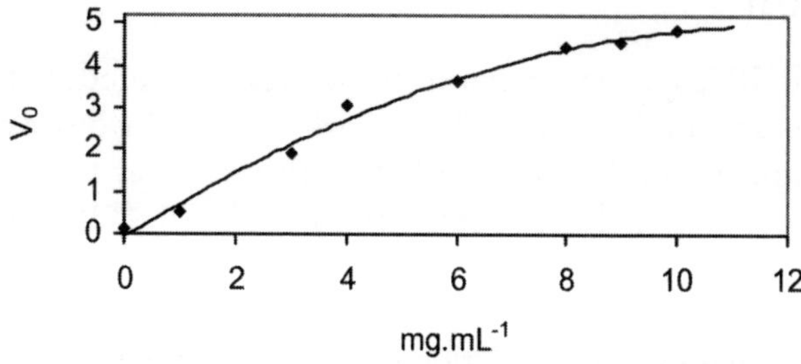

Figure 7. Hydrolysis (μmol·min^{-1}·mL^{-1}) of β-glucan by the purified 1,3-1,4-β-glucanase from Rhizopus microsporus var. microsporus, in the presence of different concentrations of 1,3-1,4-β-glucan.

Capillary Viscosimetry and Filtration Rate

The specific filtration rate and specific viscosity rate of the mash after incubation with the 1,3-1,4-β-glucanase from R. microsporus var microsporus were compared with those values calculated for two commercial β-glucanases currently used in the brewing industry. The results are shown in Tables 4 and 5. Even at lower enzyme concentration, the 1,3-1,4-β-glucanase from R. microsporus var microsporus caused a higher reduction in the filtration rate (20.4%) of the mash (table 4). Similar results were obtained for the specific viscosity of the brewer's mash after treatment with the three β-glucanases (table 5).

Table 4. Total protein in the enzyme samples, filtration time, filtration time reduction and specific filtration time reduction of the brewer's mash not treated or treated with enzymes.

	Total Protein in the enzyme sample (μg)	Filtration time (seconds)	Filtration time reduction (%)	Specific filtration time reduction (%/μg)
Control	-	274	-	-
R. microsporus var microsporus enzyme	11.0	218	20.4	1854.5 × 10^{-3}
Commercial enzyme A	394	195	29.9	75.9 × 10^{-3}
Commercial enzyme B	373	176	35.8	95.8 × 10^{-3}

Table 5. Total protein in the enzyme samples, viscosity, viscosity reduction and specific viscosity reduction of the brewer's mash not treated or treated with enzymes.

	Total Protein in the enzyme sample (μg)	Viscosity (cP)	Viscosity reduction (%)	Specific viscosity reduction (%/μg)
Control	-	1.070	-	-
R. microsporus var microsporus enzyme	110	1.020	4.7	763 × 10^{-3}
Commercial enzyme A	394	1.002	6.3	14.2 × 10^{-3}
Commercial enzyme B	373	1.001	6.4	17.3 × 10^{-3}

Conclusion

The zygomycete Rhizopus microsporus var. microsporus produced a 1,3-1,4-β-D-glucan 4-glucanhydrolase (EC 3.2.1.73) which could hydrolyze β-D-glucan substrate containing both 1,3- and 1,4-bonds. Its molecular mass as determined by both electrophoresis and mass spectrometry (MALDI-TOF) was about 33.7 kDa. Its optimum pH and temperature were found to be in the ranges of 4–5 and 50–60°C, respectively. Kinetic analysis and its capacity to reduce both the viscosity of the brewer mash and the filtration time, indicate the possibility to use this enzyme in the brewing industry.

Methods

Chemicals

Barley 1,3-1,4-β-glucan, chitin, CM-cellulose, manan, xylan, laminarin, molecular mass standard proteins and sodium dodecyl sulfate (SDS) were from Sigma Chemical Co., USA. Sephacryl S-100 and SP-Sepharose were from Pharmacia-LKT, Sweden. All other chemicals were of analytical grade.

Organism and Enzyme Production

The aerobic zygomycete microfungus Rhizopus microsporus var. microsporus was isolated from a malt silo. The fungus was maintained at 4°C, after growing for 48 hours in TLE modified solid medium [(0.5% chitin, 0.2% KH_2PO_4, 0.14%$(NH_4)_2SO_4$, 0.03% $MgSO_4.7H_2O$, 0.0152% $CaCl_2$, 0.02% glucose, 1.0 mL of 0.01% trace elements solutions (Fe^{+2}, Mn^{+2}, Zn^{+2}, Co^{+2}), 0.003% bacto-peptone, 0.003% urea, and 2% agar, pH 6.8)], at 40°C.

For enzyme production, one liter Erlemeyer flasks containing 250 mL of the liquid medium (TLE with no agar), were inoculated with 150 cm^2 blocks of solid medium taken from 2-day old R. microsporus var. microsporus cultures. Liquid cultures were then incubated for 24 hours with agitation (120 rpm) at 40°C. The culture supernatants were then separated from the mycelium by filtration, using filter paper. The supernatants were then freeze-dried and used either for enzyme assay or enzyme purification as described in the following sections.

Enzyme Assay

1,3-1,4-β-glucanase activity was assayed by the reducing-sugar method [6] with β-1,3-β-1,4-glucan as the substrate. The assay system consisted of 50 μL of 1%

(wt/vol) β-glucan dissolved in 100 mM sodium acetate buffer, pH 5.0, and 50 μL enzyme sample. The reaction was allowed to proceed for 30 min at 50°C, and was then stopped by the addition of 300 μL dinitrosalicylate reagent [6], and 5 min of boiling. The absorbance of the reaction mixture was determined at 550 nm using a Perkin Elmer mod. Lambda 11/Bio spectrophotometer. The amount of reducing sugar produced was determined using a curve constructed with glucose as standard. One unit of enzyme was defined as the amount of protein necessary to produce one μmol of reducing sugars.min-1.

The assays for xylanase, cellulase, 1,3-β-glucanase and mananase were performed as for 1,3-1,4-β-glucanase, except for the use of the substrates carboximetilcellulose, laminarin and manan, respectively. For chitinase the enzyme system consisted of 100 μL of enzyme sample, regenerated chitin 0,5% in 50 mM sodium acetate buffer, pH 5.2 [35]. The reaction was allowed to run for 12h at 37°C and stopped by addition of dinitrosalycilic reagent. The amount of reducing sugar produced was quantified using a standard curve constructed with glucose.

Purifications of the 1,3-1,4-β-glucanase from Rhizopus microsporus var. microsporus

The supernatants from cultures of R. microsporus var. microsporus grown in liquid medium containing β-glucan were concentrated by ultrafiltration [(Amicon system; 10 k-Da cut-off membrane (PM10)]. Aliquots of concentrated β-glucanase were loaded on a Sephacryl S-100 gel column (2.5 × 40 cm), equilibrated and eluted with 50 mM sodium acetate buffer, pH 5.0. Elution was performed at a flow rate of 24 mL.h^{-1}, and fractions of 4.0 mL were collected. Active fractions were pooled and applied on a SP-Sepharose ion-exchange column (3.0 × 15 cm), previously equilibrated and eluted with 50 mM sodium acetate buffer, pH 5.0, and further eluted with a linear gradient formed with 100 mL of the acetate buffer and 100 mL of the same buffer containing 1 M NaCl. Elution was carried out at a flow rate of 24 mL.h^{-1}, and fractions of 4 mL were collected. The resulting active fractions were pooled and dialyzed overnight against distilled water at 4°C, concentrated by ultrafiltration as above, and stored at -20°C until their use.

Protein Determination

Protein was determined by the Bradford method [7], with bovine serum albumin as standard.

Electrophoresis

Enzyme samples were examined by electrophoresis under denaturing conditions in polyacrylamide slab gels (SDS-PAGE) as described by Laemmli [8]. Protein bands in the gel were visualized by the silver staining method [9].

Mass Spectrometry

1,3-1,4-β-glucanase was analyzed by matrix-assisted laser desorption ionization-time of flight (MALDI-TOF) mass spectrometry with a Reflex IV mass spectrometer (Bruker Daltonik, Bremen, Germany) in linear positive mode. The purified enzyme sample (50 µg) was dissolved in 50 µL of 0.1 % (v/v) TFA, from which 1 µL was mixed with 1 µL of a saturated matrix solution of sinapinic acid dissolved in 50% (v/v) acetonitrile and 0.1% (v/v) trifluoroacetic acid, and applied to the MALDI plate. BSA was used for external mass calibration.

Effects of Ions

The effects of several metallic ions (Cu^{+2}, Mg^{+2}, Fe^{+3}, Zn^{+2}, Ca^{+2}, and Al^{+3}) on the purified 1,3-1,4-β-glucanase were tested measuring the activity of the enzyme at 50°C (see 1,3-1,4–β-glucanase assay) in the presence of the ions.

pH and Temperature Optima

The effect of temperature on the enzyme was carried out at the temperature range of from 4° to 70°C, at pH 5.0 (in 50 mM sodium acetate buffer). The optimum pH value was determined by monitoring the enzyme activity at 50°C at pH values from 3.0 to 9.0. The following buffers were used: pH 3.0—6.0, 50 mM sodium acetate; pH 7.0, 50 mM sodium phosphate; and pH 8.0—9.0, 50 mM tris-HCl.

Kinetic Parameters

For determination of kinetic parameters, the enzyme assays were performed at 50°C, using 1,3-1,4-β-glucan at concentrations varying from 0.05 to 2.0 % dissolved in 50 mM sodium acetate buffer, pH 5.0. Km, Kcat and Vmax values were obtained using a Michaelis-Menten plot with a non-linear regression data analysis program [10].

Preparation of the Mash

12.5 g of malt was triturated in a hammer mill (MAROTEC), drizzled into a sieve of 0.2 mm spacing, and dissolved in 50 mL of sodium acetate buffer (100 mM, pH 5.5), pre-heated to 45°C. Reaction started with 1.0 mL of enzyme sample taken from chromatography on a Sephacryl S-100 column, and allowed to proceed for 30 min at 45°C, followed by other periods of 10 min at 50°C, 15 min at 60°C, 60 min at 70°C, and 5 min of boiling. The reaction was then stopped by the addition of 100 mL of cold water and immediate cooling in an ice-water bath at about 20°C.

Capillary Viscosimetry

Decrease in mash viscosity was measured by capillary viscosimetry using an Oswald viscosimeter [12,20]. Samples of 30 mL of mash were filtered using filter paper and placed in a viscosimeter at 20°C. Mash viscosity in the absence of enzyme was used as a control. The specific viscosity rate was calculated using the following equations:

$$\mu_{mash} = (\mu_{water} \times T_{mash} \times \rho_{mash})/(T_{water} \times \rho_{water}) \tag{1}$$

$$\Delta\mu = (\mu_{mash\ control} - \mu_{mash}) \times 100/(\mu_{mash\ control}) \tag{2}$$

$$\Delta\mu\phi = \Delta\mu/\delta \tag{3}$$

where μ is the viscosity, T is the flow time, $\Delta\mu$ is viscosity reduction, δ is total protein, $\Delta\mu\phi$ is specific viscosity rate and ρ is the density.

Filtration Rate

The filtration rate was determined by filtration of 50 mL of mash through a filter paper [12]. Filtration rate in the absence of enzyme was used as a control. The specific filtration rate was calculated using the following equations:

$$\Delta_{\psi} = (\psi mash\ control - \psi mash) \times 100/(\psi mash\ control) \tag{4}$$

$$\Delta\psi\phi = \Delta\psi/\delta \tag{5}$$

where ψ is the flow time, $\Delta\psi$ is the filtration time of 50 mL, δ is total protein and $\Delta\psi\phi$ is the specific filtration reduction.

Authors' Contributions

KRSC conceived of the study and participated in its design, performed all experiments, data quantification and was involved in the literature search and data interpretation. RBC participated and supported the mass spectrometry analysis with a Reflex IV mass spectrometer. CRF participated in the design and coordination of the study, as well as on its supervision, and helped to draft the manuscript. All authors read and approved the final manuscript.

Acknowledgements

CRF acknowledge the scholarship awarded by CNPq (process: 305123/2005-0).

References

1. Stone BA, Clarke AE: Chemistry and Biology of 1,3-β-Glucans. Bundoora: La Trobe University Press; 1992.

2. Planas A: Bacterial 1,3-1,4-β-glucanases: structure, function and protein engineering. Biochimica et Biophysica Acta 2000, 1543:361–382.

3. Bamforth CW: β-Glucan and β-glucanases in malting and brewing: practical aspects. Brewers Digest 1994, 69:12–16.

4. Fincher GB, Stone BA: Advances in cereal science and technology. Volume 7. American Association of Cereal Chemists. St. Paul; 1986.

5. Godfrey T, West S: Industrial enzymology. New York : Stockton Press; 1996.

6. Miller GL: Use of dinitrosalycilic acid reagent for determination of reducing sugar. Analytical Chemistry 1959, 31:426–428.

7. Bradford MM: A rapid and sensitive method for the quantitation of microgram quantities of protein utilizing the principle of protein dye binding. Analitical Biochemical 1976, 72:248–254.

8. Laemmli UK: Cleavage of structural proteins during assembly of head of bacteriophage-T4. Nature 1970, 227:680–685.

9. Blum H, Beier H, Gross B: Improved silver staining of plant proteins, RNA and DNA in polyacrilamide gels. Electrophoresis 1987, 8:93–99.

10. Leatherbarrow RJ: Enzifitter: a non-linear regression data analysis program for the IBM PC. London: Biosoft; 1987:1–91.

11. McCleary BV, Nurthen E: Measurement of (1–3)(1–4)-β-D-glucan in malt, wort and beer. Journal of the Institute of Brewing 1986, 92:168–173.

12. McCleary BV, Shameer I, Glennieholmes M: Measurement of (1–3),(1–4)-β-D-glucan. Methods in Enzymology 1988, 160:545–551.

13. Kettunen A, Hamalainen JJ, Stenholm K, Pietila K: A model for the prediction of β-glucanase activity and beta-glucan concentration during mashing. Journal of Food Engineering 1996, 29:185–200.

14. Bamforth CW, Martin HL: The degradation of β-glucan during malting and mashing—the role of β-glucanase. Journal of the Institute of Brewing 1983, 89:303–307.

15. McCarthy T, Hanniffy O, Lalor E, Savage AV, Tuohy MG: Evaluation of three thermostable fungal endo-β-glucanases from Talaromyces emersonii for brewing and food applications. Process Biochemistry 2005, 40:1741–1748.

16. Bhat MK: Cellulases and related enzymes in biotechnology. Biotechnology Advances 2000, 18:355–383.

17. Jayus , McDougall BM, Seviour RJ: Purification and characterization of the (1>3)-β-glucanases from Acremonium sp IMI 383068. FEMS Microbiology Letters 2004, 230:259–264.

18. Scheffler A, Bamforth CW: Exogenous β-glucanases and pentosanases and their impact on mashing. Enzymes and Microbial Technology 2005, 36:813–817.

19. Murray PG, Grassick A, Laffey CD, Cuffe MM, Higgins T, Savage AV, Planas A, Tuohy MG: Isolation and characterization of a thermostable endo-β-glucanase active on 1,3-1,4-β-D-glucans from the aerobic fungus Talaromyces emersonii CBS 814.70. Enzyme and Microbial Technology 2001, 29:90–98.

20. Vlasenko EY, Ryan AI, Shoemaker CF, Shoemaker SP: The use of capillary viscometry, reducing end-group analysis, and size exclusion chromatography combined with multi-angle laser light scattering to characterize endo-1,4-β-D-glucanases on carboxymethylcellulose : A comparative evaluation of the three methods. Enzyme amd Microbial Technology 1998, 23:350–359.

21. Gan Q, Howell JA, Field RW, England R, Bird MR, O'Shaughnessy CL, MeKechinie MT: Beer clarification by microfiltration—product quality control and fractionation of particles and macromolecules. Journal of Membrane Science 2001, 194:185–196.

22. Wang JM, Zhang GP, Chen JX, Wu FB: The changes of β-glucan content and β-glucanase activity in barley before and after malting and their relationships to malt qualities. Food Chemistry 2004, 86:223–228.

23. Wen TN, Chen JL, Lee SH, Yang NS, Shyur LF: A Truncated Fibrobacter succinogenes 1,3-1,4-β-D-Glucanase with Improved Enzymatic Activity and Thermotolerance. Fibrobacter succinogenes 2005, 44:9197–9205.

24. Lusk LT, Duncombe GR, Kay SB, Navarro A, Ryder D: Barley β-glucan and beer foam stability. Journal of the American Society of Brewing Chemists 2001, 59:183–186.

25. Edney MJ, LaBerge DE, Langrell DE: Relationships among the β-glucan contents of barley, malt, malt congress extract, and beer. Journal of the American Society of Brewing Chemists 1998, 56:164–168.

26. Almin KE, Eriksson KE: Enzymic degradation of polymers I. viscometric method for determination of enzymic activity. Biochimica Et Biophysica ACTA 1967, 139:238–248.

27. Lima LHC, Ulhoa CJ, Fernandes AP, Felix CR: Purification of a chitinase from Trichoderma sp. and its action on Sclerotium rolfsii and Rhizoctonia solani cell walls. Journal of General and Applied Microbiology 1997, 43:31–37.

28. McCarthy TC, Lalor E, Hanniffy O, Savage AV, Tuohy MG: Comparison of wild-type and UV-mutant beta-glucanase-producing strains of Talaromyces emersonii with potential in brewing applications. Journal of Industrial Microbiology & Biotechnology 2005, 32:125–134.

29. Murashima K, Nishimura T, Nakamura Y, Koga J, Moriya T, Sumida N, Yaguchi T, Kono T: Purification and characterization of new endo-1,4-β-D-glucanases from Rhizopus oryzae. Enzyme and Microbial Technology 2002, 30:319–326.

30. Miyanishi N, Hamada N, Kobayashi T, Imada C, Watanabe E: Purification and characterization of a novel extracellular β-1,3-glucanase produced by Bacillus clausii NM-1 isolated from ezo abalone Haliotis discus hannai. Journal of Bioscience and Bioengineering 2003, 95:45–51.

31. Rana DS, Theodore K, Naidu GSN, Panda T: Stability and kinetics of β-1,3-glucanse from Trichoderma harzianum. Process Biochemistry 2003, 39:149–155.

32. Akita M, Kayatama K, Hatada Y, Ito S, Horikoshi K: A novel β-glucanase gene from Bacillus halodurans C-125. FEMS Microbiology Letters 2005, 248:9–15.

33. Erfle JD, Teather RM, Wood PJ, Irvin JE: Purification and properties of a 1,3-1,4-β-D-glucanase (lichenase, 1,3-1,4-β-D-glucan 4-glucanohydrolase, EC-3-.2.1.73) from Bacteroides succinogenes cloned in Escherichia coli. Biochemical Journal 1988, 255:833–841.

34. Schimming S, Schawarz WH, Staudenbauer WL: Properties of a thermoactive β-1,3-1,4-glucanase (lichenase) from Clostridium thermocellum expressed in Escherichia coli. Biochemical and Biophysical Research Communications 1991, 177:447–452.

35. Molano J, Duram A, Cabib E: A rapid and sensitive assay for chitinase using tritiated chitin. Anal Biochem 1977, 83:648–656.

Chemical Analysis and Risk Assessment of Diethyl Phthalate in Alcoholic Beverages with Special Regard to Unrecorded Alcohol

Jenny Leitz, Thomas Kuballa, Jürgen Rehm and
Dirk W. Lachenmeier

ABSTRACT

Background

Phthalates are synthetic compounds with a widespread field of applications. For example, they are used as plasticizers in PVC plastics and food packaging, or are added to personal care products. Diethyl phthalate (DEP) may be used to denature alcohol, e.g., for cosmetic purposes. Public health concerns of phthalates include carcinogenic, teratogenic, hepatotoxic and endocrine effects. The aim of this study was to develop and validate a method for determining phthalates in alcohol samples and to provide a risk assessment for consumers of such products.

Methodology/Principal Findings

A liquid-liquid extraction procedure was optimized by varying the following parameters: type of extraction solvent (cyclohexane, n-hexane, 1,1,2-trichlorotrifluoroethane), the ratio extraction solvent/sample volume (1:1 to 50:1) and the number of extraction repetitions (1–10). The best extraction yield (99.9%) was achieved with the solvent 1,1,2-trichlorotrifluoroethane, an extraction solvent volume/sample volume ratio of 10:1 and a double extraction. For quantification, gas chromatography/mass spectrometry with deuterated internal standards was used. The investigated samples were alcoholic beverages and unrecorded alcohol products from different countries (n = 257). Two unrecorded alcohol samples from Lithuania contained diethyl phthalate in concentrations of 608 mg/L and 210 mg/L.

Conclusions/Significance

The consumption of the phthalate-positive unrecorded alcohols would exceed tolerable daily intakes as derived from animal experiments. Both positive samples were labelled as cosmetic alcohol, but had clearly been offered for human consumption. DEP seems to be unsuitable as a denaturing agent as it has no effect on the organoleptic properties of ethanol. In light of our results that DEP might be consumed by humans in unrecorded alcohols, the prohibition of its use as a denaturing agent should be considered.

Introduction

Phthalic acid esters, commonly referred to as phthalates, are a group of industrial chemicals that have become ubiquitous environmental contaminants because of their widespread usage and high persistence in the environment [1]. They are generally colourless and odourless liquids with a low solubility in water [2]. Phthalates are used as plasticizers in many consumer products, such as household furnishings, in personal care products as vehicles for fragrance, in medical devices and children's toys, in food packaging, cleaning materials or insecticides. Since phthalates are not chemically bound to the plastic materials, they can leach out, migrate or evaporate into the air or into foodstuffs, for example. Accordingly, humans are exposed to phthalates through ingestion, inhalation and dermal contact throughout their lifetimes [3].

Another source of phthalates for human exposition is the use of diethyl phthalate (DEP) as a denaturing agent for ethyl alcohol. In Russia, alcohols denaturized by

DEP have been available on the market for human consumption [4]. The risk of these surrogate alcohols, a subgroup of the category of unrecorded alcohol (see [5] for definition), often is not known to the consumers and it was hypothesized that these products currently lead to increased mortality in eastern European countries [4].

At the present time, a thorough toxicological evaluation is not available. Nevertheless, phthalates are suspected of causing health problems. The acute toxicity of phthalates is very low (LD50 1–30 g/kg); however, the subchronic and chronic toxic effects of phthalates and their metabolites are of more importance. Toxicological investigations are now focused on carcinogenic, teratogenic and endocrine effects; some phthalates even show reproductive and developmental toxicity in animal experiments [2]. In general, the dose-response relationship of phthalates is difficult to evaluate. In Europe, there are tolerable daily intake (TDI) values only for some phthalates, such as di-2-ethylhexl phthalate (DEHP, 50 µg/kg bodyweight), di-n-butyl phthalate (DBP, 10 µg/kg bodyweight) or butylbenzyl phthalate (BBP, 500 µg/kg bodyweight) [6]–[8].

In the context of our ongoing investigation of unrecorded alcohol from different countries [9], [10], the aim of this study was to develop and validate a method for the determination of phthalates in alcoholic beverages. Especially the sample preparation should be as fast and inexpensive as possible, which is why we aimed to use a simple solvent extraction of the alcohols. Different parameters of this liquid-liquid extraction (LLE) method had to be optimized, including selection of the extraction solvent, the ratio extraction solvent volume/sample volume or the number of extraction repetitions. The results from our large collection of samples of unrecorded alcohols were used to provide a toxicological evaluation of these types of alcohols for consumers.

Materials and Methods

Chemicals and Reagents

Dimethyl phthalate (DMP), diethyl phthalate, diallyl phthalate (DAP), di-iso-butyl phthalate (DIBP), di-n-butyl phthalate (DBP), diethylhexyl adipate (DEHA), butyl-benzyl phthalate (BBP), di-2-ethylhexyl phthalate (DEHP), di-heptyl phthalate (DHP) and di-n-octyl phthalate (DNOP) were all of GC grade and were purchased from Merck (Darmstadt, Germany). 3,4,5,6-d_4-DEHP (d_4-DEHP) was synthesized according to Loupy et al. [11] and Schwetlick [12]. The solvents 1,1,2-trichlorotrifluoroethane (>99.9%), ethyl acetate (>99.8%), cyclohexane

(>99.9%) and n-hexane (>98.0%) were also purchased from Merck (Darmstadt, Germany).

A stock standard solution of the target analytes was prepared at a final concentration of about 1 g/L in cyclohexane/ethyl acetate (1:1, v:v). From this solution, a standard solution was prepared at a final concentration of 50 mg/L in cyclohexane/ethyl acetate (1:1, v:v). Calibration solutions of phthalates at nominal concentrations of 0.5, 1.0, 2.0, 4.0, 6.0, 10.0 and 20.0 mg/L were prepared by diluting the standard solution in cyclohexane/ethyl acetate (1:1, v:v). The nominal concentration of internal standard in each calibration dilution was about 7.5 mg/L.

Glassware and Reagent Control

To avoid phthalate contamination, all glassware used in this study was rinsed with acetone and dried at 220°C at least 5 h. All glassware and reagents were checked for potential phthalate contamination. The solvents cyclohexane, ethyl acetate and n-hexane were checked by gas chromatographic-mass spectrometric (GC/MS) analysis once a week. 1,1,2-Trichlorotrifluoroethane was checked with every sample measurement.

GC/MS Method

The GC/MS system used for analysis was a Trace GC in combination with a CTC Combi-PAL auto sampler and a Polaris Q mass spectrometer (Thermo Finnigan, Bremen, Germany). Data acquisition and analysis were performed using standard software supplied by the manufacturer (Xcalibur 1.3.1 and CTC Cycle Composer 1.5.3 for acquisition and Xcalibur 2.0.7 for data analysis). Substances were separated on a VF-Xms column (Factor Four, 29.3 m×0.25 mm I.D., film thickness 0.25 µm, Varian, Darmstadt, Germany). A 1 µL sample was injected into the split/splitless inlet in splitless mode (splitless for 1.5 min, split flow 10 mL/min) at 250°C. The temperature of the GC/MS transfer line was 280°C, the temperature of the ion source was 225°C. Helium with a constant flow rate of 1 mL/min was used as carrier gas. The oven temperature program was: 100°C, held for 1 min, 5°C/min up to 270°C, held for 0 min, 10°C/min up to 320°C, held for 10 min. The ion-trap mass spectrometer was operated in the electron ionization mode (70 eV) and the analytes were recorded in full-scan mode (m/z 40–300) (Table 1). For quantification, peak area ratios of the analytes to the internal standard d_4-DEHP were calculated as a function of the concentration of substances.

Table 1. Retention time and selected ions for the analysis of the target phthalates.

Compound	Retention time (min)	Quantification ions(m/z)	Identification ions (m/z)
DMP	11.6	163	164
DEP	14.7	149	177
DAP	18.0	149	189, 132
DIBP	20.3	149	223, 150
DBP	22.2	149	150, 223
DEHA	29.5	129	147, 241
BBP	29.5	149	91, 206
DEHP	31.9	149	176, 279
DHP	32.0	149	265, 150
DNOP	34.9	149	150, 279
d_4-DEHP	31.9	153	171

Samples

A total of 257 samples submitted to the CVUA Karlsruhe were analyzed for the different phthalates listed in table 1. The samplings were conducted in the context of different international projects designed to characterize the quality of alcoholic beverages, including unrecorded alcohol. Further details on samples from Nigeria (illegally produced spirits; n = 6) [13], Mexico (tequila, mezcal; n = 24) [14], Lithuania (cheap spirits and cosmetic surrogate alcohols; n = 10) [5], Hungary (cheap fruit-derived spirits; n = 15) [5], Guatemala (surgarcane spirits (cuxa), commercial and clandestine variants; n = 22) [15], [16], Poland (commercial fruit wines and unrecorded spirits (moonshine); n = 44) [17], Vietnam (commercial and homemade spirits, mainly rice-based; n = 10) [18], and Brazil (commercial cachaça; n = 24) [19] were previously published. Furthermore, samples from India (spirits; n = 2), Ukraine (predominantly homeproduced spirits, so-called samogon; n = 61), Dominican Republic (unrecorded spirits; n = 2) and Romania (fruit spirits; n = 2), as well as samples legally available on the German market (spirits, mainly vodka; n = 35) were included in the study. The samplings were not representative but risk-oriented [20] as we have specifically searched for unrecorded products, more likely to be contaminated with diethyl phthalate from possible use of denatured alcohol (see references for details on sampling strategies in the respective countries).

Sample Preparation

The sample preparation is a liquid-liquid extraction (LLE). 0.1 mL of the sample was placed in a glass test tube and 0.1 mL of internal standard (end concentration in the sample was about 7.5 mg/L) and 1 mL of 1,1,2-trichlorotrifluoroethane were added. The tube was closed with a ground-glass stopper and shaken on a Vortex mixer for 1 min. After centrifugation for 5 min (3000 rpm), the solvent phase (lower phase) was removed to a separate vial. A fresh 1 mL volume of 1,1,2-trichlorotrifluoroethane was added to the sample and the extraction was repeated. The two solvent phases were then combined and analyzed by GC/MS.

Optimization and Validation Studies

Before validation, the LLE method had to be optimized by different parameters in order to completely separate the phthalates from the sample matrix. Three extraction solvents, cyclohexane, n-hexane and 1,1,2-trichlorotrifluoroethane, were compared by extracting 1 mL of an authentic alcoholic beverage sample with different volumes of these solvents. The ratio of extraction solvent volume/ sample volume was chosen by extracting 0.1 mL of the same alcoholic beverage sample with different volumes (0.1–5 mL) of 1,1,2-trichlorotrifluoroethane. The optimum number of repetitions of the extraction procedure was determined by extracting 1 mL of the alcoholic beverage sample ten times successively with 1,1,2-trichlorotrifluoroethane.

For method validation and analytical quality assurance, we followed the demands for governmental food and alcohol control authorities [21]. Specifically, the principles outlined in ISO 17025 [22]. The method validation was conducted for DEP. For the validation, an authentic DEP-positive alcoholic beverage sample from Lithuania and a blank sample, i.e., a DEP-free vodka that was spiked with DEP-standard solution (end concentration about 25 mg/L), were extracted and analyzed several times daily (intraday, n = 6) and over several days (interday, n = 5) using the optimized procedure. The linearity of the calibration curves was evaluated between 0.1 and 20.0 mg/L. For the determination of the limit of detection (LOD) and the limit of quantitation (LOQ), a separate calibration curve in the range of LOD (0.1–1.0 mg/L) was established. The recovery rate was ascertained by adding DEP at two different concentrations (about 80 mg/L and 200 mg/L end concentration) to a blank sample (DEP-free vodka). The applicability of the procedure was proven by routine analysis of over 200 samples.

Statistics

The experimental designs and calculations were done using the Software Package Design Expert v7 (Stat-Ease Inc., Minneapolis, MN, USA). The experiments were evaluated using Analysis of Variance (ANOVA) to find the significance of variables and their interactions in the models. The models were checked for consistency by looking at the lack of fit and possible outliers. Statistical significance was assumed at below the 0.05 probability level.

Results

Parameter Optimization for the LLE Method

In the literature, many different extraction solvents, such as cyclohexane, n-hexane, ethyl acetate or dichloromethane (each solvent also in addition with NaCl), have been suggested for the extraction of phthalates from various food matrices [23]–[27]. Another solvent, 1,1,2-trichlorotrifluoroethane, was suggested for the extraction of volatile compounds from alcoholic beverages [28]. Preliminary tests showed that of the mentioned solvents, cyclohexane, n-hexane and 1,1,2-trichlorotrifluoroethane, without the addition of NaCl, offered the best extraction results from alcoholic beverages. Figure 1 shows a comparison of these three solvents. The solvent type as well as solvent volume both significantly influence the extraction efficiency (ANOVA p<0.0001 for response surface model). 1,1,2-Trichlorotrifluoroethane extracted a significantly higher amount of phthalate from the sample than the two other solvents. Figure 1 also shows that a higher extraction solvent volume led to significantly higher amounts of extracted phthalate.

For the following optimization experiments, 1,1,2-trichlorotrifluoroethane was chosen as the solvent for the LLE method. Figure 2 shows that the amount of extracted phthalate increases up to an extraction solvent volume of 1 mL, but exceeding that volume does not lead to any further significant increase in the extracted phthalate. Therefore, the optimal ratio extraction solvent volume/sample volume was 1 mL/0.1 mL. Repetition of the LLE showed that after two extractions, over 99.9% of the phthalate was extracted out of the sample (Figure 3). The optimal LLE parameters are summarized in Table 2. To improve precision and to correct for possible variability in extraction and GC/MS, deuterated DEHP was added as internal standard prior to the extraction in all further measurements.

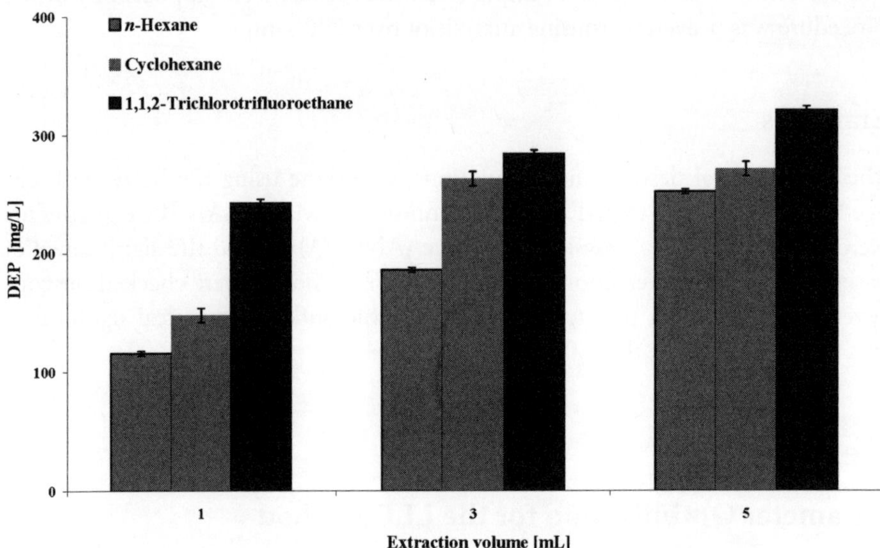

Figure 1. Comparison of the extraction of an authentic alcoholic beverage sample with three different extraction solvents.

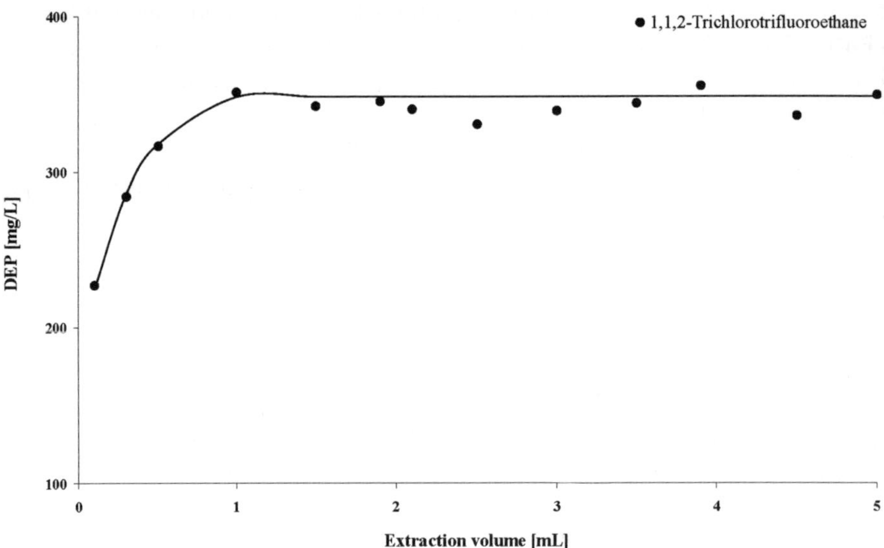

Figure 2. Evaluation of the optimal ratio extraction solvent volume/sample volume.

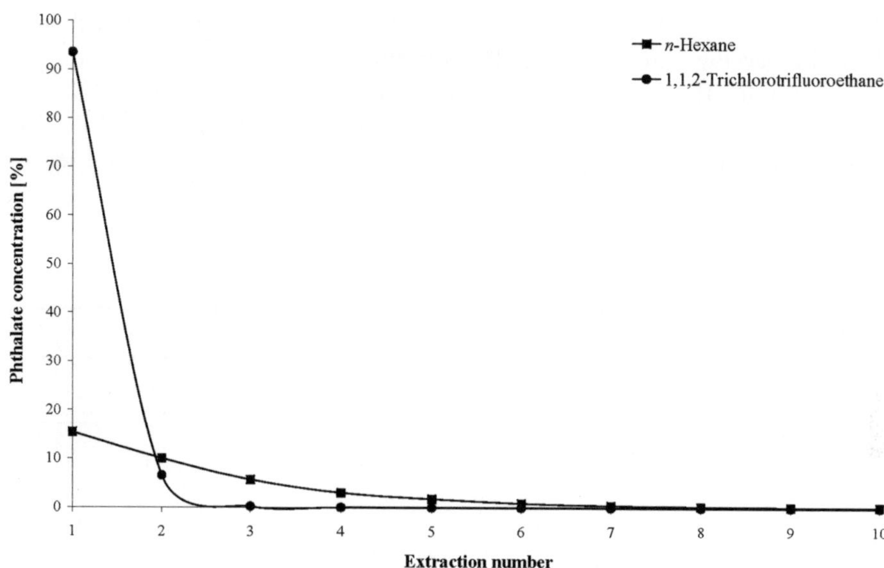

Figure 3. Repetition of the LLE and comparison of the two extraction solvents n-hexane and 1,1,2-trichlorotrifluoroethane.

Table 2. Results of LLE method optimization.

Parameter	Result
Extraction solvent	1,1,2-Trichlorotrifluoroethane
Ratio extraction solvent volume/sample volume	1 mL/0.1 mL
Repetition of LLE	1 (i.e. 2 extractions in total)

Results of Method Validation and Sample Measurement

During routine analyses of 257 authentic alcoholic beverage samples, no interfering peaks from the spirit matrix were observed. Table 3 summarizes the method validation results for DEP. The assay was linear in the required concentration range between 0.1 and 20.0 mg/L with a regression coefficient of 0.9995. When determined according to DIN 32645 [29], the limit of detection (LOD) of DEP was 0.7 mg/L, the limit of quantitation (LOQ) was 2.6 mg/L. Both values were sufficient for our purposes. The precision expressed as relative standard deviation (RSD) of the optimized method for

DEP never exceeded 9.0% (intraday) and 8.4% (interday) for the authentic alcoholic beverage sample and 8.2% (intraday) and 9.7% (interday) for the spiked sample. The method was verified by the recovery ranges of 103.9% (at 80 mg/L addition of DEP) and 110.4% (at 200 mg/L addition of DEP). The results of the method validation show that the method is precise and reproducible (Table 3).

Table 3. Results of method validation for DEP.

Parameter	Result
Linear range	0.1–20.0 mg/L
LOD[a]	0.7 mg/L
LOQ[a]	2.6 mg/L
Precision intraday[b]	9.0% (authentic sample); 8.2% (blank sample)
Precision interday[b]	8.4% (authentic sample); 9.7% (blank sample)
Recovery range	103.9% (at 80 mg/L); 110.4% (at 200 mg/L)

[a]Limit of detection (LOD) and quantitation (LOQ) were determined by establishing a separate calibration curve in the range of 0.1–1.0 mg/L. The limits were calculated from the residual standard deviation of the regression line [29].
[b]Precisions are expressed as relative standard deviation (RSD) (%), intraday ($n = 6$), interday ($n = 5$).

Only two of the 257 analysed authentic alcoholic beverage samples from different countries contained DEP, in concentrations of 608 mg/L and 210 mg/L (Figure 4). Both samples were obtained from Lithuania and were available on the market as cosmetic alcohol ("for Men Eau de Cologne" and "Cologne Syren Cupe Hb"). However, these were clearly sold for human consumption [5].

None of the other phthalates included in this study (DMP, DAP, DIBP, DBP, DEHA, BBP, DEHP, DHP, DNOP) was found in any of the 257 samples (this is also the reason why we have focused method optimization and validation purely on DEP).

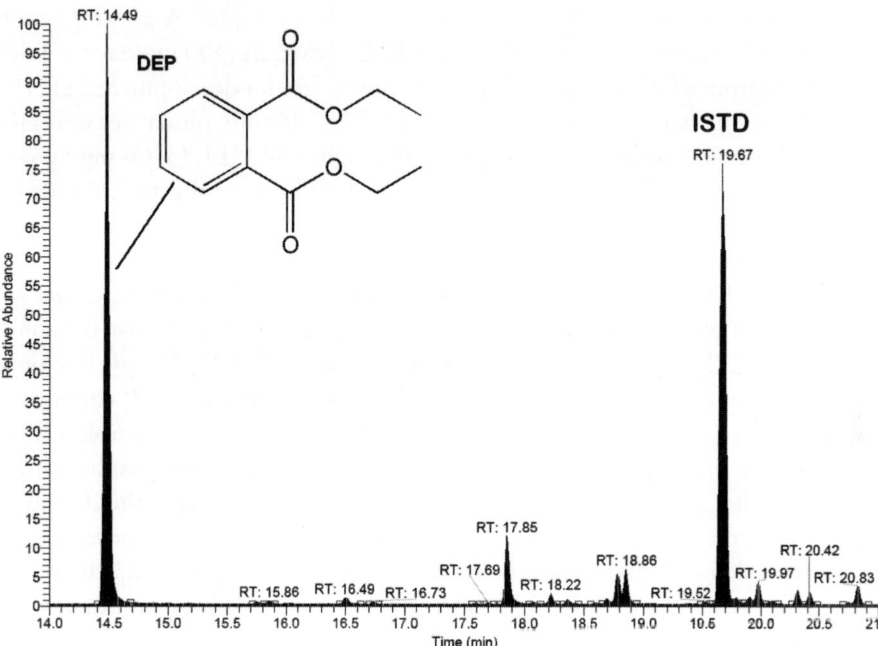

Figure 4. Chromatogram of an unrecorded alcohol sample from Lithuania containing 608 mg/L of diethyl phthalate (DEP); 3,4,5,6-d4-di-2-ethylhexyl phthalate (d4-DEHP) was used as internal standard (ISTD).

Discussion

The results of our large collective of samples suggest that the public health risk relevance of phthalates in alcohol is restricted to DEP in certain unrecorded alcohols. Recorded alcohols and migration from plastic bottles (many of the samples in the study were filled in plastic bottles) appear to be of negligible risk. The problem may also be restricted to legislations that allow the use of diethyl phthalate to denature alcohol. However, the sample size in some of the countries investigated was relatively small and does not allow a final conclusion on the worldwide scale of phthalate contamination in alcohol.

Regarding DEP, there are several reviews on its toxicological profile that raise concerns because of the ubiquitous occurrence. After inhalation, oral or dermal uptake, DEP is mainly hydrolyzed to ethanol and monoethyl phthalate, which is then excreted in the urine [30]. DEP exerts low acute toxicity; lethal doses are reported in the range of 9–31 g/kg [31]. No carcinogenic effects after dermal or oral exposure were detected in long-term studies with rodents [30], [32]. An oral reference dose (RfD) was set by the US Environmental Protection Agency [33] at 0.8 mg/kg bodyweight/day (48 mg/day for a 60-kg-human) with extrapolation from the short-term animal toxicity experiments of Brown et al. [34]. A working group of the WHO [35] estimated a TDI of 5 mg/kg bodyweight (300 mg/day for a 60-kg-human) from a NOAEL of 1600 mg/kg bodyweight for developmental effects, derived from the same study of Brown et al. [34]. In addition, phthalates were also suspected as being hepatotoxic, with a slightly higher NOAEL (3160 mg/kg/day for male rats [36]). According to the RfD and the TDI for DEP we estimated a drinking volume of the two DEP containing samples a 60-kg person could drink everyday without expecting adverse effects. Subsequently consumption of more than 79 mL–1.4 L of our positive samples could be critical (Table 4). Under assumption of a 60-kg person drinking 100 mL of phthalate contaminated alcohol with 40% vol, the alcohol could contain 480–3000 mg/L (120–750 g/hL of pure alcohol) of DEP to reach the RfD or TDI. While the use of DEP-containing products above this level might be safe for cosmetic use on the human skin [30], [37], we have major concerns about the oral use of these products. As the critical level in alcohol products appears to be in the upper mg/L-range, the detection limit of our procedure of 0.7 mg/L appears to be adequate for the purpose, and we have refrained from further sample preparation steps (e.g., solid phase extraction) that might improve the detection limit but also increase the time and cost of the assay.

Table 4. Toxicological evaluation of the DEP containing samples from Lithuania.

Product	Drinking volume for a 60-kg person to exhaust toxicological threshold	
	US EPA RfD (0.8 mg/kg bodyweight/day)	WHO TDI (5 mg/kg bodyweight/day)
Surrogate alcohol #1 (608 mg/L)	79 mL	493 mL
Surrogate alcohol #2 (210 mg/L)	229 mL	1429 mL

Further investigations are necessary to estimate the risk of DEP intake for humans. Some member states of the European Union still allow the use of DEP as a denaturing agent. In Germany, 0.5 kg DEP can be used for denaturing 100 L ethanol for the production of cosmetic agents or agents used to improve the odour [38]. In Russia, a regulation for the use of DEP as a denaturing agent for ethanol and ethanol-containing products existed [39]. However, this regulation appears to have been amended, without the inclusion of DEP [40].

The high amount of DEP in samples from Lithuania cannot be explained by the migration of DEP out of the packaging material. Instead, it documents the use of DEP as a denaturing agent for the alcohol or as an ingredient in the cosmetic as a vehicle for fragrance [37]. It can be assumed that this alcohol was only declared as perfume or eau-de-cologne to evade taxation, but in fact it was offered for human consumption.

Another aspect of the toxicological evaluation that cannot be excluded is the interaction of DEP with ethanol, which is also hepatotoxic. We also have found a third liver toxic compound, coumarin, in the same two products that contained DEP above international limits [5]. Therefore, we have three liver-toxic agents in the unrecorded alcohols (ethanol, DEP and coumarin). We currently do not have enough evidence to postulate a real public health threat, because the occurrence of DEP and coumarin in unrecorded alcohol is generally unknown. However, regions with high consumption of unrecorded alcohol also have higher incidence of liver cirrhosis, which cannot be explained by the volume and patterns of drinking alone [5]. Therefore, constituents in unrecorded alcohol, such as DEP, provide an interesting hypothesis to explain these increased effects, and are worthy of further investigation. The methodology developed in this study will be used to analyze a larger collective of unrecorded samples in the context of the Alcohol Measures for Public Health Research Alliance (AMPHORA), a collaborative project funded under the European Commission Seventh Framework Program.

Currently we can conclude that DEP seems to be unsuitable as a denaturing agent as it has no effect on the organoleptic properties of ethanol and can easily be separated by distillation [4]. For cosmetics, oils that are part of the recipe anyway or bitter agents such as denatonium benzoate (bitrex®) should be used as alternative denaturing agents [41]. Consequently, prohibition of the use of DEP as a denaturing agent should be considered, as its toxicological effects are still uncertain, but a clear potential public health risk exists.

Acknowledgements

H. Havel is thanked for excellent technical assistance. All collaborators in the corresponding projects on unrecorded alcohols are thanked for providing samples.

Authors' Contributions

Conceived and designed the experiments: TK DWL. Performed the experiments: JL. Analyzed the data: JL TK JR DWL. Contributed reagents/materials/analysis tools: TK. Wrote the paper: JL TK JR DWL.

References

1. Chou K, Wright RO (2006) Phthalates in food and medical devices. J Med Toxicol 2: 126–135.

2. Mikula P, Svobodova Z, Smutina M (2005) Phthalates: Toxicology and Food Safety—a Review. Czech J Food Sci 23: 217–223.

3. Heudorf U, Mersch-Sundermann V, Angerer J (2007) Phthalates: Toxicology and Exposure. Int J Hyg Env Health 210: 623–634.

4. Savchuk SA, Nuzhnyi VP, Kolesov GM (2006) Factors affecting the accuracy of the determination of diethyl phthalate in vodka, ethanol, and samples of illegal alcoholic products. J Anal Chem 61: 1198–1203.

5. Lachenmeier DW, Sarsh B, Rehm J (2009) The composition of alcohol products from markets in Lithuania and Hungary, and potential health consequences: A pilot study. Alcohol Alcohol 44: 93–102.

6. EFSA (2005) Opinion of the Scientific Panel on Food Additives, Flavourings, Processing Aids and Material in Contact with Food (AFC) on a request from the Commission related to Di-Butylphthalate (DBP) for use in food contact materials. EFSA J 242: 1–17.

7. EFSA (2005b) Opinion of the Scientific Panel on Food Additives, Flavourings, Processing Aids and Materials in Contact with Food (AFC) on a request from the Commission related to bis(2-ethylhexyl)phthalate (DEHP) for use in food contact materials. EFSA J 243: 1–20.

8. EFSA (2005) Opinion of the Scientific Panel on Food Additives, Flavourings, Processing Aids and Materials in Contact with Food (AFC) on a request from the Commission related to butylbenzylphthalate (BBP) for use in food contact materials. EFSA J 241: 1–14.

9. Lachenmeier DW, Rehm J (2009) Unrecorded Alcohol: a threat to public health? Addiction 104: 875–877.

10. Rehm J, Kanteres F, Lachenmeier DWUnrecorded consumption, quality of alcohol and health consequences. Drug Alcohol Rev.

11. Loupy A, Pedoussant M, Sansoulte J (1986) Solid-Liquid Phase Transfer Catalysis without Solvent: Mild and Efficient Conditions for Saponifications and Preparations of Hindered Esters. J Org Chem 51: 740–742.

12. Schwetlick K (2001) Organikum. 21st Ed. Chapter D 2.6.3, p. 241. Wiley-VCH Verlag Weinheim, Germany.

13. Ejim OS, Brands B, Rehm J, Lachenmeier DW (2007) Composition of surrogate alcohol from South-Eastern Nigeria. African J Drug Alcohol Stud 6: 65–74.

14. Lachenmeier DW, Kanteres F, Kuballa T, López MG, Rehm J (2009) Ethyl Carbamate in Alcoholic Beverages from Mexico (Tequila, Mezcal, Bacanora, Sotol) and Guatemala (Cuxa): Market Survey and Risk Assessment. Int J Env Res Pub Health 6: 349–360.

15. Kanteres F, Lachenmeier DW, Rehm J (2009) Alcohol in Mayan Guatemala: consumption, distribution, production and composition of cuxa. Addiction 104: 752–759.

16. Kanteres F, Rehm J, Lachenmeier DW (2009) Artisanal alcohol production in Mayan Guatemala: Chemical safety evaluation with special regard to acetaldehyde contamination. Sci Total Environ 407: 5861–5868.

17. Lachenmeier DW, Ganss S, Rychlak B, Rehm J, Sulkowska U, et al. Association between quality of cheap and unrecorded alcohol products and public health consequences in Poland. Alcohol Clin Exp Res 33: 1757–1769.

18. Lachenmeier DW, Anh PTH, Popova S, Rehm J (2009) The quality of alcohol products in Vietnam and its impications for public health. Int J Environ Res Public Health 6: 2090–2101.

19. Lachenmeier DW, Kuballa T, Lima MCP, Nóbrega ICC, Kerr-Corrêa F, et al. (2009) Ethyl carbamate analysis in German fruit spirits and Brazilian sugarcane spirits (cachaça): Improved sample cleanup with automated parallel evaporation. Deut Lebensm Rundsch 105: 507–512.

20. Roth M, Hartmann S, Renner R, Hörtig W (2007) Risikoorientiertes Probenmanagement in Baden-Württemberg. Deut Lebensm Rundsch 103: 45–52.

21. European Council (1993) Council Directive 93/99/EEC of 29 October 1993 on the subject of additional measures concerning the official control of foodstuffs. Off J Europ Comm L290: 14–7.

22. ISO (2005) ISO/IEC 17025:2005 General requirements for the competence of testing and calibration laboratories. Geneva Switzerland: International Organization for Standardization.

23. Aurela B, Kulmala H, So"derhjelm L (1999) Phthalates in paper and board packaging and their migration into Tenax and sugar. Food Add Contaminants 16: 571–577.

24. Del Carlo M, Pepe A, Sacchetti G, Compagnone D, Mastrocola D, et al. (2008) Determination of phthalate esters in wine using soild-phase extraction and gas chromatography-mass spectrometry. Food Chem 111: 771–777.

25. Gruber L, Wolz G, Piringer O (1998) Untersuchung von Phthalaten in Baby-Nahrung, Deut Lebensm-Rundsch 94: 177–179.

26. Kueseng P, Thavarungkul P, Kanatharana P (2007) Trace Phthalate and Adipate Esters Contaminated in Packaged Food. J Environ Sci Health B 42: 569–576.

27. Pfordt J (2004) Di(2-ethylhexyl)phthalat (DEHP) und Dibutylphthalate in einigen Lebensmitteln mit Kunststoffverpackungen und in Frauenmilch. Deut Lebensm Rundsch 100: 431–436.

28. Rapp A, Hastrich H, Yavas I, Ullemeyer H (1994) Zur einfachen, schnellen Anreicherung ("Kaltronmethode") und quantitativen Bestimmung von flüchtigen Inhaltsstoffen aus Spirituosen: Bestimmung von Thujon, Safrol, Isosafrol, β-Asaron, Pulegon und Cumarin. Branntweinwirtsch 134: 286–289.

29. DIN 32645 (1994) Chemische Analytik: Nachweis-, Erfassungs- und Bestimmungsgrenze, Ermittlung unter Wiederholbedingungen. Begriffe, Verfahren, Auswertung. Beuth Verlag, Berlin, Germany.

30. Api, AM (2001) Toxicological profile of diethyl phthalate: a vehicle for fragrance and cosmetic ingredients. Food Chem Toxicol 39: 97–108.

31. Shibko SI, Blumenthal H (1973) Toxicology of Phthalic Acid Esters Used in Food-Packaging Material. Env Health Perspect 3: 131–137.

32. NTP (1995) NTP technical report on the toxicology and carcinogenesis studies of diethyl phthalate (CAS No. 84-66-2) in F344/N rats and B6C3F1 mice (dermal studies) with dermal initiation/promotion study of diethyl phthalate and diemethyl phthtalate (CAS No 131-11-3) in male Swiss (CD-1) mice. Research Triangle Park, NC, US Department of Health and Human Services, National Institute of Health, National Toxicology Program (NTP TR 429).

33. Diethyl phthalate (CASRN 84-66-2) (2003) Integrated Risk Information System. US Environmental Protection Agency. Washington, DC. Available: http://www.epa.gov/iris/subst/0226.htm via the Internet.

34. Brown D, Butterworth KR, Gaunt IF, Grasso P, Gangolli SD (1978) Short-Term Oral Toxicity Study of Diethyl Phthalate in the Rat. Food Cosmet Toxicol 16: 415–422.

35. WHO (2003) Diethyl Phthalate. Concise International Chemical Assessment Document 52. World Health Organization, Geneva, Switzerland.

36. ATSDR (1995) Toxicological Profile for diethyl phthalate. Atlanta, GA, US Department of Health and Human Services. Agency for Toxic Substances and Disease Registry.

37. Andersen FA (2005) Annual Review of Cosmetic Ingredient Safety Assessments-2002/2003: Dibutyl Phthalate, Diethyl Phthalate and Dimethyl Phthalate. Int J Toxicol 24: 43–42.

38. Branntweinsteuerverordnung (2008) Branntweinsteuerverordnung vom 21. Januar 1994 (BGBl. I S. 104), zuletzt geändert durch Artikel 3 der Verordnung vom 19. März 2008. Bundesgesetzbl I 450–456.

39. Russian regulation (1998) Regulation of the Government of Russian Federation #732 of June 9th 1998 about implementation of state registration of denatured ethanol and spirits-containing products made of all types of substrates. Compendium of Laws of Russian Federation 28: 3365.

40. Russian regulation (2006) Regulation # 349 of June 3rd 2006 about performing of state control over the process of denaturation (use of denaturating substances) of ethanol and spirits-containing non-alimentary products and the contents of denaturating substances in them. Compendium of Laws of Russian Federation 24: 2602.

41. Lachenmeier DW, Rehm J, Gmel G (2007) Surrogate alcohol: what do we know and where do we go? Alcohol Clin Exp Res 31: 1613–1624.

Effects of Photochemical Formation of Mercuric Oxide

Evan J. Granite, Henry W. Pennline and James S. Hoffman

ABSTRACT

The photochemistry of elemental mercury and oxygen was examined using quartz flow reactors. Germicidal bulbs were used as the source of 253.7 nm ultraviolet radiation. The formation of mercuric oxide, as visually detected by yellow-brown stains on the quartz walls, was confirmed by both ICP-AES and SEM-EDX analyses. In addition, a high surface area calcium silicate sorbent was used to capture the mercuric oxide in one of the experiments. The implications of mercuric oxide formation with respect to analysis of gases for mercury content, atmospheric reactions, and direct ultraviolet irradiation of flue gas for mercury sequestration are discussed.

Introduction

The thermal oxidation of elemental mercury in air proceeds within the narrow temperature band between 350- 500EC.[1] However, ultraviolet light allows the oxidation to occur at much lower temperatures due to the formation of ozone. Mercury is a known photo sensitizer for the formation of ozone from oxygen. Ozone is a powerful oxidizer that can rapidly react with elemental mercury to form mercuric oxide.

The photochemistry of mercury has been extensively studied.[2-7] Dickinson and Sherrill found that 253.7 nm irradiation of oxygen in the presence of mercury vapor produced a decrease in pressure, which they ascribed to the formation of ozone, viz:

$$3\ O_{2(gas)} \bullet 2\ O_{3(gas)} \tag{1}$$

Dickinson and Sherrill also noted yellow-brown deposits on the walls of their quartz reactor, which they attributed to the formation of mercuric oxide via thermal oxidation of mercury by ozone, as shown below:

$$Hg_{(gas)} + O_{3(gas)} \bullet HgO_{(solid)} + O_{2(gas)} \tag{2}$$

Mercury is usually detected by the absorption or emission of 253.7 nm ultraviolet radiation. The measurement of mercury will be affected by the photosensitized ozone reaction, unless precautions are implemented, e.g., using an inert carrier gas. Many of the proposed on-line detectors (continuous emissions monitors) for mercury in flue gas are based upon atomic absorption (AAS) or atomic fluorescence spectrophotometry (AFS) utilizing 253.7 nm radiation.[8] Excited mercury atoms can revert to the ground state when colliding with a polyatomic species. Oxygen is among the most efficient quenching agents known.[9-11] Mercury excited by the absorption of 253.7 nm ultraviolet light can initiate the formation of ozone from oxygen. The ozone can then react with mercury to form mercuric oxide, which could deposit on the quartz detection cell. Furthermore, ozone is often measured by the absorption of 253.7 nm ultravioletradiation.[7] These two factors can greatly complicate the ultraviolet determination of mercury in carrier gases containing oxygen.

Ozone chemistry can effect the fate of mercury in coal-fired utilities. Electrostatic precipitators are used in a vast majority of coal-fired utilities. The corona discharge will generate both ultraviolet radiation and ozone.[7,12-17] Ozone can potentially react with elemental mercury in flue gas to form mercuric oxide, although it most likely reacts with other flue gas components whose concentrations are orders of magnitude larger than mercury.

Additionally, mercury has an impact upon ozone chemistry in the upper levels of the atmosphere. Mercuric oxide, associated with fine particulates, has been detected recently in the tropopause and is speculated to form by oxidation of elemental mercury by ozone.[18,19] Ground level mercury concentrations in the Arctic have also been found to vary with seasonal changes in sunlight, temperature, and upper atmosphere ozone level.[20,21] Ground level mercury could also impact the level of ground level ozone, where ozone is a prime constituent of smog.[22]

Experimental

The assembly used for studying the photochemical oxidation of mercury consists of an elemental mercury permeation tube, a quartz photoreactor, and an ultraviolet lamp. The reactor scheme is shown in Figure 1. A certified Dynacal permeation tube from VICI Metronics is used as the source of elemental mercury. The permeation tube has been certified by the manufacturer to release 144 ng Hg/min at 212°F. The permeation tube is located at the bottom of a Dynacal glass U-tube, which is maintained at 212°F ± 1.6°F at all times by immersing it in a Haake D8 oil bath. A flow (30 ml/min) of gas passes over the permeation tube and is maintained at all times with a thermal conductivity mass flow controller. The output of the permeation tube and the flow rate of carrier gas yields a calculated concentration of mercury of 585 ppb. The mercury output of the tube has been verified on a regular basis via weight loss measurement and has been found to be consistent (152 ng Hg/min) with the certified release.

Two quartz photoreactors were used to study the oxidation of elemental mercury. The first photoreactor is a 20 inch long by 1/4 inch outer diameter cylindrical quartz tube. The second photoreactor is a 1-3/4 inch long by 1/2 inch wide rectangular clear flow-through quartz cuvette from Brooks Rand. Quartz, unlike glass, is sufficiently transparent towards 253.7 nm ultraviolet radiation. All of the plumbing and valves which come into contact with mercury are constructed from either stainless steel or teflon. These materials have been demonstrated to have good chemical resistance and inertness towards mercury. A Raytech LS¬7 Ultraviolet lamp was used as the source of 253.7 nm light for the 20 inch quartz photoreactor. Four watt mercury germicidal bulbs (SW-6) were used as the source of 253.7 nm radiation. Similar four watt unfiltered mercury germicidal lamps (Bulbtronics bulb G4T5) were used as the source of 253.7 nm radiation when the quartz cuvette photoreactor was used. Eighty percent of the output radiation is 253.7 nm light.[23] All of the experiments were done at ambient temperature, at near ambient pressure, and with a carrier gas flow-rate of 30 cm³/sec and a mercury concentration of 585 ppb.

The quartz cuvette cell was part of a detector (Brooks Rand CVAFS-2 cold vapor atomic fluorescence spectrophotometer (AFS)) for determining elemental mercury. When used as a continuous on-line monitor for elemental mercury in argon, the detection limit is around 1 ppb. The AFS is a ultraviolet detector for elemental mercury; mercury atoms absorb 253.7 nm light and re-emit (fluoresce) this wavelength. A mercury bulb serves as the ultraviolet source, and a photomultiplier tube serves as the ultraviolet fluorescence detector. Any gas can be used as a carrier, although sensitivity can vary dramatically, due to quenching of the excited Hg atoms by collisions with polyatomic species. Therefore, maximum sensitivity is achieved with high purity argon or helium carrier gases. In comparison to argon, nitrogen was found to reduce the relative response of the AFS to elemental mercury by a factor of 10. In comparison to argon, air was found to reduce the relative response of the AFS to elemental mercury by a factor of 100. This is in good agreement with the data provided by the manufacturer of the detector.[23] However, the response of the detector decayed rapidly in oxygen containing carrier gases, due to the deposition of mercuric oxide upon the quartz cell walls.

Key process parameters were recorded with a data acquisition system. This on-line data acquisition system was used to take and store the various voltage signals from the thermocouples, flow meters, and the atomic fluorescence spectrophotometer. Data sampling occurred every 15 seconds.

Scanning electron microscopy with energy-dispersive x-ray methods (SEM-EDX) was used to confirm the formation of mercuric oxide (HgO) on the walls of the quartz photoreactor. A Leica 360i scanning electron microscope was used with a Kevex Delta EDX spectrometer. The Kevex spectrometer is equipped with a Quantum Superdry light element detector. Inductively coupled argon plasma atomic emission spectroscopy (ICP-AES) was used to quantitatively determine the mass of mercury converted to mercuric oxide. The instrument is a Perkin-Elmer Optima 3000 Radial View Spectrometer. The yellow stains on the quartz wall were extracted with warm nitric acid. A small quantity of the solution was then aspirated into the argon torch, and mercury was quantitatively determined by the intensity of the 253.7 nm emission line.

In one test, ten milligrams of a sorbent was placed in the cylindrical reactor and irradiated. The sorbent (Manville Micro-Cel) is a synthetic calcium silicate with a BET surface area given by the manufacturer as between 100 - 200 m^2/gram. The sorbent was exposed for 350 minutes to 30 ml/min of 585 ppb Hg in air while being irradiated by 253.7 nm light. The used sorbent was digested in aqua regia and the capacity was determined by ICP-AES analysis.

Results and Discussion

The results of the removal of elemental mercury in the presence of oxygen as mercuric oxide is shown in Table 1. Mercury removals are defined as the mass of mercury deposited in the reactor divided by the inlet mass of mercury. The mass of mercury within the reactor is determined by the analysis of the mercury deposits; the inlet mass of mercury was calculated from the known rate of mercury release from the permeation tube and the total time of the test. As can be seen in Table 1, a high level of removal for the empty quartz reactors was achieved by photoxidation. The empty cuvette reactor was part of the Brooks Rand AFS. The voltage signal from the AFS decayed rapidly in oxygen containing carrier gases. For the 4% oxygen carrier gas, the voltage signal decayed by over 56% due to the mercuric oxide film that formed within the quartz cuvette. Thus the tracking of mercury concentration with the AFS could not be conducted.

Table 1. Photochemical Removal of Mercury as Mercuric Oxide

Test	Carrier Gas	Time On-line (hr)	% Hg Removal
Empty Cuvette Reactor	4% O_2 in N_2	28	83
Empty Cuvette Reactor	Air	6	100
Empty Cylindrical Reactor	Air	21.3	100
Packed Bed Sorbent	Air	6	42

It should be noted that the recovery of mercury was accomplished by extraction with warm nitric acid. This may not be the optimal recovery method, but it was improved for the sorbent experiment. It is also noted that ICP-AES is not the optimal method for the quantitative determination of mercury. Nevertheless, SEM-EDX analysis clearly confirms the formation of mercuric oxide on the quartz walls. The formation of mercuric oxide on quartz walls in similar experiments was first noted by Dickinson.[2]

The removal of mercury as mercuric oxide captured on Micro-Cel was examined at room temperature. In this experiment, 585 ppb of elemental mercury in air was exposed to 253.7 nm radiation while passing over a high surface area calcium silicate sorbent for six hours. The mass of mercury captured on the silicate was found to be 22.5 micrograms out of the 53.2 micrograms passed over the sorbent. An improved procedure was used for the extraction of mercury from the used sorbent by soaking in aqua regia. However, some of the mercury appears to have adhered upon the quartz walls as mercuric oxide, as evidenced by a reddish-brown stain.

The mechanism for the removal of mercury as mercuric oxide was deduced by Dickinson and Sherrill[2] and Bamford and Tipper[5] and is shown below:

$$Hg + 253.7 \text{ nm light} \bullet Hg^* \tag{3}$$

$$Hg^* + O_2 \bullet Hg + O_2^* \tag{4}$$

$$O_2^* + O_2 \bullet O_3 + O \tag{5}$$

$$Hg + O_3 \bullet HgO + O_2 \tag{2}$$

$$O_2 + O \bullet O_3 \tag{6}$$

Reaction (3) is the excitation of elemental mercury by 253.7 nm uv radiation. Reaction (4) is the quenching of the excited mercury atom by oxygen, with the formation of an excited oxygen molecule. Step (5) is the quenching of an excited oxygen molecule, with the formation of ozone and an oxygen atom. Reaction step (2) is the thermal reaction of elemental mercury and ozone, with the formation of mercuric oxide and oxygen. Reaction (6) is the combination of an oxygen molecule with a reactive oxygen atom to form ozone.

The overall reaction is the sum of reaction steps (3), (4), (5), (2), and (6):

$$Hg + 2 O_2 + 253.7 \text{ nm light} \bullet HgO + O_3 \tag{7}$$

Biswas[14,15] found that the longer 360 nm uv light alone is ineffective at oxidizing mercury in air. The 360 nm wavelength was highly effective when used with a titanium oxide photooxidation catalyst, capturing mercury as mercuric oxide in a heterogeneous oxidation reaction.[14,15] The experiments described in the present research note involve gas phase oxidation to convert elemental mercury to mercuric oxide, with subsequent deposition of the oxide on a quartz or silicate surface.

One reason for conducting the above rudimentary tests was due to previous attempts at using the on-line AFS to measure elemental mercury in gases containing oxygen. It was found that the voltage signal from the AFS rapidly decayed, presumably due to the deposition of mercuric oxide on the quartz cell walls. The photosensitized formation of ozone can interfere with the ultraviolet measurement of elemental mercury by several ways: absorption of ultraviolet radiation by ozone; decrease in the population of mercury atoms by formation of mercuric oxide; and attenuation in the detected intensity of absorbed (AAS) or emitted (AFS) 253.7 nm radiation by elemental mercury due to the deposition of mercuric oxide on the quartz cell walls. Additionally, a reduction of the population of excited mercury atoms via energy transfer to oxygen (quenching) can influence the ultraviolet measurement of elemental mercury. However, the formation and deposition of mercuric oxide can be inhibited by heating the quartz photocell, although heating will neither stop the quenching of excited mercury atoms nor the absorption of 253.7 nm radiation by ozone. It is noted that several prototype continuous emissions monitors for mercury in flue gas, based upon absorption of 253.7 nm light, use quartz cells that are heated to over 500•C.[8]

The direct irradiation of flue gas by 253.7 nm light could be a method for the removal of elemental mercury. A typical flue gas composition from a coal-fired utility can contain the following: 4% O_2, 16% CO_2, 6% H_2O, 1000 ppm SO_2, 500 ppm NO_x, 10 ppm hydrocarbons, 1 ppb Hg, and the remainder N_2. The concentration weighted quenching effect for each gas can be obtained from the cross-sectionalarea for quenching given by Bamford and Tipper.[5] Oxygen will be an important quenching agent in flue gas, but carbon dioxide, nitrogen, and water are also significant quenching species. The quenching of excited mercury atoms by carbon dioxide, nitrogen, and water could significantly reduce the amount of ozone formed by reactions (5) and (6) by reducing the population of excited oxygen molecules. Any ozone, a powerful oxidizer, formed through reactions (5) and (6) may also oxidize other components of the flue gas, such as sulfur dioxide, nitric oxide, carbon monoxide, and unburned hydrocarbons. These oxidations will compete with the oxidation of mercury by ozone shown by reaction (2). It is also noted that these oxidizable species are present in flue gas at concentrations which are orders of magnitude greater than the concentration of mercury. These factors could adversely impact the removal of mercury as mercuric oxide from flue gas by direct 253.7 nm irradiation.

Conclusions

Based on the experimental results presented in this research note and the experience of past researchers, certain concerns must be addressed in the development of on-line ultraviolet continuous emissions monitors for the measurement of mercury in flue gas. The formation of mercuric oxide can interfere with the on-line continuous monitoring of elemental mercury in oxygen-containing gas streams. As other researchers have speculated, ozone formation can impact results since ozone can also absorb 253.7 nm radiation and can interfere with the determination of mercury even in the absence of reaction to form mercuric oxide. The quartz cells used in the ultraviolet detection of mercury could be heated above 500•C, the thermal decomposition temperature of HgO, in order to prevent the formation of mercuric oxide.

The relationship between the mercuric oxide and ozone formation through 253.7 nm irradiation warrents further attention as a technique to remove mercury from flue gas. Additionally, the presence of mercury and ozone in the upper levels of the atmosphere may have an environmental impact. Before being converted to mercuric oxide, as suggested by some researchers, elemental mercury could act as a photosensitizer for the formation of ozone in the upper atmosphere. Halogens in the upper atmosphere are known photosensitizers for the decomposition of ozone. Elemental mercury in the upper levels of the atmosphere may help mitigate some of the damaging effects of halogens upon the ozone layer. However, elemental mercury in the lower levels of the atmosphere may contribute to smog formation by acting as a photosensitizer for the formation of ground level ozone.

Acknowledgements

Evan Granite appreciates the support of a postdoctoral fellowship at the United States Department of Energy administered by Oak Ridge Institute For Science and Education. Robert Thompson of Parsons Infrastructure and Donald Martello of the United States Department of Energy provided outstanding chemical analyses.

Disclaimer

Reference in this paper to any specific commercial product, process, or service is to facilitate understanding and does not necessarily imply its endorsement by the United States Department of Energy.

Literature Cited

1. Greenwood, N.N. and Earnshaw, A., Chemistry of the Elements, Pergamon Press: New York, 1984.

2. Dickinson, R.G. and Sherrill, M.S., Formation of Ozone By Optically Excited Mercury Vapor, in Proceedings National Academy Science 1926, 12, 175.

3. Volman, D.H., Photochemical Gas Phase Reactions in the Hydrogen-Oxygen System, pp. 43-82, in Advances in Photochemistry, vol.1, John Wiley: New York, 1963.

4. Kondrat'ev, V.N., Chemical Kinetics of Gas Reactions, Pergamon: New York, 1964.

5. Bamford, C.H. and Tipper, C.F.H., editors, Comprehensive Chemical Kinetics, volume 3, The Formation and Decay of Excited Species, Elsevier: New York, 1969.

6. Maron, S.H. and Lando, J.B., Fundamentals of Physical Chemistry, Macmillan: New York, 1974.

7. Horvath, M., Bilitzky, L., and Huttner, J., Ozone, Elsevier: New York, 1985.

8. Silverberg, P., Cooper, C., and Ondrey, G., Monitoring Emissions Of Toxic Metals, Chemical Engineering 1999, 106 (4), 43.

9. Christian,G.D. and O'Reilly, J.E., Instrumental Analysis, Allyn and Bacon: Boston, 1986.

10. Willard, H.H., Merritt, L.L., Dean, J.A., and Settle, F.A., Instrumental Methods of Analysis, Van Nostrand: New York, 1981.

11. Moore, W.J., Physical Chemistry, Prentice Hall: New Jersey, 1972.

12. Loeb, L.B., Electrical Coronas, University of California Press: Los Angeles, 1965.

13. Murphy, J.S. and Orr, J.R., editors, Ozone Chemistry and Technology, Franklin Institute Press: Philadelphia, 1975.

14. Biswas, P. and Wu, C.Y., Control of Toxic Metal Emissions From Combustors Using Sorbents: A Review, Journal of the Air & Waste Management Association 1998, 48, 113.

15. Biswas, P., Wu, C.Y, Lee, T-G, Tyree, G., and Arar, E., Capture of Mercury in Combustion Systems by In Situ Generated Titania Particles with UV Irradiation, Environmental Engineering Science 1998, 15 (2), 137.

16. Perry, R.H. and Green, D., editors, Perry's Chemical Engineers' Handbook, 6th edition, McGraw Hill: New York, 1984.

17. Roberts, L.M. and Walker, A.B., Electrostatic Precipitation, in Kirk-Othmer Encyclopedia of Chemical Technology, vol.8, John Wiley: New York, 1965.

18. Murphy, D.M., The Composition of Aerosol Particles at 5-19 km Altitude. Annual Meeting of American Geophysical Union, San Francisco, CA, December 6-10, 1998.

19. Murphy, D.M., Thomson, D.S., and Mahoney, M.J., In Situ Measurements of Organics, Meteoritic Material, Mercury, and Other Elements in Aerosols at 5-19 Kilometers, Science 1998, 282, 1664.

20. Schroeder, B., Springtime Transformation of Atmospheric Mercury Vapor in the Artic. Annual Meeting of American Geophysical Union, San Francisco, CA, December 6-10, 1998.

21. Anlauf, K.G., Artic Ozone Measurements. Annual Meeting of American Geophysical Union, San Francisco, CA, December 6-10, 1998.

22. National Air Quality And Emissions Trends Report 1997, EPA 454/R-98-016, United States Environmental Protection Agency, Office of Air Quality, Research Triangle Park, NC, 1998.

23. Brooks Rand, Manual for CVAFS-2 Atomic Fluorescence Spectrophotometer, Seattle, WA, 1993.

Novel Sorbents for Mercury Removal from Flue Gas

Evan J. Granite, Henry W. Pennline and Richard A. Hargis

ABSTRACT

A laboratory-scale packed-bed reactor system is used to screen sorbents for their capability to remove elemental mercury from various carrier gases. When the carrier gas is argon, an on-line atomic fluorescence spectrophotometer (AFS), used in a continuous mode, monitors the elemental mercury concentration in the inlet and outlet streams of the packed-bed reactor. The mercury concentration in the reactor inlet gas and the reactor temperature are held constant during a test. For more complex carrier gases, capacity is determined off-line by analyzing the spent sorbent with either a cold vapor atomic absorption spectrophotometer (CVAAS) or an inductively coupled argon plasma atomic emission spectrophotometer (ICP-AES). The capacities and breakthrough times of several commercially available activated carbons, as well as novel sorbents, were determined as a function of various parameters. The mechanisms of mercury removal by the sorbents are suggested by combining the results of the packed-bed testing with various analytical results.

Introduction

Over 32% of anthropogenic mercury emissions in the United States are from coal-burning utilities. This percentage will increase over the next few years due to the mandated control of mercury emissions from municipal solid waste and medical waste incinerators. A low concentration of mercury, on the order of 1 ppbv, exists in flue gas when coal is burned. The primary forms in the flue gas are elemental mercury and mercuric chloride.[1]

Control technologies for removing mercury from flue gas include scrubbing solutions and activated carbon sorbents. Mercuric chloride is soluble in water; elemental mercury is not. Dry sorbents have the potential to remove both elemental and oxidized forms of mercury. Activated carbons have been successfully applied for the control of mercury emissions from incinerators.[1,2]

Several sorbents, such as activated carbons, can remove mercury from flue gas produced by the combustion of coal. However, there are problems associated with the use of activated carbons for mercury removal from flue gas. Activated carbons are general adsorbents; most of the components of flue gas will adsorb on carbon, with some in competition with mercury. Carbon sorbents operate effectively over a limited temperature range, typically working best at temperatures well below 300°F. The projected annual costs for an activated carbon cleanup process are high, not only because of the high cost of the sorbent, but also because of its poor utilization/selectivity for mercury. Carbon-to-mercury weight ratios of 3,000:1 to 100,000:1 have been projected.[1,3-5] In addition, activated carbons can only be regenerated a few times before exhibiting an unacceptably low activity for mercury removal. Therefore, the development of improved activated carbons, as well as novel sorbents, merits further research.

A sorbent can capture mercury via amalgamation, physical adsorption, chemical adsorption, and/or chemical reaction. The noble metal sorbents[6-14] can capture mercury via amalgamation. Unpromoted activated carbons and aluminosilicates[15] physisorb elemental mercury. Both amalgamation and physisorption are low temperature processes, typically occurring below 300°F. Chemically promoted (with sulfur, iodine, or chlorine) activated carbons [16-21], selenium [22,23], and manganese dioxide or hopcalite[24,25] are examples of sorbents which chemisorb or chemically react with mercury. Chemisorption and chemical reaction can occur over a wider range of temperatures than physical adsorption and amalgamation. The enthalpy and activation energies of chemisorption/chemical reaction are typically larger than those for physical adsorption.

In this work, which is sponsored by the Advanced Research and Environmental Technology Power Subprogram of the U.S. Department of Energy's Fossil Energy Program, various sorbents were examined for the removal of elemental mercury from argon. It was realized that elemental mercury in flue gas would be more difficult to remove than oxidized mercury, and thus the thrust was to initially identify sorbents that could remove the less reactive elemental mercury. Very few techniques can be used to make an on-line and continuous determination of elemental mercury down to ppb levels, and the exact mechanism by which most sorbents remove mercury is unresolved. The atomic fluorescence spectrophotometer can be used to measure the concentration of elemental mercury in argon on a continuous basis[26] and was used in determining the breakthrough curves of sorbents in a packed bed. When more complex carrier gases were used, capacity was determined off-line via ICP-AES or CVAAS. The capacities of several commercially available activated carbons, as well as metal oxides, a halide salt, metal sulfides, silicates, chlorinated sorbents, a noble metal, and fly ashes were determined.

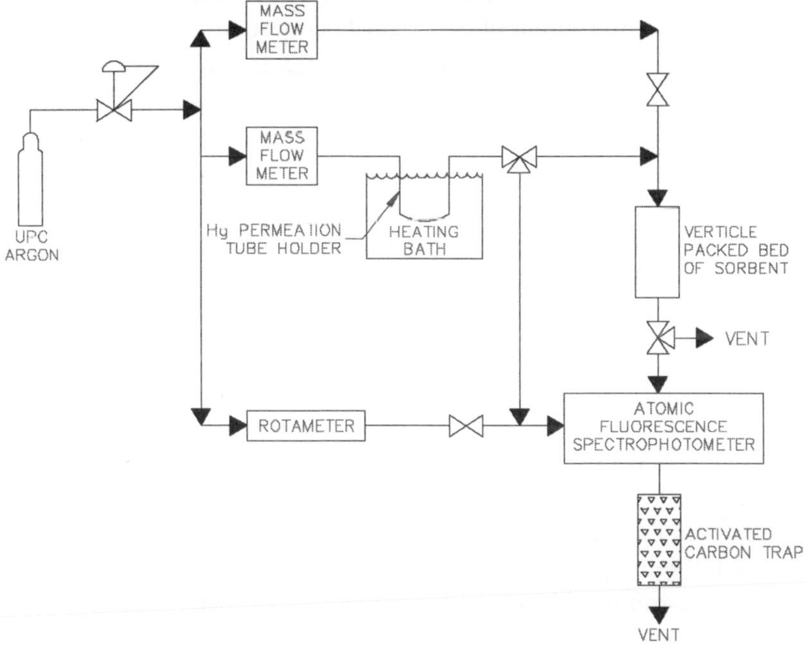

Figure 1. Schematic of Sorbent Screening Unit

Experimental Procedures

The assembly used for measuring sorbent capacities consists of an elemental mercury permeation tube, a packed-bed reactor, an on-line atomic fluorescence spectrophotometer, and a data acquisition system. The reactor scheme is shown in

Figure 1. A certified Dynacal permeation tube from VICI Metronics is used as the source of elemental mercury. The permeation tube has been certified by the manufacturer to release 144 ng Hg/min at 212•F. The permeation tube is located at the bottom of a Dynacal glass U-tube, which is maintained at 212•F ± 1.6•F at all times by immersing it in a Hacke L oil bath. A flow (30-ml/min) of ultra-high purity carrier grade (99.999%) argon gas passes over the permeation tube and is maintained at all times with a thermal conductivity mass flow controller. The output of the permeation tube and the flow rate of argon yields a calculated concentration of mercury in argon of 585 ppb. The mercury output of the tube has been verified on a monthly basis via weight loss measurement and has been found to be consistent (155 ng Hg/min) with the certified release. After a year in service, the output of the permeation tube dropped to 139 ng Hg/min and was replaced with a second certified tube rated for 119 ng Hg/min. The output of this tube has also been verified by weight loss measurement and has been found to be consistent (107 ng/min) with the certified release. Sorbent capacities have all been normalized to reflect the output of the original permeation tube.

The reactor (adsorber) is a quartz tube (20 inches in length with an outer diameter of 1/4 inch and inner diameter of 1/6 inch) held in a vertical position. All of the plumbing and valves which come into contact with mercury are constructed from either stainless steel or Teflon. These materials have been demonstrated to have good chemical resistance and inertness towards mercury. The packed bed of sorbent is surrounded by a large clam-shell furnace. A Self-tune Plus 300 PID controller is used to maintain the bed at the desired temperature. The temperature at the top of the bed has been determined to be within 1.8•F of the temperature at the bottom of the bed.

The detector for elemental mercury is a Brooks Rand CVAFS-2 cold vapor atomic fluorescence spectrophotometer (AFS). When used as a continuous on-line monitor for elemental mercury in argon, the detection limit is below 0.1 ppb. The AFS is a ultraviolet (UV) detector for elemental mercury; mercury atoms absorb 253.7 nm light, and re-emit (fluoresce) this wavelength. A mercury bulb serves as the UV source, and a photomultiplier tube serves as the UV fluorescence detector. Any gas can be used as a carrier, although sensitivity varies dramatically with inertness, due to quenching of the excited Hg atoms by collisions with polyatomic species. Maximum sensitivity (ppt) is achieved with high purity argon or helium carrier gases. When the AFS was used as an on-line detector for elemental mercury, argon was used as the carrier gas. For the more complex carrier gases, sorbent capacity was determined off-line by analyzing the spent sorbent with ICP-AES or CVAAS. The ICP-AES is a Perkin-Elmer Optima 3000 Radial View Spectrometer. The CVAAS is a Cetac M-6000A unit dedicated for the analysis of elemental mercury.

Key process parameters were recorded with a data acquisition system . This on-line data acquisition system was used to take and store the various voltage signals from the thermocouples, flowmeters, and the atomic fluorescence spectrophotometer. Data logging occurred every 15 seconds.

Typically, ten milligram (mg) of 200/325 mesh (45-75 micron) sorbent is placed in the center of the tube and is supported by about 50 mg of quartz wool. The quartz wool and reactor tube have been demonstrated to be inert toward elemental mercury. Separate argon gas streams flow through the bed and through the permeation tube holder. The latter flow is sent to the AFS to determine a baseline for the mercury concentration. Once thermal stability is reached in the reactor, the mercury/argon mixture is diverted to flow through the reactor. Breakthrough curves were generated by plotting the atomic fluorescence spectrophotometer voltage signal at the reactor exit versus time. Sorbent capacities were determined by integration under the breakthrough curve.

Sorbent Preparation

The sorbents examined in this study and their characterization are listed in Table 1. The sorbents I-AC, S-AC, AC-1, and AC-2 are commercially available activated carbons. I-AC is an iodine promoted activated carbon, containing both elemental iodine and potassium iodide. S-AC is a sulfur promoted activated carbon. AC-1 and AC-2 are unpromoted carbons from Calgon and CarboChem respectively. Some typical mercury control applications for AC-1 include municipal waste combustors, hazardous waste combustors, and hospital waste incinerators.[2] AC-2 is a food grade activated carbon used commercially for decolorizing corn syrup.

Cl-AC-1 is a chlorine promoted activated carbon, prepared by boiling AC-1 in 37% hydrochloric acid. Cl-BPL-AC is also a chlorine treated activated carbon prepared by MacDonald.[27] It was formulated by treating the commercially available activated carbon BPL-AC from Calgon with chlorine gas. The chlorine treatment took place in a sealed stainless steel reaction vessel maintained at a pressure of about one half an atmosphere of chlorine gas, at 330EF for 30 minutes. HNO_3-AC-1 is prepared by boiling AC-1 in 70% nitric acid. S-BPL-AC is a sulfur promoted activated carbon prepared by Vidic.[17]

Five novel sorbents were prepared for investigation with chemicals that were typically analytical reagent grade or ACS grade. The vanadium pentoxide dispersed on celkate, a magnesium silicate ($MgSiO_3$) support, was initially prepared by thermally decomposing a mixture of ammonium vanadate with the support to obtain 8 weight percent vanadia. In a later preparation, the supported sorbent $V_2O_5/MgSiO_3$ was prepared by the incipient wetness technique according to the procedure outlined elsewhere [28] using vanadium oxalate solution and the celkate (a synthetic magnesium silicate with surface area 180 ± 25 m^2/g from Manville Products Corp.) support material. Water was added to ammonium meta-vanadate, NH_4VO_3 (J.T. Baker Inc.), and oxalic acid, (Mallincrodt). Reaction occurred immediately and the resultant solution was used to impregnate the celkate support followed by drying at 572°F for 2 hours and calcining at 932°F in an oven with air flow. Incipient wetness occurred at about 0.9 ml/g of celkate. Also, a potassium superoxide promoted vanadium pentoxide (KO_2-V_2O_5) celkate-supported sorbent, whose preparation was similar to the preceding sorbent, was fabricated as well.

Table 1. Characterization of Sorbents

Sorbent	Composition	BET Surface Area m^2/g	
I-AC	3.5% I	750	
Cl-AC-1	6.0% Cl	550	
Cl-BPL-AC	6.7% Cl	1000	
HNO$_3$-AC-1	---	575	
S-BPL-AC	5.9% S	790	
S-AC	7.6% S	690	
AC- 1	0.9% S	650	
AC- 2	0.4% S	900	
Celkate	MgSiO$_3$	160	
Alumina	Al$_2$O$_3$	82	
MoO$_3$/MgSiO$_3$	46% MoO$_3$	70	
MnO$_2$/Al$_2$O$_3$	7% MnO$_2$	65	
V$_2$O$_5$-MgSiO$_3$-1	8% V$_2$O$_5$	91	
V$_2$O$_5$-MgSiO$_3$-2	50% V$_2$O$_5$	60	
KO$_2$-V$_2$O$_5$	3.4% K, 1.4% V	85	
Cr$_2$O$_3$/Al$_2$O$_3$	13% Cr$_2$O$_3$, 11% C	156	
Fe$_2$O$_3$	100% Fe$_2$O$_3$	250	
TS-7	3.5% S	450	
Cl-Celkate	15.0% Cl	80	
MoS$_2$	87% MoS$_2$	50	
FeS	57.8% Fe, 22.6% S	32	
FeS$_2$	81.3 % FeS$_2$	1	
CERF-FA-#2	59.3% C	37	
CERF-FA-#4	37% C		24
FA-1	5% C	5	
WCFA-1	64% C		32
WCFA-1-air-750F	50% C		127
Cl-WCFA-1	----	12	
DCFA-1	29% LOI	16	
DCFA-2	52% LOI	25	
DCFA-3	82% LOI	34	
HFA	23% Ca, 1.1 ppm Hg	2	
CaCl$_2$/Al$_2$O$_3$	10% CaCl$_2$	41	
Pt/wool	40% Pt	20	

The supported sorbent $MoO_3/MgSiO_3$ was prepared by the incipient wetness technique by dissolving ammonium molybdate, $(NH_4)_6Mo_7O_{24} \cdot 4H_2O$ (Fisher Scientific Company), with ammonium hydroxide in distilled water and then contacting the celkate. The solution pH was 8. Impregnation was followed by drying at 248°F for 24 hours and then calcining at 932°F for 6 hours.[29]

The alumina supported MnO_2 sorbent was prepared by the incipient wetness technique using an aqueous solution of manganese nitrate, $Mn(NO_3)_2 \cdot 4H_2O$ (Sigma Chemical Corp.), with alumina, Al_2O_3 (Catalox SCFA 90 with surface area 82 ± 25 m^2/g from Condea Vista). Incipient wetness occurred at about 0.6 ml/g of alumina. Impregnation was followed by the thermal decomposition of manganese nitrate in air at 261°F as outlined elsewhere.[30] Preliminary X-ray diffraction data did not show a MnO_2 diffraction pattern, indicating that the MnO_2 phase was well dispersed over the alumina and that the crystallite size was below 5nm.

The chromium oxide sorbent Cr_2O_3/Al_2O_3 that was obtained from Cadus was prepared by impregnation of alumina by chromic acid at room temperature for 30 minutes.[31] Immediately after impregnation, the sorbent was dried overnight to 122•F in a vacuum oven and then calcined at 1202•F in air for 7 hours.

MACH I Inc. supplied the ferric oxide sorbent Fe_2O_3. It was the Nanocat super fine iron oxide,which is a dark brown amorphous powder. The particle size is 3 nm.

The platinum sorbent Pt/wool was prepared by deposition of Engelhard metallo organic platinum ink upon quartz wool. The ink was fired in air at red heat to form a platinum film.

Pacific Northwest National Laboratory provided a novel self-assembled monolayer thiol promoted aluminosilicate sorbent (TS-7). The sorbent has been successfully applied to purify mercury contaminated water streams. This sorbent has a high BET surface area and is 3.5% sulfur by weight.

A chlorine treated celkate sorbent (Cl-celkate) was prepared by boiling celkate in 37% hydrochloric acid. The slurry is boiled in air until it is thoroughly dry. The boiling hydrochloric acid turns green, indicating the evolution of chlorine.

Themolybdenum sulfide sorbent (MoS_2) is a hydrodesulfurization catalyst prepared in-house at the National Energy Technology Laboratory (NETL). Bulk analysis by ICP-AES indicates a composition of 87% by weight molybdenum sulfide. Surface analysis of the fresh sorbent by x-ray photoelectron spectroscopy (XPS) also indicates a fairly pure sample of MoS_2.

The iron sulfides FeS and FeS_2 (marcasite) were prepared in-house at NETL. FeS contains 57.8% iron and 22.6% sulfur by weight. This suggests iron enrichment as in a non-stoichiometric compound or multi-phase mixture. The FeS_2 sorbent contains 81.3% FeS_2 by weight.

CERF-FA-#2 and CERF-FA-#4 are fly ashes obtained from a 35-lb/hr pulverized coal combustion unit located at NETL. The fly ash samples were derived from the combustion of Pittsburgh #8 coal and were extracted from the furnace at high temperatures, having short residence times for combustion of the coal. The resulting fly ash samples are atypical and extraordinarily high in unburned carbon.

Sorbents were prepared from fly ash in an effort to utilize unburned carbon from the fly ash. The starting material, FA-1, is fly ash obtained from the combustion of Blacksville coal in the 500-lb/hr pilot scale coal combustion unit located at NETL. FA-1 contains 5% carbon and has a BET surface area of 5 m^2/gram. WCFA-1 is a unburned carbon separated from fly ash obtained from the 500-lb/hr combustion unit. The carbon is concentrated from the fly ash through a wet separation technique. WCFA-1 contains 64% carbon and has a BET surface area of 32 m^2/gram.

Cl-WCFA-1 is a chlorine treated carbon derived from fly ash. It is prepared by soaking WCFA-1 in aqua regia for 24 hours and drying in air. Also, WCFA-1-air-750F is prepared by heating the carbon WCFA-1 in air at 750•F for two hours. This is done to increase the BET surface area of the carbon.[32] The thermal oxidation in air increases the microporosity of carbon due to the chemical reaction. WCFA-1 has a BET surface area of 32 m^2/gram, whereas WCFA-1-air-750F has a higher surface area of 127 m^2/gram. The oxidation in air decreases the carbon content from the original 64% down to 50%. DCFA-1 is a fly ash that is high in carbon content due to poor combustion at a commercial utility.

DCFA-2 and DCFA-3 are unburned carbon fractions separated from the DCFA-1 fly ash. The two carbon samples are obtained from the fly ash by a dry separation method (triboelectrostatic) where the first sample is a one-pass separation, and the second is a two-pass separation. The elements present in these sorbents were determined via ICP-AES and are silicon, aluminum, iron, titanium, potassium, calcium, magnesium, phosphorus, and sodium. Sulfur, chlorine, and several other elements were not determined by the ICP-AES, suggesting that the mass balances obtained (near 90%) are reasonable. Silicon and aluminum accounted for 70 to 80 weight percent of these sorbents (excluding the carbon). These carbons were subsequently treated with chlorine by soaking in hydrochloric acid to form Cl-DCFA-1, Cl-DCFA-2, and Cl-DCFA-3.

Additionally, a supported halide salt was prepared. A 10% $CaCl_2/Al_2O_3$ sorbent was fabricated by the incipient wetness technique using an aqueous solution of $CaCl_2 \cdot 2H_2O$ (Mallincrodt) with Al_2O_3 (Catalox SCFA 90 from Condea Vista). Incipient wetness occurred at about 0.6 ml/g of alumina. Impregnation was followed by heating at 392°F overnight to remove the moisture.

Various analytical techniques were used to characterize the fresh and spent sorbents. A review of literature pertinent to surface analyses and of reports pertaining to Hg detection was conducted to determine the best available analytical techniques. These methods included BET surface areas and pore size distributions determined with a Coulter Omnisorp 100 CX apparatus; an x-ray photoelectron spectrometer (XPS) for Hg speciation and surface concentration; bulk chemical analyses; and x-ray diffraction (XRD) for supported oxide sorbent phase identifications.

Results and Discussion

A rigorous evaluation of the experimental setup was initially conducted in an attempt to identify, quantify, and eliminate, if possible, experimental artifacts that could exist in the system. Quantities that were used to characterize the behavior of the sorbent toward elemental mercury removal were the capacity and breakpoint. Capacity was defined as the amount of elemental mercury removed by the sorbent after 350 minutes on stream. When the continuous on-line AFS monitor for elemental mercury was used, breakpoint was defined as the time when the outlet concentration of mercury emerging from the reactor bed equaled 10% of the inlet mercury concentration.

The reproducibility of the experimentally determined 350 minute capacity and the 10% breakpoint was determined for the baseline sorbent, iodine-promoted activated carbon. Ten milligrams of 200/325 mesh iodine-promoted activated carbon was exposed to 585 ppb of elemental mercury in a 30 cc/min flow of argon at 350°F in order to generate the breakthrough curves. This sorbent was used in this exercise since it represented the most reactive sorbent to date. The experiment was replicated with good results. The capacity determined via the on-line AFS in argon was reproducible to within ± 0.2 mg/gram and the breakpoint time to within ± 25%.

The capacity determined with the on-line AFS was compared with the results obtained from analyzing the spent sample with cold vapor atomic absorption spectrophotometry (CVAAS) and the inductively coupled argon plasma atomic emission spectrophotometer (ICP-AES). The on-line AFS is the most reliable technique for determining sorbent capacity. Unfortunately, this technique is primarily limited to argon (or other noble gas) or nitrogen carrier gas streams.[33] The AFS also has a detection limit for mercury which is an order of magnitude less than the detection limit for the atomic absorption spectrophotometer (AAS).

CVAAS is the next most reliable method for capacity determination and is the preferred analytical technique for the quantitative determination of trace levels of mercury in solids because of the elimination of the background matrix. Great care is taken to transfer the mercury into a noble carrier gas, providing good reproducibility. In CVAAS, the solid is dissolved into solution. Mercury is reduced from solution with tin chloride, aerated onto a gold trap, thermally desorbed from the gold trap, and swept into an argon stream to a ultraviolet (AAS) detector. Chemical (tin chloride reduction) and physical (amalgamation) steps are taken to separate the mercury. The detection limit of CVAAS is 10 ng/g.[34] The typical precision for measurement of mercury concentrations in solids is 5-10% relative standard deviation.[34] A comparison of capacity determinations via the on-line AFS and CVAAS shown in Tables 2 and 3 for I-AC in argon at 350•F, MoS_2 in argon at 280•F, CERF-FA #2 in argon at 280•F, and CERF-FA #4 in argon at 280•F, shows a fair agreement.

The ICP-AES is the least reliable of the three techniques for trace level mercury determinations. It can be seen from Table 2 that the ICP-AES yields capacities which are high by a factor of two. The ICP-AES is the most versatile tool for multielement analysis, but it is not the preferred method for trace level mercury measurements in solids. No steps are taken to separate the mercury from the other elements present in the solid sample. Other elements could interfere in the detection of mercury. For example, the cobalt emission line at 253.649 nm could interfere in the determination of mercury by the 253.652 nm emission line.[35] The concentration of the interfering element (in this case cobalt), monochromator slit width, and relative intensity of the shared emission line are factors in determining the extent of spectral interference. Experiments with the Perkin-Elmer Optima 3000 Radial View ICP-AES confirmed that cobalt will interfere in the trace level determination of mercury in solids. Additionally, because of their emission lines close to 253.7 nm, iron and manganese[36] will also interfere in the determination of mercury by ICP-AES.

The effect of intraparticle mass transfer resistance due to the diffusion of mercury within the pores was determined by carrying out the same experiment but with various size fractions of the baseline iodine-promoted activated carbon. For a sub 400 mesh size fraction of the same carbon, the 350 minute capacity was 4.9 mg Hg/gram, and the breakpoint was 405 minutes. This is in good agreement with the data for the larger size fraction (see I-AC in Table 2), suggesting that mass transfer resistance due to the diffusion of mercury into the sorbent at the sizes used in the testing is negligible. Calculations further indicated that bulk mass transfer effects, heat transfer effects, channeling, and pressure drop would not be significant in the experimentation.

Table 2. Sorbent Experimental Results: Argon Carrier Gas Sorbent Capacity (mg/g)** Breakpoint (min) Temperature (EF)

Sorbent	Capacity (mg/g)**	Breakpoint (min)	Temperature (• F)
I-AC	3.1		350
I-AC	4.8	330	350
S-AC	0.4	4	350
S-AC	3.5	7	280
S-BPL-AC	1.9		280
Cl-AC-1	4.0	70	280
Cl-BPL-AC	2.6		280
HNO_3-AC-1	1.2	3	280
AC-1	0.37		280
AC-2	0.4	4	280
Celkate	0.5	2.5	350
Alumina	0.6	2.5	140
MnO_2/Al_2O_3	2.2	3	350
MnO_2/Al_2O_3	2.4	11	280
MnO_2/Al_2O_3	2.4	40	140
V_2O_5-$MgSiO_3$-1	0.4	3	350
V_2O_5-$MgSiO_3$-2	0.1	2	350
$MoO_3/MgSiO_3$	0.2	3	350
Cr_2O_3/Al_2O_3	1.2	2	140
Cr_2O_3/Al_2O_3	3.1		280
Cr_2O_3/Al_2O_3	3.3	9	350
Fe_2O_3	0.1		280
TS-7	0.01 *icp-aes*		150
Cl-Celkate	0.8	0.5	280
MoS_2	8.8 *icp-aes*		280
MoS_2	3.9		280
MoS_2	3.6	17	280
MoS_2	4.5	194	140
FeS	cap < 0.01		280
FeS_2	0.2		280
CERF-FA-#2	1.7	4	280
CERF-FA-#2	1.4		280
CERF-FA-#4	2.2	20	280
CERF-FA-#4	1.7		280
FA-1	0.02		280
WCFA-1	0.1		280
WCFA-1	0.04		350

Table 2. *(Continued)*

Cl-WCFA-1	2.5	280
Cl-WCFA-1	0.64	350
DCFA-1	0.03	280
DCFA-2	0.12	280
DCFA-3	0.15	280

Sorbent	Capacity (mg/g)**	Breakpoint (min)	Temperature ($\bullet$F)
Cl-DCFA-1	0.24		280
Cl-DCFA-2	0.30		280
Cl-DCFA-3	0.41		280
Pt/wool	5.0		280
$CaCl_2/Al_2O_3$	0.6	2	140

**Capacity determined via on-line AFS when a breakpoint time is given; otherwise capacity determined by CVAAS, except as noted.

Most of the experiments used a gas feed of 585 ppb elemental mercury in argon. This is dramatically different than the composition of a typical flue gas from a coal-fired utility. Most of the components in a typical flue gas (e.g. acid gases, etc.) can adsorb on an activated carbon and could possibly hinder or help the adsorption of mercury on carbon. As pointed out above, the ultra-high purity argon carrier gas was selected to maximize the sensitivity of the AFS for elemental mercury. However, the capacity of the sorbents in argon can be quite different from the capacity in flue gas. Also, the temperatures at which sorbent capacities were typically determined are 140$\bullet$F, 280$\bullet$F, and 350EF. These temperatures were chosen because of their potential relevance to coal-fired utilities. If a sorbent were contacted with the flue gas by injection into the duct work of a coal-fired utility after the air preheater but before the particulate collection device, it would experience temperatures in the range of 350$\bullet$F to 280$\bullet$F. If a sorbent was placed downstream of a wet scrubber, it would encounter a temperature near 140$\bullet$F.

The 350 minute capacities and the 10% breakpoint times for the sorbents are listed in Table 2. The baseline sorbent—the iodine-promoted activated carbon— exhibited both the largest most reliably determined (on-line AFS) capacity and longest breakpoint time. A typical breakthrough curve for the iodine-promoted activated carbon is shown in Figure 2.

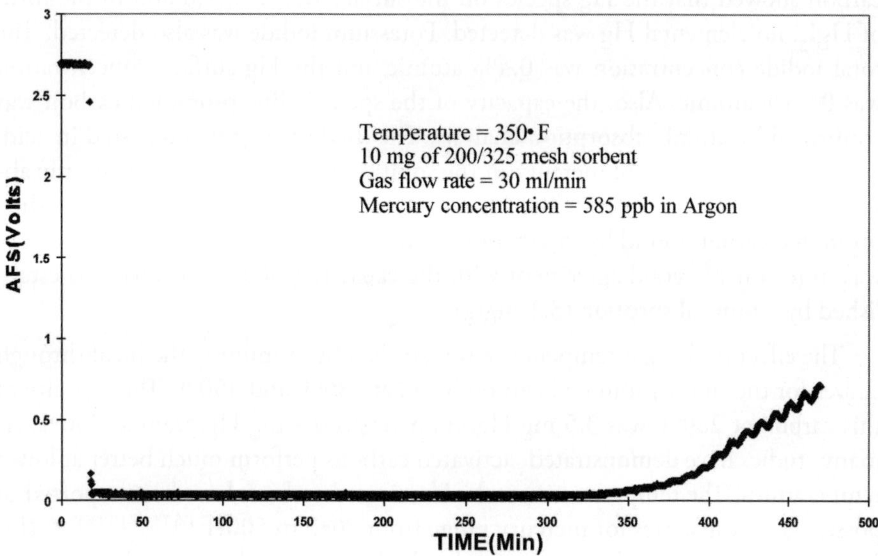

Figure 2. Breakthrough Curve for I-AC Using the AFS for Mercury Detection

Activated Carbons

With the activated carbons, the effect of the chemical promoter on the capacity for elemental mercury was determined by comparing the capacities of the commercially available unpromoted carbon with the capacities of the sulfur, iodine, chlorine, and nitric acid-treated carbons at the same temperature. The sulfur and iodine promoted carbons are available commercially. The carbons, when chemically promoted, exhibited a far greater capacity for elemental mercury. An unpromoted carbon primarily captures elemental mercury via physical adsorption. Chemically promoted carbons capture elemental mercury by both physical adsorption and chemisorption/chemical reaction, where mercuric sulfide, mercuric iodide, etc. formation enables the promoted carbons to remove more elemental mercury.

Various analyses were performed on the spent baseline iodated carbon sorbent to elucidate the role of the promoter. A 3-day run in the packed bed was performed on 35 mg of iodated activated carbon so that gross differences, if any, between the fresh sorbent and spent sorbent could be differentiated by the BET-surface analysis. Results indicate a reduction in surface area from 780 to 300 m^2/g. Additionally, after determining that the vacuum treatment would not impact the mercury concentration, XPS studies with the spent iodated activated carbon showed that the Hg species on the surface was oxidized and in the form of HgI_2; no elemental Hg was detected. Potassium iodide was also detected. The total iodide concentration was 0.4% atomic and the Hg surface concentration was 0.13% atomic. Also, the capacity of the spent iodine promoted carbon was confirmed by atomic absorption analysis. The used sorbent was digested in acid, and the concentration of mercury in the solution was measured by an atomic absorption spectrophotometer. The capacity (gaseous determination) of the iodine promoted carbon found by integration under the breakthrough curve (4.8 mg/g) was in reasonably good agreement with the capacity (solid determination) established by atomic absorption (3.1 mg/g).

The effect of flue gas temperature was studied by examining the breakthrough curves for the sulfur promoted carbon S-AC at 280ºF and 350ºF. The capacity of this carbon at 280ºF was 3.5 mg Hg/gram versus 0.4 mg Hg/gram at 350ºF. As many studies have demonstrated, activated carbons perform much better at lower temperatures. The temperatures at which activated carbons have been reported to possess good capacities for mercury range from 70•F to 500•F.[1-6,13,16-19,23,37-46] This suggests that physical adsorption may be the first step in the removal of mercury for both unpromoted and promoted carbons. Physical adsorption, analogous to condensation, is a low temperature process. For a chemically promoted carbon, such as sulfur impregnated carbon, chemisorption/reaction between the physically adsorbed mercury and sulfur promoter to form mercuric sulfide could be the second step in the mechanism of mercury removal.

The hydrochloric acid treated activated carbon Cl-AC-1 exhibited a large capacity of 4.0 mg Hg/gram when tested in argon at 280EF, making it one of the most active sorbents studied to date. Additionally, the chlorine gas-treated activated carbon, Cl-BPL-AC, exhibited a modest capacity for elemental mercury removal. One previous study suggests that hydrochloric acid treatment yields activated carbons which have chemisorbed chlorine.[47] Quimby demonstrated that HCl treated activated carbon will adsorb mercuric chloride from air at 300•F.[21] Mercury is known to primarily form the tetrachloromercury complex $HgCl_4^{2-}$ on the surface of activated carbons used for the removal of mercuric chloride from waste water; little mercuric chloride was found on the surface of these carbons.[48] Other prior studies have shown that HCl treatment of silica increases its capacity for mercury. It can be speculated that elemental mercury reacts with chemisorbed chlorine to form the tetrachloromercury complex on the surface of the carbon.

A nitric acid treated activated carbon HNO_3-AC-1 was examined as a sorbent for the removal of elemental mercury from argon at 280•F. The untreated carbon AC-1 is a relatively inactive sorbent. The treated carbon exhibited a small capacity of 1.2 mg Hg/gram versus 0.4 mg Hg/gram for the untreated carbon. Previous studies suggest that nitric acid treatment of carbon oxidizes the surface[20,47], and increases its capacity for the removal of mercury from nitrogen at 86EF by a factor of twenty.[20] Only a modest increase in capacity was observed in our lab at 280•F. It can be speculated that oxygen surface complexes such as carboxyl groups which are formed by nitric acid treatment of carbon are destroyed at the higher temperatures.

The unpromoted carbons AC-1 and AC-2 were found to possess relatively small capacities for elemental mercury, whether from argon or 4% oxygen in nitrogen. Oxygen will adsorb on carbon. This could either help (by promoting the carbon surface by oxidation) or hinder (by competitive adsorption) the removal of elemental mercury by an unpromoted carbon. The latter effect was probably observed in the packed bed experiments. Also, oxygen may dramatically reduce the capacity of the sulfur promoted carbon as capacity dropped from 3.5 mg/g in argon for S-AC to 0.5 mg/g in air at 280•F for S-BPL-AC. S-AC and S-BPL-AC both exhibit high capacities for elemental mercury from inert carrier gases.[17] This suggests that oxygen competitively adsorbs on sulfur, reducing capacity.

The results obtained from the packed bed unit require judicious interpretation when attempting to extrapolate their relevance to activated carbon sorbent duct injection as a mercury control technique for industrial size combustors. AC-1 was also studied in the FETC 500-lb coal/hr pilot-scale combustor unit forthe removal of mercury from the flue gas.[49] When introduced at a large sorbent to mercury ratio of around 5,000 to 1, AC-1 used in the 500-lb/hr unit achieved a high level of mercury removal. However, the used AC-1 recovered from the baghouse had mercury levels of less than 300 ppm (0.3 mg Hg/gram), but a high level of mercury removal is achieved. Unpromoted activated carbons sequester elemental mercury via physical adsorption, and therefore exhibit small capacities. Duct injection at large sorbent to mercury ratios of 5,000/1 to 100,000/1 allows them to, nevertheless, achieve high levels of removal of mercury from flue gas.

Metal Oxides

Metal oxides are proposed as novel alternatives to activated carbon sorbents. It is noted that there are many binary oxides of mercury, such as mercury vanadates, mercury molybdates, and mercury manganates.[50-52] Vanadium pentoxide, molybdenum trioxide, and manganese dioxide are all partial oxidation oxide catalysts for the oxidation of various hydrocarbons.[26,53] In the oxidation of various hydrocarbons, lattice oxygen serves as the oxidant in a Mars-Maessen mechanism. This suggests that lattice oxygen of partial oxidation oxides could also serve as the oxidant of mercury. The reaction mechanism for the capture of mercury by oxide catalysts can be written as:

$$Hg_{(g)} + surface \bullet Hg_{(ad)} \tag{1}$$

$$Hg_{(ad)} + M_xO_y \bullet HgO_{(ad)} + M_xO_{y-1} \tag{2}$$

$$HgO_{(ad)} + M_xO_{y-1} + 1/2\ O_{2(g)} \bullet HgO_{(ad)} + M_xO_y \tag{3}$$

$$HgO_{(ad)} + M_xO_y \bullet HgM_xO_{y+1} \tag{4}$$

The overall reaction in the presence of gas phase oxygen is the sum of reaction steps one through four:

$$Hg_{(g)} + 1/2\ O_{2(g)} + M_xO_y \bullet HgM_xO_{y+1} \tag{5}$$

where: M_xO_y is the sorbent metal oxide and HgM_XO_{y+1} is the binary oxide.

Step (1) is the collision of elemental mercury with the surface of the sorbent, resulting in adsorption. Step (2) is the reaction of adsorbed mercury with the metal oxide, forming adsorbed mercuric oxide and reducing the surface of the sorbent. Step (3) is the reoxidation of the sorbent by gas phase oxygen. Step (4) is the reaction of adsorbed mercuric oxide with the sorbent to form the binary oxide.

Note that mercury can be captured in the absence of gas phase oxygen by reactions (1) and (2), as demonstrated by the modest capacity for elemental mercury displayed by manganese oxide in argon, shown in Table 2. There are many potential rate limiting factors which can impact oxide capacity for mercury, including surface area, activity of sorbent as oxidation catalyst, stability of lower oxides, oxygen partial pressure, and tendency to form the binary oxide. Mercury is a semi-noble metal with a standard reduction potential similar to palladium. Mercury may not be easily oxidized by the metal oxide sorbent. An oxide's tendency to form sulfates is a critical factor for sorbent performance in flue gas because sulfur dioxide is present at concentrations orders of magnitude greater than mercury.

Alumina (Al_2O_3) or celkate ($MgSiO_3$), which were used as supports for some of the novel sorbents, were examined as sorbents for the removal of elemental mercury from argon. Both exhibit small capacities, demonstrating their inertness towards elemental mercury. The role of the alumina or celkate support is to provide a high surface area substrate for maximizing the number of collisions between mercury and the sorbent.

Supported vanadium pentoxide and supported molybdenum trioxide exhibited low capacities for the removal of elemental mercury from argon at 350°F. Preparation of the V_2O_5 supported sorbent either via the thermal decomposition of ammonium vanadate or via incipient wetness did not impact the sorbent capacity. Manganese dioxide supported on alumina was examined as a sorbent for the removal of elemental mercury from argon at 350°F, 280°F, and 140°F. Manganese dioxide has been reported to remove elemental mercury from both air and argon at room temperature.[23,24] MnO_2 exhibited modest 350 minute capacities of 2.2 at the higher temperature and 2.4-mg Hg/gram, at both 140 and 280°F.

In the Mars-Maessen mechanism, gas phase oxygen can serve to reoxidize the reduced oxide. Oxygen was absent from the gas phase in these experiments. Manganese dioxide is the most powerful oxidation catalyst[53] of the oxide oxidation catalysts examined and exhibited the largest capacity for mercury. A Mars-Maessen redox mechanism for the removal of mercury has been proposed above for partial oxidation oxide sorbents. The capacity of the manganese dioxide sorbent was observed to be larger in air than in argon at 280°F (see Table 3).

Table 3. Sorbent Experimental Results: Air Carrier Gas

Sorbent	Capacity (mg/g)	Analysis Method	Temperature (• F)
KO_2-V_2O_5	0.02	CVAAS	280
KO_2-V_2O_5	0.04	CVAAS	350
MnO_2/Al_2O_3	3.50	CVAAS	280
TS-7	0.00	ICP-AES	140
S-BPL-AC	0.53	ICP-AES	280
S-BPL-AC	0.28	ICP-AES	350
MoS_2	5.6	ICP-AES	140
MoS_2	5.2	ICP-AES	280
MoS_2	1.1	ICP-AES	350
AC-1*	0.04	CVAAS	280
AC-1*	0.19	AFS	280

*4% O_2 in N_2 Carrier Gas

Nanoscale iron oxide was examined as a sorbent for mercury removal from argon at 280°F. Each particle contains about 600 iron atoms and 900 oxygen atoms. A surface will always be more reactive than the bulk lattice because of the dangling bonds and availability for collision with a reactant species. A nanoscale particle has a significant proportion of its atoms exposed on the surface whereas a larger particle has most of its atoms contained within the crystalline lattice. The chemical and physical properties of nanoscale particles will, therefore, often differ dramatically from those exhibited by larger particles. Nevertheless, the ferric oxide displayed a poor capacity, despite the unusually small (3 nm) particle size and high surface area.

A potassium superoxide promoted vanadium pentoxide sorbent exhibited a miniscule capacity for elemental mercury from air at both 280 and 350°F, as seen in Table 3. The potassium superoxide (KO_2) is a powerful oxidizing agent, and was expected to oxidize elemental mercury to mercuric oxide. The mercuric oxide could then chemisorb/react with vanadium pentoxide to form mercury vanadate (HgV_2O_6). Chromium oxide was found to exhibit modest capacities for elemental mercury. Cr_2O_3 is a fairly strong oxidation catalyst, with a catalytic activity for the deep oxidation of methane comparable to manganese dioxide.[53] A crude correlation was found between catalytic activity for deep oxidation exhibited by the oxide and sorbent capacity for elemental mercury removal. Sorbents that are active catalysts for the deep oxidation of methane Cr_2O_3/Al_2O_3 and MnO_2/Al_2O_3 exhibit large capacities, whereas the inactive oxide catalysts Fe_2O_3, MoO_3/Al_2O_3, and V_2O_5 –$MgSiO_3$–1 show small capacities.

Promotion of metal oxide supports was also investigated. The chlorine promoted magnesium silicate Cl-Celkate exhibited a small capacity for the removal of elemental mercury from argon at 280°F. Braman demonstrated that HCl treated Chromosorb-W, a diatomite chromatographic packing, will adsorb mercuric chloride vapors at 70°F.[7] Additionally, the novel thiol promoted aluminosilicate sorbent (TS-7) exhibited very small capacities for elemental mercury in both argon and in air. Thiols are the sulfur analogs of alcohols. Thiols are also called mercaptans, from the Latin, mercurium captans, meaning "capturing mercury."[54] Mercaptans react with mercuric ions and the ions of other heavy metals to form precipitates. The sorbent was developed for the removal of oxidized mercury from contaminated water. Elemental mercury is insoluble in water. Oxidized forms of mercury are known to react efficiently with thiols. The low decomposition temperatures of thiols, as well as the lack of reactivity with elemental mercury, suggest that thiols are not practical promoters for the removal of elemental mercury from flue gas.

Metal Sulfides

Molybdenum sulfide (MoS_2) displayed a large capacity for the removal of elemental mercury from argon and air. This sorbent was originally developed as a hydrodesulfurization catalyst for the conversion of thiopene and mercaptans to hydrogen sulfide and alkanes. A possible mechanism of mercury capture is chemisorption/chemical reaction to form mercuric sulfide. XPS analysis of the used sorbent run in argon at 280°F confirms the presence of mercury on the surface. Elemental mercury was not detected on the surface of the used MoS_2 sorbent. This rules out physical adsorption of elemental mercury as the primary means of sequestration. The x-ray excited photoelectron spectra suggests the presence of mercuric sulfide on the surface of the sorbent. The sorbent exhibits a much lower capacity at 350°F in air versus the capacities in air at 140°F and 280°F. This suggests that physical adsorption of elemental mercury is the first step in the sequestration mechanism, and/or the physical-chemical degradation of the sorbent at the higher temperature. Molybdenum disulfide is known to decompose in air at elevated temperatures.[36] The sorbent removed nearly all of the mercury entering the packed bed at 140°F in argon.

Less expensive sulfides, such as iron sulfides, were also examined as sorbents. The iron sulfides FeS and FeS_2 exhibited poor capacity for elemental mercury from argon at 280°F. The FeS_2 lost sulfur during the sorption of elemental mercury from argon at 280°F, as evidenced by a yellow film which formed at the bottom of the packed bed reactor.

Unburned Carbons from Fly Ash

The atypical high carbon flyashes CERF-FA-#2 and CERF-FA-#4 exhibited modest capacities for the removal of elemental mercury from argon at 280°F. These capacities are, however, significantly higher than those exhibited by the unpromoted carbon and the alumina and celkate supports. Further characterization of these flyash sorbents is needed to determine the mechanism of mercury capture. These carbons were extracted from the combustor at high temperatures of around 2300°F. It is speculated that novel forms of carbon present in these samples could positively impact capacity.

The flyash obtained from the combustion of Blacksville coal, FA-1, exhibited a miniscule capacity for elemental mercury at 280°F. The carbon separated from this flyash, WCFA-1, exhibited a small capacity for the removal of elemental mercury from argon at 280°F. Nevertheless, WCFA-1 does exhibit a larger capacity than the parent flyash, FA-1. The capacity of WCFA-1 was smaller at 350°F, as expected. The unpromoted activated carbons show similarly low capacities. The chlorine promoted carbon extracted from flyash, Cl-CFA-1, exhibited a much larger capacity for elemental mercury, much like the chlorine promoted activated carbons. Capacity was lower at the higher temperature, as expected.

DCFA-2 and DCFA-3 are carbons separated from the parent fly ash, DCFA-1, by a dry separation method and exhibit small capacities for elemental mercury. Capacity increases with increasing carbon content. The chlorine treated materials Cl-DCFA-1, Cl-DCFA-2, and Cl-DCFA-3 showed significantly larger, but still small capacities. Capacity again increases with increasing carbon content.

Halide salts are also proposed as an alternative to carbon sorbents. There are many binary halides of mercury such as calcium chloromercurate and potassium iodomercurate.[50] These are double salts of calcium chloride and mercuric chloride, and potassium iodide and mercuric iodide, respectively. Potassium iodide is used as a chemical promoter in some of the commercially available activated carbons[18,19], such as the baseline sorbent in this study. However, the thermal stability of the binary halides of mercury is poor, as evidenced by their low decomposition temperatures.[50] Mercuric chloride was absent from the gas phasein these experiments. The absence of mercuric chloride could explain the small capacity exhibited by the calcium chloride sorbent. Mercuric chloride can be present in the flue gas obtained from the combustion of coal, municipal waste, and medical waste.[1]

Noble Metals

The platinum sorbent Pt/wool exhibited a large capacity for elemental mercury from argon at 280°F. Breakthrough was not observed. After the absorption experiment, the used Pt/wool sorbent was slowly heated in argon to 770°F over a seventy minute period, with the effluent sent directly to the AFS. Over 99.4% of the mercury remained sequestered on the platinum. A minor desorption spike of mercury was observed at 320°F, likely due to unburned carbon from the organometallic platinum paint precursor. The noble metals are often used for small-scale sampling of gases for mercury, i.e., mercury is often collected on gold, thermally desorbed, and sent to a UV detector for its analytical determination. Thermal desorption of the mercury is accomplished by heating the noble metal to 1470°F[13], greater than the 770°F maximum temperature in the desorption experiment.

Conclusions

A packed-bed reactor system was used to screen sorbents for the removal of elemental mercury from a carrier gas. An on-line atomic fluorescence spectrophotometer was used to measure elemental mercury in argon on a continuous basis. For more complex carrier gases, sorbent capacities were determined off-line via CVAAS or ICP-AES. Chemically promoted activated carbons exhibit a far greater capacity for mercury than unpromoted carbons. The activated carbons possess higher capacities at lower temperatures. Chlorine could be a cost-effective chemical promoter for carbon sorbents for the removal of mercury.

Metal oxides and sulfides are proposed as a possible alternative to activated carbon sorbents, with MnO_2, Cr_2O_5, and MoS_2 exhibiting moderate capacities for mercury removal among the candidates investigated. Unburned carbon sorbents from fly ash typically showed poor performance towards mercury removal, although promotion of these increases the activity for elemental mercury removal.

Future work will concentrate on testing inexpensive chlorine promoted carbons, as well as metal oxides and sulfides, in a simulated flue gas matrix which includes acid gases, oxygen, water, and mercuric chloride. Promising sorbent candidates will be further evaluated on a pilot-scale system.[49,55]

Acknowledgements

Evan Granite acknowledges the support of a postdoctoral fellowship at the United States Department of Energy administered by Oak Ridge Institute For Science and Education (ORISE). George Haddad, also through the ORISE program, provided project support. Michael Hilterman, Edward Zagorski, Dennis Stanko, Sheila Hedges, John Baltrus, Mac Gray, Yee Soong, Rod Diehl, Murphy Keller, Mark Freeman, Anthony Cugini, James Hoffman, Karl Schroeder of the Department of Energy, Robert Thompson and Joseph Niedzwicki of Parsons Infrastructure, Donald Floyd of EG&G Technical Services, and Mark Wesolowski of SSI Services, Inc., provided outstanding technical assistance.

Several individuals and organizations provided sorbent samples for examina tion by DOE. Professor Radisav Vidic of The University of Pittsburgh supplied a sulfur promoted carbon sorbent. Professor Jayne MacDonald of the Royal Military College of Canada provided a chlorinated carbon sample. CarboChem Inc. donated an unpromoted activated carbon sorbent. Dr. Shas Mattigod of Pacific Northwest National Laboratory provided the thiol promoted aluminosilicate. Professor Luis Cadus of INTEQUI Instituto de Investigaciones en Tecnologia Quimica, San Luis, Argentina supplied the chromium oxide sorbent. Dr. John Stencel of the Kentucky Center For Applied Energy Research provided carbons derived from flyash. MACH I Inc. of King of Prussia Pennsylvania contributed the iron oxide sample. ADA Technologies Inc. of Englewood Colorado supplied the high calcium flyash sorbent.

Disclaimer

Reference in this paper to any specific commercial product, process, or service is to facilitate understanding, and does not necessarily imply its endorsement by the United States Department of Energy.

Literature Cited

1. Mercury Study Report to Congress, EPA - 452/R-97-003; United States Environmental Protection Agency, Office of Air Quality Planning and Standards, Washington, D.C., December 1997.

2. Flue Gas Treatment Reference Bulletin, Calgon Carbon Corporation: Pittsburgh, PA, November 1994.

3. Ghorishi, B.; Jozewicz, W.; Gullet, B.K.; Sedman, C.B. Mercury Control: Overviewof Bench-Scale Research at EPA/APPCD. Presented at the First Joint Power & Fuel Systems Contractors Conference, Pittsburgh, PA, July 1996.

4. Waugh, E.G.; Jensen, B.K.; Lapatnick, L.N.; Gibbons, F.X.; Haythornthwate, S.; Sjostrom, S.; Ruhl, J.; Slye, R.; Chang, R. Mercury Control on Coal-Fired Flue Gas Using Dry Carbon-Based Sorbent Injection: Pilot-Scale Demonstration. Presented at the Fourth International Conference on Managing Hazardous Air Pollutants, Washington, D.C., November 1997.

5. Brown, T.D.; Smith, D.N.; Hargis, R.A.; O'Dowd, W.J. Mercury Measurement and Its Control: What We Know, Have Learned, and Need to Further Investigate. J. Air Waste Manage. Assoc. 1999, 6, 1.

6. Attari, A.; Chao, S. Trace Constituents in Natural Gas; Sampling and Analysis. Oper. Sect. Proc. Am. Gas Assoc., 1994, pp.28-32.

7. Braman, R.S.; Johnson, D.L. Selective Absorption Tubes and Emission Technique for Determination of Ambient Forms of Mercury in Air. Env. Sci. & Technology, 1974, 8(12), 996.

8. Williston, S.H. Mercury in the Atmosphere. J. Geophysical Research, 1968, 73(22), 7051.

9. Henriques, A.; Isberg, J.; Kjellgren, D. Collection and Separation of Metallic Mercury and Organo-mercury Compounds in Air. Chemica Scripta, 1973, 4(3), 139.

10. Henriques, A.; Isberg, J. A New Method for Collection and Separation of Metallic Mercury and Organomercury Compounds in Air. Chemica Scripta, 1975, 8(4), 173.

11. Roberts, D.L.; Stewart, R.M. Novel Process for Removal and Recovery of Vapor Phase Mercury. Presented at First Joint Power & Fuel Systems Contractors Conference, Pittsburgh, PA, July 1996.

12. Yan, T.Y. A Novel Process for Hg Removal from Gases. Ind. Eng. Chem. Res. 1994, 33, 3010.

13. Dumarey, R.; Dams, R.; Hoste, J. Comparison of the Collection and Desorption Efficiency of Activated Charcoal, Silver, and Gold for the Determination of Vapor-Phase Atmospheric Mercury. Anal. Chem. 1985, 57, 2638.

14. Roberts, D. Novel Process for Removal and Recovery of Vapor-Phase Mercury. Presented at Advanced Coal-Based Power and Environmental Systems '97 Conference, Pittsburgh, PA, July 1997.

15. Livengood, C.D.; Huang, H.S.; Mendelsohn, M.H.; Wu, J.M. Enhancement of Mercury Control in Flue Gas Cleanup Systems. Presented at the First Joint Power & Fuel Systems Contractors Conference, Pittsburgh, PA, July 1996.

16. Walker, P.L.; Sinha, R.K. Removal of Mercury by Sulfurized Carbons. Carbon 1972, 10, 754.

17. Korpiel, J.; Vidic, R. Effect of Sulfur Impregnation Method on Activated Carbon Uptake of Gas-Phase Mercury. Environ. Science and Tech. 1997, 31(8), 2319.

18. Material Safety Data Sheet for Activated Carbon 717; Barnebey & Sutcliffe: Columbus, OH, September 1996.

19. Carbon Product Bulletin HGR-LH; Calgon Carbon: Pittsburgh, PA, February 1994.

20. Matsumura, Y. Adsorption of Mercury Vapor on the Surface of Activated Carbons Modified by Oxidation or Iodization. Atmospheric Environment 1974, 8, 1321.

21. Quimby, J.M. Mercury Emissions Control From Combustion Systems. Proceedings of Annual Meeting, Air Waste Management Association, 1993.

22. Hogland, W.K.H. Usefulness of Selenium for the Reduction of Mercury Emissions From Crematoria. J. Environ. Qual. 1994, 23, 1364.

23. Tsuji, K.; Shiraishi, I.; Olson, D.G. EPRI Report no. TR-105258, vol.3, conf-950332, pp.66.1¬66.15, 1995.

24. Janssen, J.; Van Den Enk, J.; De Groot, D. Determination of Total Mercury in Air by Atomic Absorption Spectrometry After Collection on Manganese Dioxide. Analytica Chimica Acta 1977, 92, 71.

25. Rathje, A.; Marcero, D.; Dattilo, D. Personal Monitoring Technique for Mercury Vapor in Air and Determination by Flameless Atomic Absorption. J. Am. Ind. Hyg. Assoc. 1974, 35, 571.

26. Granite, E.J.; Pennline, H.W.; Hargis, R.A. Sorbents For Mercury Removal From Flue Gas. Topical Report DOE/FETC/TR-98-01, National Energy Technology Laboratory, Pittsburgh, PA, January 1998.

27. MacDonald, J.; Evans, M.; Halliop, E.; Liang, S. The Effect of Chlorination On Surface Properties Of Activated Carbon. Carbon 1998, 36(11), 1677.

28. Oyama, S.T.; Went, G.T.; Lewis, K. B.; Bell, A.T.; Somarjai, G. A. J. Phys. Chem 1989, 93, 6786.

29. Mahipal Reddy, B.; Padmanabha Reddy, E.; Srinivas,S.T. J. Catal. 1992, 136, 50.

30. Maltha, A.; Favre, T.L.F.; Kist, H.F.; Zuur, A.P.; Ponec, V. J. Catal. 1994, 149, 364.

31. Mentasty, L.; Gorriz, O.; Cadus, L. Chromium Oxide Supported on Different Al OSupports.23 Industrial & Engineering Chemistry Research 1999, 38, 396.

32. MacDonald, J.A.F.; Halliop, E.; Evans, M.J.B. The Production of Chemically-Activated Carbons. Carbon 1999, 37, 269.

33. Granite, E.J.; Pennline, H.W.; Hoffman, J.S. Effects of Photochemical Formation of Mercuric Oxide. Industrial & Engineering Chemistry Research 1999, 38, 5034.

34. Golightly, D.W.; Simon, F.O. editors. Methods For Sampling and Inorganic Analysis of Coal. U.S. Geological Survey Bulletin 1823, USGS, Reston, VA, 1995.

35. Reeves, R.D.; Brooks, R.R. Trace Element Analysis of Geological Materials, John Wiley: New York, 1978.

36. Weast, R.C. editor. CRC Handbook of Chemistry and Physics, 63rd edition; CRC Press: Boca Raton, Florida, 1982.

37. Brochure on Sorbalit; Dravo Corporation: Pittsburgh, PA, June 1994.

38. Teller, A.J.; Hsieh, J.Y. Control of Hospital Incineration Emissions-Case Study. Presented at 84th Annual Meeting of Air & Waste Management Association, Vancouver, Canada, 1991.

39. Hsieh, J.Y.; Confuorto, N. Operation of a Medical Waste Incinerator and the Emission Control System. Proc. SPIE-Int. Soc. Opt. Eng. 1993, 1717 (Industrial, Municipal, and Medical Waste Incineration Diagnostics and Control), 46.

40. Tsuji, K.; Shiraishi, I. ACS Division of Fuel Chemistry Preprints 1996, 41(1), 404.

41. Tsuji,K.; Shiraishi, I. EPRI Report no. TR-101054, vol.3, conf-911226, 8A.1, 1992.

42. Dunham, G.E.; Miller, S.J. Evaluation of Activated Carbon For Control of Mercury From Coal-Fired Boilers. Presented at the First Joint Power & Fuel Systems Contractors Conference, Pittsburgh, PA, July 1996.

43. Livengood,C.D.; Huang, H.S.; Wu, J.M. Experimental Evaluation of Sorbents for the Capture of Mercury in Flue Gases. Presented at the 87th Annual Meeting Air & Waste Management Association, Cincinnati, Ohio, June 1994.

44. McNamara, J.D.; Wagner, N.J. Gas Sep. Purif. 1996, 10 (2), 137.

45. Roberts, D.L.; Stewart, R.M. Characterization of Sorbents for Removal of Vapor Phase Mercury. Presented at the EPRI Mercury Sorbent Workshop, Grand Forks, ND, July 1996.

46. Van Wormer, M. Control of Mercury Emissions From Combustion Applications. Presented at 1992 AIChE National Meeting, August 1992.

47. Castilla, C.M.; Marin, F.C.; Hodar, F.J.M.; Utrilla, J.R. Effects Of Non-Oxidant And Oxidant Acid Treatments On The Surface Properties Of An Activated Carbon With Very Low Ash Content. Carbon 1998, 36(1-2), 145.

48. Carrott, P.J.M.; Carrott, M.M.L.R.; Nabais, J.M.V. Influence of Surface Ionization On The Adsorption Of Aqueous Mercury Chlorocomplexes By Activated Carbons. Carbon 1998, 36(1-2), 11.

49. Hargis, R.A.; O'Dowd, W.J.; Pennline, H.W. Mercury Investigations in a Pilot-Scale Unit. Presented at AIChE National Meeting, Houston, Texas, March 1999.

50. Mellor, J.W. A Comprehensive Treatise on Inorganic and Theoretical Chemistry; Longmans Green and Company: New York, 1952.

51. Angenault, J. Revue de Chimie Minerale 1970, 7, 651.

52. Wessels, A.L.; Jeitschko, W. The Mercury Vanadates with the Empirical Formulas HgVO3 and Hg VO . 4 Journal of Solid State Chemistry 1996, 125(2), 140.2

53. Golodets, G.I. Heterogeneous Catalytic Reactions Involving Molecular Oxygen; Elsevier: New York, 1983.

54. Solomons, T.W. Graham Organic Chemistry; John Wiley: New York, 1980.

55. Hargis, R.A.; Pennline, H.W. Trace Element Distribution and Mercury Speciation in A Pilot-Scale Combustor Burning Blacksville Coal. Paper No.97-WP72B.04 Presented at the 90th AWMA Annual Meeting, Toronto, Canada, June 1997.

Toward a New U.S. Chemicals Policy: Rebuilding the Foundation to Advance New Science, Green Chemistry, and Environmental Health

Michael P. Wilson and Megan R. Schwarzman

ABSTRACT

Objective

We describe fundamental weaknesses in U.S. chemicals policy, present principles of chemicals policy reform, and articulate interdisciplinary research questions that should be addressed. With global chemical production projected to double over the next 24 years, federal policies that shape the priorities of the U.S. chemical enterprise will be a cornerstone of sustainability. To date, these policies have largely failed to adequately protect public health or the environment or motivate investment in or scientific exploration of cleaner chemical technologies, known collectively as green chemistry. On this trajectory, the United States will face growing health, environmental, and economic problems related to chemical exposures and pollution.

Conclusions

Existing policies have produced a U.S. chemicals market in which the safety of chemicals for human health and the environment is undervalued relative to chemical function, price, and performance. This market barrier to green chemistry is primarily a consequence of weaknesses in the Toxic Substances Control Act. These weaknesses have produced a chemical data gap, because producers are not required to investigate and disclose sufficient information on chemicals' hazard traits to government, businesses that use chemicals, or the public; a safety gap, because government lacks the legal tools it needs to efficiently identify, prioritize, and take action to mitigate the potential health and environmental effects of hazardous chemicals; and a technology gap, because industry and government have invested only marginally in green chemistry research, development, and education. Policy reforms that close the three gaps—creating transparency and accountability in the market—are crucial for improving public and environmental health and reducing the barriers to green chemistry. The European Union's REACH (Registration, Evaluation, Authorisation and Restriction of Chemicals) regulation has opened an opportunity for the United States to take this step; doing so will present the nation with new research questions in science, policy, law, and technology.

Key Words: chemicals policy, data gap, environmental health, green chemistry, innovation, REACH, safety gap, sustainability, technology gap, TSCA

Introduction

In the United States, the primary legal framework for managing industrial chemicals used in processes and products, the Toxic Substances Control Act (TSCA 1976), is now over 30 years old and is widely recognized as having failed to meet the intent of Congress. Analyses of TSCA by the National Research Council (NRC 1984), U.S. Government Accountability Office (U.S. GAO 1994, 2005a, 2007, 2009), Congressional Office of Technology Assessment (OTA 1995), U.S Environmental Protection Agency (U.S. EPA 1998, 2003), Environmental Defense Fund (Denison 2009; Roe et al. 1997), and a former U.S. EPA assistant administrator (Goldman 2002) have concluded that the statute has all but prevented government, businesses, and the public from a) assessing the hazard traits of the great majority of chemicals in commerce; b) controlling chemicals of significant concern; and c) motivating broad industry investment in cleaner chemical technologies and safer alternatives, known collectively as green chemistry.

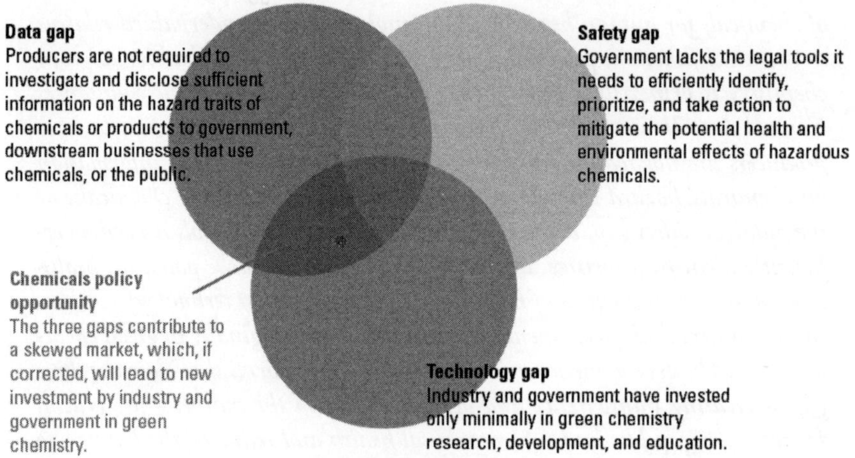

Figure 1. The three gaps in U.S. chemicals policy. Policy measures that address the gaps will promote sustainable innovation in the chemical enterprise while improving human health and the environment.

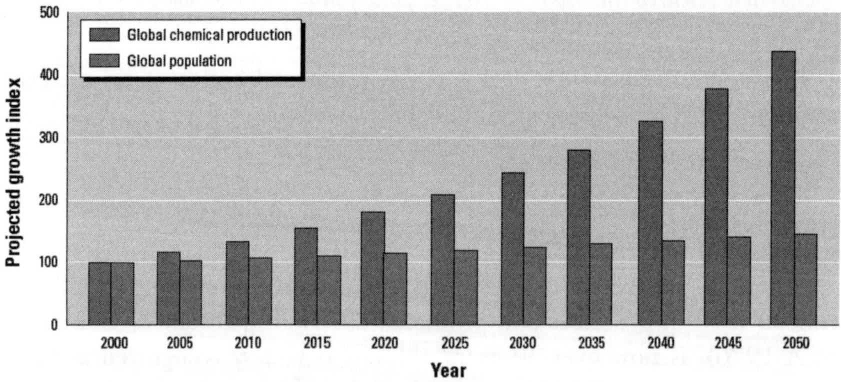

Figure 2. Global chemical production is projected to grow at a rate of 3% per year, rapidly outpacing the rate of global population growth, estimated at 0.77% per year. On this trajectory, chemical production will double by 2024, indexed to 2000 (American Chemistry Council 2003; OECD 2001; United Nations 2004).

These outcomes are the consequence of weaknesses in TSCA that have produced three overarching gaps in the U.S. chemicals policy (Figure 1) (Wilson et al. 2006, 2008):

Table 1. Numbers of industrial chemicals and pollutants governed by U.S. law, excluding TSCA.

Table 1. Numbers of industrial chemicals and pollutants governed by U.S. law, excluding TSCA.

Federal statute	No. of substances
Clean Water Act (1966)	148
Resource Conservation and Recovery Act (1976)	502
Clean Air Act (1963)	189
Occupational Safety and Health Act (1970)	453
Emergency Planning and Community Right-to-Know Act; Toxics Release Inventory (1986)	600

Although TSCA applies to tens of thousands of substances, only 1,134 chemicals and pollutants are listed under five major federal statutes (with overlap) (Dernbach 1997).

- Data gap: Producers are not required to investigate and disclose sufficient information on the hazard traits of chemicals to government, the public, or businesses that use chemicals.

- Safety gap: Government lacks the legal tools it needs to efficiently identify, prioritize, and take action to mitigate the potential health and environmental effects of hazardous chemicals.

- Technology gap: Industry and government have invested only marginally in green chemistry research, development, and education.

Over the last 30 years, the three gaps have given rise to a U.S. chemicals market that undervalues the safety of chemicals relative to their function, price, and performance. Chemicals and products are bought and sold primarily on the basis of how much work they perform per unit cost, with very little attention given in the market to their potential impacts on human health and ecosystems throughout the chemical lifecycle. Hazardous chemicals have thus remained competitive, and broad industrial investment in green chemistry has lagged, despite efforts of some leading companies. Reflecting these market conditions, the nation's research and education agendas have neither prioritized green chemistry development nor adequately prepared the next generation of scientists to lead the chemical enterprise toward sustainability. Collectively, the three gaps—and the market conditions they have engendered—present a formidable barrier to the scientific, technical, and commercial success of green chemistry in the United States.

Facing a similar set of chemical management problems—and their consequences for human health and the environment—the European Union in 2006 enacted a sweeping new chemicals regulation known as the Registration, Evaluation, Authorisation and Restriction of Chemicals (REACH) (European Commission 2008; REACH 2007; U.S. Department of Commerce 2008). REACH responds to the barriers described by the data gap and safety gap by requiring producers to disclose some hazard and exposure information on an estimated 30,000 industrial chemicals. Chemical manufacturers must also gain government authorization to use certain "substances of very high concern." These new requirements—both for data and for proving safe use—are expected to promote the development and use of safer chemical substances, closing the technology gap by fueling new investment in green chemistry science, technology, and education (Black 2008; Buckland 2008; Henzelmann et al. 2007).

Because REACH applies equally in most aspects to manufacturers in the European Union and foreign importers, it is forcing change among chemical and product manufacturers far beyond Europe's borders (Schapiro 2007). Manufacturers worldwide cannot afford the losses in market share that would result should they fail to comply with REACH.

As U.S. producers prepare hazard and exposure information for the European Union Chemicals Agency (ECHA), the United States has a unique opportunity both to make use of the data and to revisit its own chemicals policy. In doing so, the United States should consider a portfolio of measures that simultaneously close the data, safety, and technology gaps. This approach will most effectively—and with minimal delay—instill within the chemicals market a more appropriate set of incentives and disincentives that are a precondition to motivating broad investment in green chemistry.

As with REACH in the European Union, a new chemicals policy in the United States has the potential to fuel global demand for safer substances and processes, increasing the incentive for research and development in green chemistry while improving human and environmental health. It also could move the United States into a position of greater collaboration in international sustainability efforts and position the country as a global leader in green chemistry innovation. Environmental health scientists have an essential role in identifying and addressing the research questions that will arise with the development of a new U.S. chemicals policy.

Background

A Key Industry

Over the last 150 years, the U.S. chemical industry has contributed significantly to both the national and global economy (Aftalion 2001; Arora et al. 1998). The industry's contributions to economic growth, employment, and improvements in life expectancy, health, and living conditions in Western-style societies are widely recognized (NRC 1992; Spitz 2003). The industry's products are ubiquitous; in roughly the last 50 years, synthetic chemicals have become integrated into nearly all industrial processes and commercial products and now constitute the primary material base of society (Geiser 2001).

The scale of chemical production is correspondingly enormous: Data from the TSCA Inventory Update Rule (IUR) show that the United States produced or imported about 15 trillion pounds of chemical substances during the 2002 reporting cycle, or about 42 billion pounds per day (U.S. EPA 2005). For the 2005 reporting period, chemical manufacturers reported producing or importing about 27 trillion pounds of 6,200 chemicals at more than 25,000 pounds per site per year, or about 74 billion pounds per day. The IUR data include substances used in industrial processes and products and do not include fuels, pesticide products, pharmaceuticals, or food products. There is no clear explanation for the 80% increase in volume between 2002 and 2005, and the U.S. EPA has obscured the categories of the products in which each chemical is used (U.S. EPA 2009). The TSCA inventory now lists about 83,000 substances that have been for sale in the United States at some point since the inventory was first published in 1979 (U.S. EPA 2008a). Of these, approximately 62,000 were in commercial use at the time TSCA was passed in 1976, and about 20,000 new substances have entered commercial use since that time (U.S. GAO 2005a).

Global chemical production is projected to continue growing—about 3% per year, with a doubling rate of 24 years, rapidly outpacing the rate of global population growth (Figure 2) [American Chemistry Council 2003; Organisation for Economic Cooperation and Development (OECD) 2001; United Nations 2004]. This growth will distribute globally both the benefits and the health and environmental consequences of industrial chemical technologies.

Human and Environmental Health Consequences

Bioaccumulative Chemicals

Because of their wide distribution throughout the economy and environment, many industrial chemicals come in contact with people: in the workplace, in homes, through the use of products, and via air, water, food, and waste streams. Ultimately, at some point in their life cycle, all industrial chemicals will enter the earth's ecosystems. Biomonitoring studies are demonstrating widespread human exposure to certain industrial chemicals and pollutants. In 2001–2002, the U.S. Centers for Disease Control and Prevention (CDC) looked for, and found, 148 synthetic chemicals and pollutants in the blood and urine of a representative sample of the U.S. civilian population (CDC 2005). The 2008 assessment is anticipated to include testing for about 250 substances for participants in the 2003–2004 National Health and Nutrition Examination Survey period.

Early Life Exposures

Evidence that many xenobiotic chemicals pass through the placenta, entering and, in some cases, accumulating in the fetus, suggests they could pose significant risks to human development (Barr et al. 2007; Doucet et al. 2008; Grandjean et al. 2007). Rising incidence of some cancers, asthma, and developmental disorders may be due in part to chemical exposures, particularly those that occur during development (Hertz-Picciotto and Delwiche 2009; Sharpe and Irvine 2004; Surveillance, Epidemiology, and End Results Program 2008). A variety of male reproductive abnormalities may also be linked to prenatal exposures to certain pesticides or endocrine-disrupting chemicals (Bay et al. 2006; Main et al. 2006; Skakkebaek 2002; Swan et al. 2000, 2005). The data gap limits investigators' ability to establish these links, and TSCA has all but prevented the U.S. EPA from instituting more than voluntary measures to act on these early indicators of harm. In assessing the state of the science, the Faroes Statement of the International Conference on Fetal Programming and Developmental Toxicity concluded that efforts to prevent exposures to hazardous chemicals should focus on protecting the embryo, fetus, and small child as highly vulnerable populations. The United States, however, lacks both the information and the regulatory mechanisms necessary to accomplish this goal (Grandjean et al. 2007).

Occupational Disease

Workers are at particular risk from chemical exposures because, depending on their occupation, they can be more highly exposed to hazardous substances than the general public (LaDou 1997). Although the burden of all-cause occupational disease is enormous, resulting in over 60,000 deaths annually (Leigh et al. 1997) national estimates of the proportion specifically attributable to chemical exposures have not yet been compiled (Herbert and Landrigan 2000). Occupational disease reporting is generally suspected to underestimate true disease rates, given both the underdiagnosis of, and difficulty of accounting for, noninjury work-related illnesses (Herbert and Landrigan 2000). Immigrants, minorities, and lower-income communities typically bear a disproportionate burden of occupational chemical exposures and associated diseases (California EPA 2004; Pastor et al. 2002; Robinson 1991).

Occupational health data are more readily available in Europe. During the development of REACH, European Union research on potential occupational health benefits afforded by the regulation projected that the improved safety of workplace chemicals could prevent up to 40,000 cases of asthma annually (50% of occupationally related cases), an equal number of dermatitis cases, and 10,000 cases of COPD each year (Pickvance et al. 2005). The European Commission further estimated that REACH would prevent about 4,300 occupational cancers per year. In all, the commission estimated that the REACH regulation would save €50 billion ($60 billion) over a 30-year period in total occupational disease prevention (European Commission 2003).

Hazardous Waste

The management and cleanup of hazardous waste is another externalized cost largely attributable to existing chemical technology choices. Each year, the United States spends more than $1 billion managing Superfund sites, and future costs are estimated at $250 billion (U.S. EPA 2007; U.S. GAO 2005b). On the current trajectory, the U.S. EPA anticipates the need for 217,000 new hazardous waste sites over the next 20 years (U.S. EPA 2004, 2007). This "end-of-pipe" approach to hazardous waste management can fail. In California, 70% of legacy sites are leaking directly into groundwater, having breached their containment, and are now posing what the state's Department of Toxic Substances Control (2007) calls a major threat to human health or the environment. That half of the most prevalent chemicals at existing sites are known teratogens, neurotoxicants, and/ or carcinogens raises serious concerns for the health of residents in communities surrounding these sites (Agency for Toxic Substances and Disease Registry 2003; Ostrowski et al. 1999).

Toward a New Chemicals Policy

The cases noted above illustrate the expanding health and environmental problems attributable to an antiquated U.S. chemicals policy that has failed to keep pace with developments in either green chemistry or the environmental health sciences. Given the size of the chemical enterprise, the extent to which it is woven into the fabric of society, and the backlog of unexamined chemicals, a new approach is needed that does not rely on resource-intensive, chemical-by-chemical risk assessments in which government, at great public expense, bears the burden of proof. An integrated chemicals policy is needed that enables identification and prioritization of chemicals of concern and employs market and regulatory tools sufficient to motivate investment by industry in the design and production of safer chemicals and materials, based on the principles of green chemistry. Making this transition will require a chemicals policy that departs markedly from the federal policies of the last 30 years, of which TSCA is emblematic.

The Link to Green Chemistry

The principles of green chemistry offer an upstream solution to many of the health, environmental, and economic problems related to industrial chemicals (Anastas and Warner 1998; DeVito and Garret 1998; Lancaster 2002; Lempert et al. 2003; NRC 2005). Implementation of these principles is critical if the chemical enterprise is to achieve sustainability. Specifically, green chemistry products are designed to be inherently less toxic and more readily broken down in the environment. Green chemistry processes use safer materials, operate more efficiently, and produce much less hazardous waste.

To date, producers have not invested in green chemistry at a level commensurate with the scale and pace of chemical production (Cone 2008a, 2008b; Iles 2006; LaMonica 2007; Lempert et al. 2003; NRC 2005). This is, to a large extent, a consequence of the U.S. chemicals market, which has been shaped over the last 30 years by the combined effects of the data, safety, and technology gaps. The European Union, meanwhile, has built a case for green chemistry in the European market (Black 2008; Henzelmann et al. 2007), raising the twin specters of the U.S. chemicals industry lagging behind its European counterpart in green chemistry innovation, and the United States becoming a market for hazardous chemicals and products that are prohibited for sale in the European Union and elsewhere (Cone 2006; Energy & Enviro Finland 2008).

The laws governing the chemical enterprise help define the incentives and disincentives that guide economic behavior in the market (Guth 2008). We use the term green chemistry in this context: as an analytical framework that encompasses both the science of safer chemistry and the laws and policies that will motivate its development and adoption by society. Although there are a variety of financial, technical, organizational, and cultural barriers to the widespread adoption of green chemistry practices by industry (Matus et al. 2007), addressing these barriers alone will not be sufficient to transform the chemical enterprise. Removing these barriers will require policy and regulatory reforms to improve the structure of incentives in the chemicals market.

Mapping the Policy Gaps

Of the federal environmental statutes, TSCA is the only U.S. law intended to enable regulation of industrial chemicals both before and after they enter commerce. Other federal laws that pertain to chemicals are essentially "end-of-pipe" statutes that aim to control chemical emissions and exposures but do not permit pre-market review of chemicals. Whereas TSCA potentially applies to some 83,000 chemical compounds, five other major U.S. statutes combined currently apply to only 1,134 chemicals and pollutants (Dernbach 1997) (Table 1).

TSCA was the U.S. response to conditions in which, before 1976, tens of thousands of chemicals entered markets without any form of accountability or oversight (Goldman 2002). Congress articulated the statute's objectives in TSCA §2 (TSCA 1976):

- Chemical producers should develop adequate data on the health and environmental effects of chemical substances and mixtures.

- Government should have adequate authority to regulate chemical substances that present an unreasonable risk to health or the environment, and to take action on imminent hazards.

- The government's authority over chemical substances should not create unnecessary economic barriers to technological innovation.

Based on these goals, TSCA promised to be an important step forward in the regulation of industrial chemicals. In practice, however, its legal and procedural requirements have largely thwarted these objectives.

The Data Gap

Logical Paralysis

When it was passed in 1976, TSCA grandfathered the 62,000 chemical substances that were in commercial circulation at that time; that is, except on a case-by-case basis, chemical producers were not required to generate and disclose any information about the uses or hazard traits of these products (U.S. GAO 1994). In essence, this body of existing chemicals was assumed to be safe unless the EPA could prove otherwise. To gather the necessary hazard and exposure data from producers, however, TSCA §4 required the U.S. EPA to establish, on a chemical-by-chemical basis, a) that a substance may present an unreasonable risk to human health or the environment, or b) that there is either significant human exposure potential or substantial quantities of the chemical are produced, imported, and released into the environment (TSCA 1976). This created a logical paralysis for the U.S. EPA: To systematically assess risks of existing chemicals, the EPA needed hazard and exposure data that producers were under no obligation to provide, unless the EPA could first show that such an unreasonable risk might in fact exist or that high exposures were already occurring. This has proved all but paralyzing for the U.S. EPA. In the first 15 years under TSCA, the agency was able to review the risks of about 1,200 (2%) of the 62,000 existing chemicals, despite the fact that the agency estimated that about 16,000 (26%) were potentially of concern based on their production volume and chemical properties (U.S. GAO 1994). TSCA §8(e) does require that chemical manufacturers, processors, and distributors notify the U.S. EPA of any new or unpublished chemical hazard information, and the EPA receives about 300 such submissions each year (U.S. EPA 2003). Perversely, however, this requirement creates a disincentive for manufacturers to voluntarily investigate the hazard properties of their products.

Existing Chemicals

Although the TSCA inventory has grown to about 83,000 substances, the body of 62,000 existing chemicals still constitutes nearly all chemicals in commercial use in the United States. In the 2002 reporting year, 3,000 high production volume (HPV) chemicals—those produced or imported at more than one million pounds per year—made up > 99% by volume of the 15 trillion pounds of chemicals in commerce. Although these HPV chemicals constitute only one-third of existing chemicals by count, their high volume raises concern about the lack of basic hazard information on existing chemicals. Although the U.S. EPA's New Chemical Program requires some minimal data on chemicals introduced since 1976, these chemicals make up < 1% of the production volume of the substances under the jurisdiction of TSCA (U.S. EPA 2005).

In tacit acknowledgement of its constraints, the EPA has turned to voluntary initiatives to close the data gap on existing chemicals. The HPV Challenge (U.S. EPA 2008b) is one such effort, begun in 1997 as an effort to gather screening-level data on HPV chemicals. It has been substantially limited, however, by late, incomplete, and poor- quality data submissions by chemical producers (Denison 2007; Denison and Florini 2004). In 2008, the U.S. EPA announced the Chemical Assessment and Management Program, in which the agency plans to conduct risk-based prioritizations for about 6,750 chemicals (U.S. EPA 2008a). These assessments, however, rely on incomplete data and information from TSCA inventory updates, much of which is obscured by trade secret claims and a lack of transparent assessment methods (Denison 2008).

The Safety Gap

Barriers to Action

Given that the U.S. EPA bears the burden of proof, the agency is further constrained by the level of evidence required to take regulatory action. To regulate a chemical, TSCA §6 requires the U.S. EPA to provide substantial evidence of all of the following conditions: The chemical presents or will present an unreasonable risk to health and the environment; the benefits of regulation outweigh both the costs to industry of the regulation and the lost economic and social value of the product; the U.S. EPA has chosen the least burdensome means of addressing the source of unreasonable risk; and that no other statute could adequately address the risk (TSCA 1976).

Under this evidentiary burden, the U.S. EPA has been able, since 1976, to use its formal rule-making authority to partially regulate five existing chemicals (or chemical classes): polychlorinated biphenyls (PCBs), chlorofluorocarbons, dioxins, asbestos, and hexavalent chromium (U.S. GAO 1994). Of these, an amendment by Congress to TSCA required regulation of PCBs, and the U.S. EPA's asbestos regulation, promulgated after the agency spent 10 years building its case, was overturned in its most significant aspects by the 5th Circuit Court of Appeals, which concluded that the U.S. EPA had failed to meet its burdens of proof under TSCA (Goldman 2002).

New Chemicals

TSCA does grant the U.S. EPA relatively more authority to regulate new chemicals introduced since 1976, as well as new uses of existing chemicals. Producers must submit premanufacturing notices (PMNs) before marketing a new chemical. However, there is no required minimum data set beyond information already in the possession of the producer at the time they file the PMN. This has created a disincentive for manufacturers to investigate their products' potential hazards because, although known hazards must be reported, ignorance of hazard is not penalized. Not surprisingly, an EPA evaluation found that 85% of PMNs lacked data on chemicals' health effects, and 67% lacked health or environmental data of any kind (U.S. EPA 2003). Hindered by limited data and the short time period permitted for the agency's review, the U.S. EPA has taken some form of action on < 10% of the 36,600 chemicals that producers have proposed for commercial use between 1979 and 2004 (U.S. EPA 2007). Once new chemicals have entered commercial use, the U.S. EPA may regulate them only under the standards and burdens it carries for existing chemicals, as described above.

Trade Secrets

Extensive trade secret claims permitted under TSCA have exacerbated both the data and safety gaps. Although some protection of proprietary information is necessary, in practice the statute's allowances for confidential business information (CBI) claims have severely limited access to basic information on chemical identity and use. In 2005, the U.S. EPA reported that 95% of PMNs submitted by producers contained some information claimed as confidential (U.S. GAO 2005a). One assessment found that 90% of the CBI claims in PMNs hid the identity of the chemical (U.S. EPA 2003). CBI allowances under TSCA have thus contributed to a pervasive lack of supply-chain transparency about chemical hazards, despite other regulations intended to facilitate hazard communication via Material Safety Data Sheets (Sattler et al. 1997; Occupational Safety and Health Administration 2004; U.S. Government Accountability Office 1991).

The Technology Gap

Minimal Investment

By requiring less hazard information for existing substances than for new chemicals (data gap) and by stymieing effective regulation of well-known hazards (safety gap), TSCA has produced a chemicals market that favors existing chemicals over newer and potentially less toxic substances. As a result, the statute provides little motivation for industry investment in green chemistry research and development. The vast majority of chemical products manufactured in the United States rely on technologies developed 40–50 years ago, a fact that led the Council for Chemical Research to call for new technologies that incorporate economical and environmentally safer processes, use less energy, and produce fewer harmful by-products (Council for Chemical Research 1996). Twelve years after the Council's Vision 2020 report, the Web sites of the 50 largest U.S. chemical companies all state their commitment to reaching sustainability goals, but their spending on research and development has remained level or declined since about 2000 (Lempert et al. 2003; NRC 2005). A National Research Council report concluded that this trend makes it difficult to advance the science and technology needed to support such sustainability goals (NRC 2005).

Public investment is also critical to the development of green chemistry. The U.S. experience in other sectors illustrates that investment by government, federal laboratories, and research universities helps spur innovation of new technologies (Block and Keller 2008).

Education and the Market

It is a reflection of the priorities of the U.S. chemicals market that, with rare exception, students can earn an undergraduate or graduate degree in chemistry at universities throughout the United States without demonstrating an understanding of the principles of toxicology, ecotoxicology, or green chemistry (Cone 2008c; NRC 2007). Educational priorities have largely matched those of the chemicals market, such that the hazard traits of a substance, for example, are undervalued in the chemistry classroom relative to chemical function, price, and performance. Without a policy strategy that will favor green chemistry in the market, and without a corresponding research and educational effort, the United States risks lagging behind the European Union and other industrialized regions in the scientific and technical development of green chemistry.

Effects of the Three Gaps

The most striking implication of the data gap is that the U.S. EPA lacks the information it needs to identify potential threats to public health and the environment. Perhaps equally significantly, the data-poor market makes safer alternatives difficult to distinguish from hazardous chemicals, distorting market signals (Guth et al. 2007). Furthermore, chemical hazard information generated by producers is often asymmetrically distributed, with inadequate communication to the market. Given better information, many downstream businesses, as well as governments, consumers, and workers, would be better able to express a preference for safer chemicals and products. The ability to identify safer substances could potentially lower the business costs of handling hazardous substances, estimated at 7-10 times the purchase cost of the chemical (Chemical Strategies Partnershp 2004).

The safety gap compounds these problems by allowing the commercial circulation of hazardous substances. The high level of evidence TSCA requires of the U.S. EPA makes it difficult for the agency to act on information it does have and impose restrictions on chemical use. With the burden of proof on government, lack of information increases the likelihood that hazardous chemicals will not be regulated. Some have argued that what begins as a regulatory disincentive for producers to generate or disclose hazard information has become an incentive to create misleading information that casts doubt on scientific evidence (Michaels 2008). Given the high evidentiary threshold, doubt, in and of itself, effectively impedes the agency's ability to take action (Guth et al 2007).

Finally, as a result of the incentives created by TSCA, chemical producers are rationally motivated to defend existing chemicals (Ashford 2000; Echikson 2006). Those invested in existing chemicals have a strong commercial interest in resisting policies that could improve market transparency and the commercial viability of safer substitutes, as evidenced by the efforts of the American Chemistry Council to influence REACH negotiations (Brown 2003; Loewenberg 2006; U.S. House of Representatives 2004).

Although large "sunk" investments in existing chemicals and processes could make it difficult to transform the industrial system to one based on the principles of green chemistry, industry will have to make this transition if the United States is to meet the challenge of economic and environmental sustainability.

Closing the Gaps: Implications for the Environmental Health Sciences

The transition to a sustainable chemical enterprise in the United States will require a fundamental reform of TSCA that meets three overarching objectives:

- Close the data gap: Provide for the effective operation of the chemicals market by requiring that chemical producers generate, disclose, distribute, and effectively communicate sufficient information to stakeholders on the hazard properties of chemicals.

- Close the safety gap: Provide government with the legal tools necessary to identify, prioritize, and take action to reduce chemical hazards and exposures.

- Close the technology gap: Build capacity in cleaner chemicals and processes by incorporating scientific, technical, legal, and policy-related elements of green chemistry into the nation's education and research infrastructure.

Accomplishing these objectives will require both supply-side and demand-side strategies (Green Chemistry Initiative 2008). Supply-side strategies address the technology gap. They are intended to improve the supply of the science, technology, and commercial applications of green chemistry through advancements in education, research, and development. Demand-side strategies address the data and safety gaps, primarily through public policies to drive data generation and disclosure and to regulate known hazards, a combination that ultimately improves the structure of incentives in the chemicals market. Demand-side policies also include, for example, laws that extend the scope of producer responsibility to include the complete product life cycle.

The two approaches operate in tandem: Demand-side strategies provide the requisite drive for supply-side solutions by generating the market need for new science and technology. That is, demand stimulates the private and public investments necessary to advance green chemistry innovation. The importance of market demand as a driver for industrial innovation is well established in the environmental sector. A survey of the executives of 90 leading companies operating in 13 European Union countries identified public policy and market demand as the two most important factors necessary for motivating environmental innovation in their companies (Henzelmann et al. 2007). Recent reports by the U.S. GAO (2009) and the Environmental Defense Fund (Denison 2009) are among others that have called for policy changes to improve the incentive structure in the chemicals market by, for example, strengthening the U.S. EPA's authority to obtain chemical hazard information from producers and shifting more of the burden to producers to demonstrate the safety of their products.

Undertaking meaningful chemicals policy reform in the United States will engender new research questions that must be informed by the environmental health sciences. Through the lens of the three policy gaps, these questions include the following:

The Data Gap

- What is the most useful and attainable body of standardized hazard and exposure information that should be generated for chemicals, and how should a chemical's sales volume and inherent hazard traits drive the scope of data requirements?

- What is the proper role of producers in generating these data, and how can the credibility, standardization, and quality of these data be assured?

- In what ways can emerging predictive toxicity testing and exposure methods be applied to meet these data needs?

- What information on chemical hazards, exposures, and uses if made publicly available would most effectively protect public and environmental health and motivate the development of safer alternatives?

- What are the most effective means of communicating chemical hazard and exposure information to stakeholders, including product formulators, downstream businesses, communities, workers, consumers, and government agencies?

The Safety Gap

- On what measures of hazard and exposure (such as environmental persistence, bioaccumulative potential, toxicity, or presence in consumer products) should chemicals be prioritized and safer alternatives defined?

- To what extent (and in what ways) should producers carry the burden of proof of chemical safety?

- What level of evidence of potential harm to health or the environment is sufficient to trigger government action?

- What portfolio of actions should government employ to efficiently address identified hazards?
- What are the appropriate bounds of producer responsibility over the life cycle of chemicals and products, and what policies can best ensure producer responsibility within these bounds?

The Technology Gap

- How can developments in the environmental health sciences (such as biomonitoring findings or the science of endocrine-disrupting chemicals) best be communicated to the chemical enterprise so as to drive continuous improvement in chemical design?
- What are the high-priority chemicals and processes that warrant publicly funded research into green chemistry alternatives?
- What are the scientific, technical, and practical barriers to implementing these alternatives, and how are these barriers best addressed?
- How should green chemistry inform the development of next-generation environmental technologies, such as alternative energy and building materials, to reduce their health and environmental impacts and improve their overall sustainability?
- How should green chemistry education be designed to better prepare scientists, engineers, and decision makers to respond to the challenges of sustainability?

Conclusion

Although some leading businesses have adopted green chemistry methods, the vast potential of green chemistry remains untapped. This primarily reflects the priorities of the U.S. chemicals market, in which chemical safety is undervalued relative to function, price, and performance. Hazardous chemicals have thus remained competitive, despite the many costs society bears as a result of their production, use, and eventual disposal. These market conditions are a consequence of the chemical data gap, safety gap, and technology gap that have grown out of weaknesses in the language and implementation of TSCA. A new U.S. chemicals policy has the potential to address the societal costs—human, environmental, and economic—that have accompanied advancements in the chemical enterprise.

New chemical and product laws in the European Union have opened an opportunity for chemicals policy reform in the United States. A fundamental restructuring of TSCA will need to simultaneously correct the data, safety, and technology gaps using strategies that improve both the demand for and supply of green chemistry technologies. The attendant research questions demand engagement from the environmental health sciences, and their solutions offer the possibility of improving human health and environmental protection while moving the United States to a position of global leadership in green chemistry innovation.

References

1. Aftalion F. 2001. A History of the International Chemical Industry. Philadelphia: Chemical Heritage Press.

2. Agency for Toxic Substances and Disease Registry. 2003. 2003 CERCLA Priority List of Hazardous Substances. Atlanta, GA:Agency for Toxic Substances and Disease Registry. Available: http://www.atsdr.cdc.gov/cercla/ [accessed 30 January 2009].

3. American Chemistry Council. 2003. Guide to the Business of Chemistry. Arlington, VA: American Chemistry Council.

4. Anastas P, Warner J. 1998. Green Chemistry: Theory and Practice. New York:Oxford University Press.

5. Arora A, Landau R, Rosenberg N. 1998. Chemicals and Long-Term Economic Growth: Insights from the Chemical Industry. New York:John Wiley & Sons, Inc.

6. Ashford N. 2000. An innovation-based strategy for a sustainable environment. In: Innovation-Oriented Environmental Regulation: Theoretical Approach and Empirical Analysis (Hemmelskamp J, Rennings K, Leone F, eds). New York:Springer-Verlag.

7. Barr D, Bishop A, Needham L. 2007. Concentrations of xenobiotic chemicals in the maternal-fetal unit. Reprod Toxicol 23(3):260–266.

8. Bay K, Asklund C, Skakkebaek N, Andersson A. 2006. Testicular dysgenesis syndrome: possible role of endocrine disrupters. Best Pract Res Clin Endocrinol Metab 20:77–90.

9. Black H. 2008. Chemical reaction: the U.S. response to REACH. Environ Health Perspect 116:A127. Available: http://www.ehponline.org/members/2008/116-3/spheres.html [accessed 30 January 2009].

10. Block F, Keller M. 2008. Where Do Innovations Come From? Transformations in the U.S. National Innovation System, 1970–2006. Washington, DC:Information Technology and Innovation Foundation. Available: http://www.itif.org/index.php?id=158 [accessed 30 January 2009].

11. Brown VJ. 2003. REACHing for chemical safety. Environ Health Perspect 111:A766–A769. Available: http://www.ehponline.org/members/2003/111-14/spheres.html [accessed 30 January 2009].

12. Buckland D. 2008. Within REACH: Under REACH, new environmentally-friendly substances will become commercially more attractive. Industry Week, 21 July. Available: http://www.industryweek.com/PrintArticle. aspx?ArticleID=16841 [accessed 30 January 2009].

13. Cal EPA. 2004. Intra-Agency Environmental Justice Strategy. Sacramento: California Environmental Protection Agency. Available: http://www.calepa. ca.gov/EnvJustice/Documents/ 2004/Strategy/Final.pdf [accessed 30 January 2009].

14. CDC. 2005. National Report on Human Exposure to Environmental Chemicals: Third Report. Atlanta, GA:Centers for Disease Control and Prevention. Available: http://www.cdc.gov/exposurereport/report.htm [accessed 30 January 2009].

15. Chemical Strategies Partnership. 2004. Chemical Management Services Industry Report 2004: Creating Value Through Service. San Francisco:Chemical Strategies Partnership. Available: http://www.chemicalstrategies.org/resources_ industry.htm [accessed 30 January 2009].

16. Clean Air Act of 1963. 1963. Public Law 88-206.

17. Clean Water Act of 1966. 1966. Public Law 89-753.

18. Cone M. 2006. U.S. rules allow sale of products others ban. Chemical-laden good outlawed in Europe and Japan are permitted in the American market. Los Angeles Times (Los Angeles, CA) 6 October: A1.

19. Cone M. 2008a. A Greener Future: Part 1. Products derived from natural, nontoxic ingredients—once seen as fringe—are now mainstream. Los Angeles Times (Los Angeles, CA) 14 September. Available: http://www.latimes.com/ news/local/la-me-greenchem14-2008sep14,0,5897268,full.story [accessed 30 January 2009].

20. Cone M. 2008b. A Greener Future: Part 2. Most industries remain dependent on hazardous substances. Los Angeles Times (Los Angeles, CA) 19 September. Available: http://www.latimes.com/news/local/la-me-greenchem19-2008sep19,0,4450885,full.story [accessed 30 January 2009].

21. Cone M. 2008c. Potential environmental risks aren't part of chemical engineers' training. Los Angeles Times (Los Angeles, CA) 19 September. Available: http://articles.latimes.com/2008/sep/19/local/me-greenchemside19 [accessed 30 January 2009].

22. Council for Chemical Research. 1996. Technology Vision 2020: The U.S. Chemical Industry. Washington, DC:American Chemical Society, American

Institute of Chemical Engineers, Chemical Manufacturers Association, Synthetic Organic Chemical Manufacturers Association. Available: http://www.chemicalvision2020.org/whatis.html [accessed 30 January 2009].

23. Denison R. 2007. High Hopes, Low Marks: A Final Report Card on the HPV Challenge. Washington, DC:Environmental Defense Fund. Available: http://www.edf.org/documents/6653_HighHopesLowMarks.pdf [accessed 30 January 2009].

24. Denison R. 2008. Environmental Defense Fund's Comments on ChAMP: EPA's Recent Commitments and Possible New Initiatives for Existing Chemicals. Docket EPA-HQ-OPPT-2008-0319. Available: http://www.edf.org/documents/ 7871_Comments_ChAMP_May08.pdf [accessed 30 January 2009].

25. Denison R. 2009. Ten Essential Elements in TSCA Reform. Available: http://www.edf.org/documents/9070_Denison_10_%20Elements_TSCA_Reform.pdf [accessed 30 January 2009].

26. Denison R, Florini K. 2004. Facing the Challenge: A Status Report on the U.S. HPV Challenge Program. Washington, DC:Environmental Defense. Available: http://www.environmentaldefense.org/documents/3816_HPVorphansExec.pdf [accessed 30 January 2009].

27. Department of Toxic Substances Control. 2007. Failed Land Disposal Units: A Twenty Year Retrospective. Sacramento, CA:California Environmental Protection Agency, Department of Toxic Substances Control.

28. Dernbach J. 1997. The unfocused regulation of toxic and hazardous pollutants. Harvard Law Review 21(1):1–57.

29. DeVito S, Garrett R. 1998. Designing Safer Chemicals: Green Chemistry for Pollution Prevention. Washington, DC:American Chemical Society.

30. Doucet J, Tague B, Arnold DL, Cooke GM, Hayward S, Goodyear C. 2009. Persistent organic pollutant residues in human fetal liver and placenta from greater Montreal, Quebec: a longitudinal study from 1998 through 2006. Environ Health Perspect 117:605–610.

31. Echikson W. 2006. Europe fines chemical firms $489.8 million for bleach cartel. Wall Street Journal (New York, NY). 4 May: A2.

32. Energy & Enviro Finland. 2008. Russia harmonizes its chemical regulation with EU REACH. Energy and Enviro Finland, 17 July. Available: http://www.energy-enviro.fi/index.php?PAGE=1832&NODE_ID=1832&LANG=1 [accessed 30 January 2009].

33. Emergency Planning and Community Right to Know Act of 1986. 1986. Toxics Release Inventory [section 313]. Public Law 99-499.

34. European Commission. 2008. What is REACH? Available: http://ec.europa.eu/environment/chemicals/reach/reach_intro.htm [accessed 30 January 2009].

35. European Commission, Environment Directorate General. 2003. Assessment of the Impact of the New Chemicals Policy on Occupational Health—Final Report. Norfolk, UK:Risk and Policy Analysts Limited. Available: http://ec.europa.eu/environment/chemicals/reach/background/docs/finrep_occ_health.pdf [accessed 30 January 2009].

36. Geiser K. 2001. Materials Matter. Cambridge, MA:MIT Press.

37. Goldman L. 2002. Preventing pollution? U.S. toxic chemicals and pesticides policies and sustainable development. Environ Law Rev 32:11018–11041.

38. Grandjean P, Bellinger D, Bergman A, Cordier S, Davey-Smith G, Eskenazi B, et al. 2007. The Faroes Statement: human health effects of developmental exposure to chemicals in our environment. Basic Clin Pharmacol Toxicol 102:73–75. Available: http://www.blackwell-synergy.com/doi/pdf/10.1111/j.1742-7843.2007.00114.x [accessed 30 January 2009].

39. Green Chemistry Initiative. 2008. Green Chemistry Option for the State of California. Report from the Green Chemistry Science Advisory Panel to Department of Toxic Substances Control Director Maureen Gorsen. Sacramento, CA:California Environmental Protection Agency, Department of Toxic Substances Control. Available: http://www.dtsc.ca.gov/PollutionPrevention/GreenChemistryInitiative/upload/SAP_Report.pdf [accessed 3 January 2009].

40. Guth J. 2008. Law for the ecological age. Vermont J Environ Law 9(3):431–512.

41. Guth J, Denison R, Sass J. 2007. Require comprehensive safety data for all chemicals. New Solut 17(3):233–258.

42. Henzelmann T, Mehner S, Zwelt T. 2007. Innovative environmental growth markets from a company perspective. Research project on behalf of the Federal Environment Agency (UFOPLAN 206 14-132/04). Available: http://www.rolandberger.com/expertise/industries/environmental_technology/2007-11-01-rbsc-pub-Innovative_environmental_growth_markets.html [accessed 30 January 2009].

43. Herbert R, Landrigan PJ. 2000. Work-related death: a continuing epidemic. Am J Public Health 90(4):541–545.

44. Hertz-Picciotto I, Delwiche L. 2009. The rise in autism and the role of age at diagnosis. Epidemiology 20(1). Available: http://www.vetmed.ucdavis.edu/cceh/Epipaper1208.pdf [accessed 30 January 2009].

45. Iles A. 2006. Shifting to green chemistry: the need for innovations in sustainability marketing. Business Strategy and the Environment, 21 July. Available: http://www3.interscience.wiley.com/journal/112728523/abstract [accessed 30 January 2009].

46. LaDou J. 1997. Occupational and Environmental Medicine. 2nd ed. Stamford, CT:Appleton and Lange.

47. LaMonica M. 2007. Green chemistry's 'race to innovation.' CNET News, 12 November. Available: http://news.cnet.com/Green-chemistrys-race-to-innovation/2100-13838_3-6217663.html [accessed 30 January 2009].

48. Lancaster M. 2002. Green Chemistry: An Introductory Text. Cambridge, UK:Royal Society of Chemistry.

49. Leigh J, Markowitz S, Fahs M, Shin C, Landrigan P. 1997. Occupational injury and illness in the United States: Estimates of costs, morbidity and mortality. Arch Intern Med 157:1557–1568.

50. Lempert R, Norling P, Pernin C, Resetar S, Mahnovski S. 2003. Next Generation Environmental Technologies: Benefits and Barriers. Arlington, VA:RAND Corporation.

51. Loewenberg S. 2006. U.S. chemical companies leave their mark on EU law. Lancet 367(9510):556–557.

52. Main K, Mortensen G, Kaleva M, Boisen K, Damgaard I, Chellakooty M, et al. 2006. Human breast milk contamination with phthlates and alterations in endogenous reproductive hormones in infants three months of age. Environ Health Perspect 114:270–276.

53. Matus K, Anastas P, Clark W, Itameri-Kinter K. 2007. Overcoming the Challenges to the Implementation of Green Chemistry. Harvard Center for International Development Working Paper No. 155. Available: http://www.cid.harvard.edu/cidwp/155.htm [accessed 30 January 2009].

54. Michaels D. 2008. Doubt Is Their Product: How Industry's Assault on Science Threatens Your Health. New York:Oxford University Press.

55. NRC (National Research Council). 1984. Toxicology Testing: Strategies to Determine Needs and Priorities. Washington, DC:National Academies Press. Available: http://www.nap.edu/openbook.php?isbn=0309034337 [accessed 30 January 2009].

56. NRC (National Research Council). 1992. Critical Technologies: The Role of Chemistry and Chemical Engineering. Washington, DC:National Academies Press.

57. NRC (National Research Council). 2005. Sustainability in the Chemical Industry: Grand Challenges and Research Needs—A Workshop Report. Washington, DC:National Academies Press.

58. NRC (National Research Council). 2007. Exploring Opportunities in Green Chemistry and Engineering Education: A Workshop Summary to the Chemical Sciences Roundtable. Washington, DC: National Academies Press.

59. Occupational Safety and Health Act of 1970. 1970. Public Law 91-596.

60. Occupational Safety and Health Administration. 2004. Hazard Communication in the 21st Century Workplace. Washington, DC: U.S. Department of Labor, Occupational Safety and Health Administration. Available: http://www. osha.gov/dsg/hazcom/finalmsdsreport.html [accessed 30 January 2009].

61. OECD. 2001. Environmental Outlook for the Chemicals Industry. Paris:Organisation for Economic Co-operation and Development. Available: http://www.oecd.org/dataoecd/7/45/2375538.pdf [accessed 30 January 2009].

62. Office of Technology Assessment. 1995. Screening and Testing of Chemicals in Commerce: Background Paper. Washington, DC:Congressional Office of Technology Assessment. Available: http://www.princeton.edu/~ota/ disk1/1995/9553/9553.PDF [accessed 30 January 2009].

63. Ostrowski S, Wilbur S, Chou C, Pohl H, Stevens Y, Allred P, et al. 1999. Agency for Toxic Substances and Disease Registry's 1997 priority list of hazardous substances. Latent effects—carcinogenesis, neurotoxicology, and developmental deficits in humans and animals. Toxicol Ind Health 15:602–644.

64. Pastor M, Morello-Frosch R, Sadd J. 2005. The air is always cleaner on the other side: race, space and ambient air toxics exposures in California. J Urban Aff 27(2):127–148.

65. Pickvance S, Karnon J, Peters J, El-Arifi K. 2005. The Impact of REACH on Occupational Health, with a Focus on Skin and Respiratory diseases. Sheffield, UK:University of Sheffield. Available: http://hesa.etui-rehs.org/uk/newsevents/ files/reach-sheffield-complet.pdf [accessed 30 January 2009].

66. REACH. 2007. Registration, Evaluation, Authorisation, and Restriction of Chemicals. Official Journal of the European Union. Available: http://www. reach-compliance.eu/ english/legislation/docs/launchers/launch-2006-1907- EC.html [accessed 30 January 2009].

67. Resource Conservation and Recovery Act of 1976. 1976. Public Law 94-580.

68. Robinson J. 1991. Toil and Toxics: Workplace Struggles and Political Strategies for Occupational Health. Berkeley, CA:University of California Press.

69. Roe D, Pease W, Florini K, Silbergeld E. 1997. Toxic Ignorance: The Continuing Absence of Basic Health Testing for Top-Selling Chemicals in the United States. Washington, DC:Environmental Defense Fund. Available: http://www. environmentaldefense.org/pdf.cfm?ContentID= 243&FileName=toxicignorance. pdf [accessed 30 January 2009].

70. Sattler B, Lippy B, Jordan T. 1997. Hazard Communication: A Review of the Science Underpinning the Art of Communication for Health and Safety.

Available: http://www.osha.gov/SLTC/hazardcommunications/hc2inf2.html [accessed 30 January 2009].

71. Schapiro M. 2007. Exposed: The Toxic Chemistry of Everyday Products and What's at Stake for American Power. White River, VT:Chelsea Green Publishing.

72. Sharpe R, Irvine D. 2004. How strong is the evidence of a link between environmental chemicals and adverse effects on human reproductive health? BMJ 328:447–451.

73. Skakkebaek N. 2002. Endocrine disrupters and testicular dysgenesis syndrome. Hormone Res 57(suppl 2):43.

74. Spitz P. 2003. The Chemical Industry at the Millennium. Philadelphia:Chemical Heritage Press.

75. Surveillance, Epidemiology, and End Results Program. 2003. National Cancer Institute, Surveillance Research Program, Cancer Statistics Branch. Available: http://seer.cancer.gov/ [accessed 30 January 2009].

76. Swan S, Elkin E, Fenster L. 2000. The question of declining sperm density revisited: an analysis of 101 studies published 1934–1996. Environ Health Perspect 108:961–966.

77. Swan S, Main K, Liu F, Stewart S, Kruse R, Calafat A, et al. 2005. Decrease in anogenital distance among male infants with prenatal phthalate exposure. Environ Health Perspect 113:1056–1061.

78. TSCA. 1976. Toxic Substances Control Act of 1976. Public Law 94-469.

79. United Nations. 2004. World Population to 2300. New York: United Nations. Available: http://www.un.org/esa/population/publications/longrange2/WorldPop2300final.pdf [accessed 30 January 2009].

80. U.S. Department of Commerce. 2008. The Latest on REACH. Washington, DC:U.S. Department of Commerce, U.S. Commercial Service, U.S. Mission to the European Union. Available http://www.buyusa.gov/europeanunion/reach.html [accessed 30 January 2009].

81. U.S. EPA. 1998. Chemical Hazard Data Availability Study: What Do We Really Know about the Safety of High Production Volume Chemicals? Washington, DC:U.S. Environmental Protection Agency, Office of Pollution Prevention and Toxics. Available: http://www.epa.gov/hpv/pubs/general/hazchem.pdf [accessed 30 January 2009].

82. U.S. EPA. 2003. Overview: Office of Pollution Prevention and Toxics Programs. Draft Version 2.0. Washington, DC:U.S. Environmental Protection Agency. Available: http://www.chemicalspolicy.org/downloads/TSCA10112-24-03.pdf [accessed 30 January 2009].

83. U.S. EPA. 2004. Cleaning up the Nation's Waste Sites: Markets and Technology Trends. Washington, DC:U.S. Environmental Protection Agency. Available: http://www.clu-in.org/download/market/2004market.pdf [accessed 30 January 2009].

84. U.S. EPA. 2005. How Can EPA More Efficiently Identify Potential Risks and Facilitate Risk Reduction Decision for Non-HPV Existing Chemicals? Washington, DC:U.S. Environmental Protection Agency, National Pollution Prevention and Toxics Advisory Committee, Broader Issues Work Group. Available: http://epa.gov/oppt/npptac/pubs/ finaldraftnonhpvpaper051006.pdf [accessed 30 January 2009].

85. U.S. EPA. 2007. New Report Projects. Number, Cost and Nature of Contaminated Site Cleanups in the U.S. over the Next 30 Years. Washington, DC:U.S. Environmental Protection Agency, Superfund Program. Available: http://epa.gov/superfund/accomp/news/30years.htm [accessed 30 January 2009].

86. U.S. EPA. 2008a. Chemical Assessment and Management Program (ChAMP). EPA Chemical Efforts Under ChAMP. Washington, DC:U.S. Environmental Protection Agency. Available: http://www.epa.gov/champ/ [accessed 30 January 2009].

87. U.S. EPA. 2008b. High Production Volume (HPV) Challenge Program. Washington, DC:U.S. Environmental Protection Agency, Office of Pollution Prevention and Toxics. Available: http://www.epa.gov/chemrtk/index.htm [accessed 30 January 2009].

88. U.S. EPA. 2009. Inventory Update Reporting: Non-confidential 2006 IUR Records by Chemical, including Manufacturing, Processing and Use Information. Washington, DC:U.S. Environmental Protection Agency, Office of Pollution Prevention and Toxics. Available: http://cfpub.epa.gov/iursearch/ [accessed 30 January 2009].

89. U.S. GAO (U.S. Government Accountability Office). 1991. OSHA Action Needed to Improve Compliance with Hazard Communication Standard. GAO/HRD-92-8. Washington, DC:U.S. General Accounting Office. Available: http://archive.gao.gov/t2pbat7/145328.pdf [accessed 30 January 2009].

90. U.S. GAO. 1994. Toxic Substances Control Act: Legislative Changes Could Make the Act More Effective (GAO/RCED-94-103). Washington, DC:U.S. Government Accountability Office. Available: http://archive.gao.gov/t2pbat2/152799.pdf [accessed 30 January 2009].

91. U.S. GAO. 2005a. Chemical Regulation: Options Exist to Improve EPA's Ability to Assess Health Risks and Manage Its Chemicals Review Program (GAO 05-458). Washington, DC:U.S. Government Accountability Office. Available: http://www.gao.gov/new.items/d05458.pdf [accessed 30 January 2009].

92. U.S. GAO. 2005b. Hazardous Waste Programs: Information on Appropriations and Expenditures for Superfund, Brownfields, and Related Programs. GAO-05-746R. Washington, DC:U.S. Government Accountability Office. Available: http://www.gao.gov/new.items/d05746r.pdf [accessed 30 January 2009].

93. U.S. GAO. 2007. Chemical Regulation: Comparison of U.S. and Recently Enacted European Union Approaches to Protect against the Risks of Toxic Chemicals (GAO-07-825). Washington, DC:U.S. Government Accountability Office. Available: http://www.gao.gov/new.items/d07825.pdf [accessed 30 January 2009].

94. U.S. GAO. 2009. High Risk Series: An Update. GAO-09-271. Washington, DC:U.S. Government Accountability Office. Available: http://www.gao.gov/products/GAO-09-271 [accessed 22 January 2009].

95. U.S. House of Representatives. 2004. The Chemical Industry, the Bush Administration, and European Efforts to Regulate Chemicals. U.S. Congressional Committee on Oversight and Government Reform, 1 April. Available: http://oversight.house.gov/story.asp?ID=427 [accessed 30 January 2009].

96. Wilson M. 2006. Testimony before the U.S. Senate Committee on Environment and Public Works, Oversight Hearing on the Toxic Substances Control Act and the Chemicals Management Program. Washington, DC, 2 August. Available: http://epw.senate.gov/hearing_statements.cfm?id=260748 [accessed 30 January 2009].

97. Wilson M, Chia D, Ehlers B. 2006. Green Chemistry in California: A Framework for Leadership in Chemicals Policy and Innovation. Special Report to the California Senate Environmental Quality Committee and the Assembly Committee on Environmental Safety and Toxic Materials. Berkeley, CA:California Policy Research Center, UC Office of the President. Available: http://coeh.berkeley.edu/docs/news/06_wilson_policy.pdf [accessed 30 January 2009].

98. Wilson M, Schwarzman M, Malloy T, Fanning E, Sinsheimer P. 2008. Green Chemistry: Cornerstone to a Sustainable California. Special Report to the Secretary, California Environmental Protection Agency. Berkeley and Los Angeles, CA:University of California, Centers for Occupational and Environmental Health. Available: http://coeh.berkeley.edu/docs/news/green_chem_brief.pdf [accessed 30 January 2009].

Biofilm Reactors for Industrial Bioconversion Processes: Employing Potential of Enhanced Reaction Rates

Nasib Qureshi, Bassam A. Annous, Thaddeus C. Ezeji,
Patrick Karcher and Ian S. Maddox

ABSTRACT

This article describes the use of biofilm reactors for the production of various chemicals by fermentation and wastewater treatment. Biofilm formation is a natural process where microbial cells attach to the support (adsorbent) or form flocs/aggregates (also called granules) without use of chemicals and form thick layers of cells known as "biofilms." As a result of biofilm formation, cell densities in the reactor increase and cell concentrations as high as 74 gL⁻¹ can be achieved. The reactor configurations can be as simple as a batch reactor, continuous stirred tank reactor (CSTR), packed bed reactor (PBR), fluidized bed reactor (FBR), airlift reactor (ALR), upflow anaerobic sludge blanket (UASB) reactor, or any other suitable configuration. In UASB granular biofilm particles are used. This article demonstrates that reactor productivities in these reactors have been superior to any other reactor types. This article describes production of ethanol, butanol, lactic acid, acetic acid/vinegar, succinic acid, and fumaric acid in addition to wastewater treatment in the biofilm reactors. As the title suggests, biofilm reactors have high potential to be employed in biotechnology/bioconversion industry for viable economic reasons. In this article, various reactor types have been compared for the above bioconversion processes.

Introduction

Biochemical reactors play an important role in the biochemical industry as the rate of reaction, ease, and length of reactor operation affect reactor productivities and hence process economics [1,2]. In order to employ a most appropriate reactor for an industrial operation, reaction rate should be high and the reactor configuration should be simple. Under optimized parameters such as pH, temperature, substrate, and medium components, reaction rate can be increased by increasing cell mass concentration in the reactor. There are two methods commonly used for increasing cell mass concentration inside the reactor; first, use of a permeable membrane to retain cells; and the other, use of immobilized cell technique. Membrane reactors allow passing of liquid, substrate, and product out of the reactor while retaining the cells. In these reactors, high cell concentrations can be achieved [3]. Unfortunately, for some processes such as waste water treatment, these reactors are not preferred due to their high cost and problems with fouling. Other processes where the relatively high cost of these reactors does not allow their use include production of large volume, low cost chemicals such as vinegar or acetic acid.

Other types of reactors that offer high reaction rates are immobilized cell reactors [4]. In these reactors, high cell concentrations are achieved by fixing them on various supports. Cells can be immobilized by three different techniques; namely, adsorption, entrapment, and covalent bond formation. Entrapment and covalent bond formation require use of chemicals that add to the cost of production and perhaps restrict further propagation or increase in cell concentration inside the reactor. The third technique is of natural origin as cells "adsorb/and adhere" to the support naturally and firmly [4-6]. This technique is called "adsorption" and has been used extensively in the literature to adsorb microbial cells. Table 1 shows a comparison of these techniques with the membrane reactors. It should be noted that some microbial cells leach out from immobilized cell reactors which require separation (leached out cells in reactor effluent) prior to product removal possibly by centrifugation.

Table 1. A comparison of different types of reactors with biofilm reactors

Reactor Type	Comments
Membrane reactor	
Advantages	High productivities, high cell concentration can be achieved inside the reactor, clear permeates for further separation
Disadvantages	Fouling with cells, cost prohibits their use in low cost large volume chemical production
Immobilized cell reactors	
Covalent bond formation	
Advantages	High cell concentration may be achieved, high productivity
Disadvantages	Cell growth inside matrix may be restricted, cells leach out of the matrix and hence centrifugation of effluent may be required, chemical may affect the cells
Entrapment	
Advantages	High cell concentration may be achieved, high productivity
Disadvantages	Matrix often starts disintegration with time, cells leach out of matrix, centrifugation of reactor effluents is required for further separation
Biofilm	
Advantages	Comparatively high reactor productivities and high cell concentrations are achieved, reactors run longer and are economic to operate
Disadvantages	Effluent centrifugation is required

In addition to being a natural process, adsorption can be performed in place, and economical adsorbents are available. Additionally, these reactors are simple in concept and construction and the immobilization process is economical. Adsorbed cells form cell layers on the support and cell mass grows inside the reactor over time [7]. These layers of cells are called "biofilms." Biofilms can be used in various types of reactors such as continuous stirred tank reactors (CSTRs), packed bed reactors (PBRs), fluidized bed reactors (FBRs), airlift reactors (ALRs), upflow anaerobic sludge blanket (UASB) reactors, and expanded granular sludge bed (EGSB) reactors etc. [4,7-11]. In these reactors, reaction rates are usually high as compared to the other types of reactors. On the laboratory, pilot plant, and industrial scale (some), these reactors have been very successful and examples include waste water treatment [12] and vinegar or acetic acid [13] production. In addition to these, other processes that have employed these biofilm reactors include ethanol, butanol, lactic acid, fumaric acid, and succinic acid production. Since they offer high reaction rates and are economical, this review becomes their subject matter. In the authors' view, this natural process of biofilm formation can be employed to economize production of various chemicals by fermentation on a large scale [13]. In biofilm reactors, cell concentrations as high as 74 gL^{-1} can be achieved [7]. In addition, the cell layers in bioparticles become highly active, thus contributing to the high reactor productivities. Within fluidized bed reactors, the biofilm particles are of various shapes (including spherical and irregular shapes) and these reactors can be operated for long periods of time. The amount of adsorbent that is used in these reactors is low, which also reduces the cost of the cell support. In biofilm reactors, reactor configurations can vary from a simple packed bed reactor to fluidized bed, UASB, and airlift reactors as described in this article.

Biofilm Formation

Various Types of Biofilms

In nature, biofilms exist primarily as complex multi-species communities of bacteria in which each species fills an ecological niche within the biofilm depending on its metabolism and morphology [14]. The nature of mixed culture biofilms is dependent on which species are present and what role each species fills. For instance, a single species may utilize anaerobic fermentation deep within one biofilm in one environment, but may utilize an aerobic metabolism in another environment in the presence of different neighboring biofilm species. Multi-species biofilms are important clinically as well as industrially. Clinically, biofilms are important as the source of persistent infections. They are responsible for dental caries and nosocomial infections, as well as a variety of other infections and diseases [15]. Industrially, biofilms are detrimental in many cases and beneficial in many others. For instance, natural biofilms can reduce heat transfer in heat exchangers and cooling towers [16], foul reverse osmosis membranes [17], and contaminate food processing equipment [18]. Multi-species biofilms are used industrially to achieve several aims including the treatment of wastewater for removal of organics [19,20] and heavy metals [21]. The presence of multiple species allows for the treatment of waste streams that are diverse in composition and that fluctuate in component concentration.

Single species biofilm are used to produce industrially important chemicals [22,23]. Such biofilms can exist in some situations and are important industrially, although in nature they are not the norm. For example, in nature, an immature biofilm that is resultant from the attachment and growth of a single cell may exist as a single species biofilm before incorporating other species. For chemical production, single species biofilms are important because they allow for control and maximization of desired products. In this case, a single species is inoculated into a sterile environment and allowed to form a biofilm before being used to produce a particular chemical product.

In industrial applications including wastewater treatment, usually two types of biofilms are employed, namely, biofilms that grow onto supports such as charcoal, resin, bonechar, concrete, clay brick, or sand particles, and biofilms that are formed as a result of flocs and aggregate formation. On the above supports, biomass grows all around the particles and the size of the biofilm particles grows with time usually to several mm in diameter. The density of the support particles is usually higher than the fermentation broth and for this reason bioparticles tend to remain in the lower section of the reactor. Another type of biofilm is where no support is used and cells form biomass granules and flocs that also grow in size with time. This type of biofilm is called granular biofilm and the reactor where this biofilm is used is called granular biofilm reactor. Granule formation may take from several weeks to several months. The cells produce extracellular polymeric substances (EPS) that binds the cells firmly in the form of flocs and aggregates. The most commonly used bioreactors that fall in this category are upflow anaerobic sludge blanket (UASB) reactors that are used to treat domestic and industrial wastewater anaerobically. Sponza [24] examined anaerobic granulation process in a UASB to remove tetrachloroethylene. In some cases expanded bed biofilm reactors have been used with granular biofilm particles that are called expanded granular sludge bed (EGSB) reactors.

Mechanism of Biofilm Formation

A biofilm is defined as a structured community of bacterial cells enclosed in a self-produced polymeric matrix and adherent to an inert or living surface [15]. In general, there are four stages to the development of a mature biofilm: initial attachment, irreversible attachment by the production of extracellular polymeric substances (EPS), early development, and maturation of biofilm architecture [14].

The life of a biofilm starts with the planktonic or free floating cell. In order for a planktonic cell to attach to a surface, it must first interact with the surface. Surfaces immersed in an aqueous solution usually acquire a surface charge which attracts and concentrates inorganic solutes, and charged or highly polar organic molecules. The concentration of cations, glycoproteins, proteins, and organic molecules at the surface can provide a relatively nutritious zone for bacteria compared to the bulk aqueous environment [25]. In addition, fluid flow in the boundary region near the surface can be considered negligible which allows bacteria to approach the surface. Once near the surface, it will either approach the surface by Brownian motion or move by chemotaxis towards the surface in response to the chemical concentration gradient [25]. When at the interface, the cell will form a temporary association with the surface or microbes already present on the surface [26].

After initial association with the surface, a planktonic bacterial cell can dissociate from the surface and resume the planktonic state or become irreversibly attached to the surface. Irreversible attachment involves the production of EPS. EPS serves to bind the cell to the surface and to protect it from the surrounding environment. EPS can be composed of polysaccharides, proteins, nucleic acids, or phospholipids. A common EPS produced by bacterial cells in biofilms is the ex-opolysaccharide alginate. In biofilm associated cells of Pseudomonas aeruginosa, transcription of algC, the gene involved in alginate production, was fourfold that in planktonic cells [27]. EPS provide protection to biofilm cells by providing a diffusive barrier to any toxic compounds that could harm the cells as well as a barrier to phagocytes and bacteriocides. The EPS can also represent a barrier to nutrients necessary for cell growth. Cells in the interior of a biofilm often show a much reduced rate of growth and cell division rate may be near zero [26,28]. The reduced growth rate is itself protective because uptake of toxic substances is also reduced. The presence of the EPS matrix may also serve as a spatial restrictor of cell growth and division.

Water and nutrient diffusion into the interior of a biofilm is highly limited. As biofilms mature, water channels can develop that allow water and nutrient access deeper into the biofilm. These channels partially relieve the diffusion limitation within the biofilm. The architecture of the biofilm develops in response to shear forces. In low shear environments, biofilms can form as thick mushroom-like masses. In high shear environments, biofilms may be flatter or form long strands [29].

A final stage that may occur in the life of a biofilm is reversion of part of the cells to the planktonic state. When cells living in biofilm take up nutrients, they channel much of that energy towards production of EPS rather than to cell growth and division. When nutrients become scarce, cells must escape the EPS matrix or be trapped in an unfavorable environment. Biofilm associated cells are able to produce enzymes capable of breaking down the EPS matrix in times of nutrient starvation. Pseudomonas fluorescens is able to produce an exopolysaccharide lyase under starvation conditions [30]. The enzyme serves not only to break down the polysaccharide matrix allowing cells to find nutrients elsewhere, but the degraded EPS can often be used as a food source for the nutrient deprived cells. In addition to cell detachment due to starvation or nutrient deficiency, there are other detachment processes such as abrasion, shear stress, sloughing and graz-ing. Providing detailed accounts of these processes is considered beyond the scope of this article.

Factors Enhancing Biofilm Formation

Several parameters affect how quickly biofilms form and mature, including sur-face, cellular, and environmental factors. The surface onto which cells will attach has an important impact on biofilm formation. Rough surfaces tend to enhance biofilm formation [31]. Shear forces are lower near a rough surface, and there is a larger surface area to which cells can adhere. Porous materials also work well for biofilm formation. Shear forces are very low inside pores even under conditions where bulk fluid velocity is high. Pores provide a protected environment for cells to attach and grow. Porous materials such as brick and bonechar have been used to immobilize Clostridium cells used in biofilm reactors [22,32]. Biofilm formation also tends to increase with the hydrophobicity of the surface material [28]. Biofilms form much more rapidly on Teflon and other plastics than glass or metal. Possibly this is due to differences in hydrophobicity of the surfaces and ionic charges [28].

The amount of nutrients present in the medium can affect the rate of biofilm formation. Biofilms tend to form more readily in the presence of ample nutrients [33]. One function of the biofilm is to anchor cells in a friendly, nutrient rich environment. Phosphorus is a particularly important nutrient. Cells saturated with phosphate have a higher tendency to flocculate and adhere due to their increased hydrophobicity, while those cells depleted in phosphate are more hydrophilic and less likely to adhere [34].

Temperature can have an effect on biofilm formation. Temperatures at the high end of a culture's growth range can enhance biofilm formation. Depending upon the species involved, high temperature increases the rate of cell growth, EPS production, and surface adhesion, all of which enhance biofilm formation [25].

Cellular factors may affect biofilm formation. A hydrophobic cell will be more able to overcome the initial electrostatic repulsion with the solid surface and adhere more readily. The presence of fimbriae, proteinaceous bacterial appendages high in hydrophobic amino acids, can increase cell surface hydrophobicity [28,35]. Flagellated cells show increased ability to attach to surfaces. Flagellar motility may serve to overcome initial electrostatic surface repulsion.

Calculations and Data Presentation

In a continuous process, productivity ($gL^{-1}h^{-1}$) is calculated as the product concentration in gL^{-1} liquid multiplied by the dilution rate (h^{-1}). In a batch process, productivity is calculated as the product concentration in gL^{-1} liquid divided by the fermentation time (h). Specific productivity (h^{-1}) is calculated as productivity ($gL^{-1}h^{-1}$) divided by cell or protein concentration (gL^{-1}). Dilution rate (feed flow per reactor volume per h) can be based on total volume of the continuous reactor or void volume. In fully or partially packed bed reactors, void volume is total reactor volume minus the volume occupied by the cell support. For a particular flow rate, dilution rate based on void volume is higher than based on the total reactor volume. In this article, reactor productivities based on both total reactor volume and void volume have been reported as mentioned by the different authors. The reader is advised that it is difficult to correlate/compare the two productivities (based on total reactor volume or void volume) unless void volume fraction (void volume/total volume) is given along with the flow/feed rate. Residence time (h) in the reactor can be calculated by inversing the dilution rate (h^{-1}).

Types of Biofilm Reactors

Biofilm reactors can be assembled in a number of configurations including batch, continuous stirred tank (CSTR; including agitating continuous reactors, and rotary continuous reactors), packed bed (PBR), trickling bed (TBR), fluidized bed (FBR), airlift reactors (ALR), upflow anaerobic sludge blanket (UASB), and expanded bed reactors. The operation of these reactors changes from reactor to reactor. In a batch biofilm reactor, the immobilized cells have to be utilized for repeated batches. However, it is likely that during the late stationary phase of chemical production the culture would experience inhibition thus reducing productivity. Also, in a batch reactor, productivity would be reduced due to downtime necessary to fill and empty the reactor. If the reactor is packed with biofilm particles, some cells may die or become inactive due to lack of feed during emptying and filling of the reactor. As a result, it is viewed that batch reactors are not practical for biofilms.

In a CSTR feed medium is fed to the reactor and product is withdrawn at the same rate as feed. They are stirred using a mechanical device such as impeller. CSTRs cannot be packed with the adsorbent support covered by biofilms as no agitation can be provided in that case. However, they can be used if fibrous bed support is used for adsorption of cells. In that case, cells can grow and form a biofilm on the fibrous bed. In such a case agitation can be provided. This type of system was used for the production of butanol [36] and lactic acid [37] in continuous operation with a constant feed and a constant effluent from the reactor. In some cases, there may be excessive growth on the surface of the fibrous bed, and the cell layers may be sheared off the support. This type of fibrous bed biofilm CSTRs are called as agitating continuous reactors. Another type of CSTRs called rotating CSTRs have same length/diameter (L/D) ratio as in CSTRs. The rotating CSTRs are placed horizontally (lengthwise) and are rotated along the horizontal axis.

PBRs are different types of reactors as they are packed with suitable support material followed by inoculation with the culture to form biofilm. The reactor is supplied with a feed that is not deficient in nutrients. Depending on the culture, nutrients, and support, biofilm formation may take a few to several days. Such reactors are usually fed at the bottom, thus getting product at the top of the reactor. However, these reactors are prone to blockade due to excessive cell growth. In C. acetobutylicum/C. beijerinckii biofilm packed bed reactors, reaction rates up to 45 times that of the batch (control) reactors have been obtained [38,39].

TBRs are different from PBRs as they (TBRs) are fed at the top of the reactor thus obtaining product at the bottom. However, in such reactors some of the biofilms may not get sufficient feed thus affecting reactor efficiency/productivity adversely. Also, in gaseous fermentations gas may occupy significant space in the reactor and may form stagnant pockets. This also may affect the efficiency of the reactor. In anaerobic waste water treatment and acetic acid production, these reactors have been used at large scale successfully.

FBRs have played a successful role in the degradation of toxic phenolic chemicals [40-43] and butanol production [39,44]. In these reactors, cell growth occurs around the adsorbent particles. Formation of active biofilms around the particles and accumulation of sufficient biomass in the reactor may take from 2 to 4 weeks. A major advantage in these reactors is that they can be operated for much longer periods than PBR or CSTRs (with fibrous bed). These reactors do not block due to excessive growth. In these reactors butanol production was increased by approximately 40–50 times that of the batch reactors. These reactors have been operated successfully for longer than 4 months in continuous operation (Unpublished data, Qureshi and Maddox).

Airlift reactors contain two concentric tubes, a riser (an inner tube) and a downcomer (an outer tube). In these reactors, mixing is achieved by circulating essentially air at the bottom of the reactor. As a result of force applied by the air (at the bottom of the inner tube), the liquid in the inner tube moves up which then overflows (the inner tube) downward thus creating eddies to mix the liquid. In some of the airlift reactors downcomer is replaced with an external loop to circulate fermentation broth. Such reactors where air is replaced by an anaerobic gas are called gaslift reactors.

Upflow anaerobic sludge blanket (UASB) reactors (contain granular biofilm particles) are used for anaerobic treatment of wastewater/industrial effluents. As the name suggests, the flow in these reactors is in upward direction. At the top of the reactor provisions are made for gas/es to escape and sludge particles to settle to the bottom part of the reactor. Reactor effluent is removed from the top of the reactor. UASB reactor was developed by Lettinga et al. [9]. Fig. 1 shows a schematic diagram of various reactors and biofilm particles.

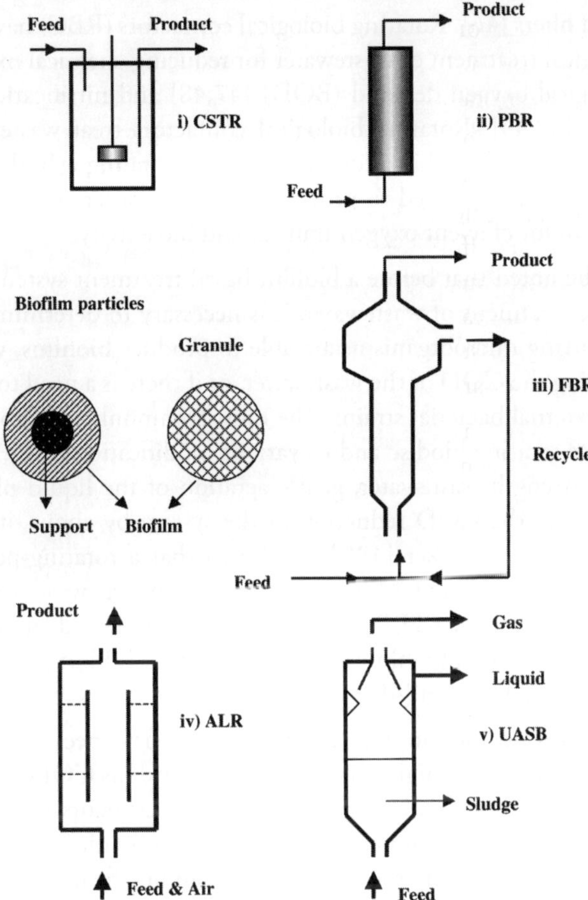

Figure 1. Schematic diagrams of various types of biofilm reactors and biofilm particles.

Biofilm Reactors in Biological Wastewater Treatment

The application of biofilm technology in wastewater treatment originated from the industrial operation of trickling filters in the early 1880s in Wales, Great Britain [45]. Biofilm processes in wastewater treatment can be divided into two categories: namely (1) the fixed-medium systems where the biofilm media are static in the reactors and the biological reactions take place in the biofilm developed on the static media, and (2) the moving-medium systems where the biofilm media are kept continually moving by means of mechanical, hydraulic, or air forces [46]. The moving-medium systems include rotating biological contactors, moving-bed biofilm reactors, vertically moving biofilm reactors, and fluidized bed biofilm reactors; while the fixed-medium systems include trickling filters and biological aerated filters [46]. Rotating biological contactors (RBC) have been widely used in biological treatment of wastewater for reducing chemical oxygen demand (COD)/biological oxygen demand (BOD) [47,48] and nitrification/denitrification purposes [48,49]. Rotating biological contactors treat wastewater streams using a thin biofilm of aerobic microorganisms on rotating cylinders or biodiscs. The rate of rotation is selected to provide optimum contact of the waste stream with the biofilm for efficient oxygen transfer and bioactivity.

It should be noted that before a biofilm-based treatment system is to be considered for the treatment of wastewater, it is necessary to determine whether the naturally occurring microorganisms are able to produce biofilms, while simultaneously reducing the COD of the wastewater, or if there is a need to inoculate the reactor with external bacterial strains. The most commonly used rotating biofilm contactor is the rotating biodisc and its various modifications [48]. For the treatment of high strength wastewater, gentle aeration of the liquid phase has been shown to improve the COD reduction of the system by about 40% [50]. In a related study, Kargi and Eker [48] have shown that a rotating-perforated-tube biofilm reactor is effective in COD removal from synthetic wastewater composed of diluted molasses, urea, KH_2PO_4, and $MgSO_4$. The liquid phase in the tank was not aerated, (the total biofilm surface area (A) was 1.34 m^2), and the rotation speed of the tubes was 5 rpm [48].

In some instances, thermophilic aerobic systems have been employed to biodegrade the wastewaters of high strength, and tremendous COD reductions have been reported in both laboratory and pilot scale experiments [51,52]. However, the thermophilic systems exhibited poor bacterial flocculation characteristics due to the dispersed growing microorganisms (no biofilm formation) which made bacterial separation from the treated effluent difficult [53,54].

The fluidized bed biofilm reactors (FBBR; also called as FBR) (in which particles move up and down within the expanded bed in the well defined zone of the reactor) have been used for more than two decades for treating industrial wastewater [55,56]. Immobilized bacterial systems configured as fluidized bed biofilm reactors (FBBRs) offer some technical advantages. Since chemical wastes are injected into the recycle, toxic chemicals are immediately diluted, which make the microorganisms more resistant to direct chemical toxicity than many conventional treatment systems. In addition, since FBBRs are usually oxygenated by supplying air into the recycle loop, a high level of microbial activity may be supported with minimal air stripping of volatile chemicals.

Jesis and Owen [57] studied FBBR, and they found that the use of small, fluidized media enabled the FBBR to retain high biomass concentrations and, thereby, operate at significantly reduced hydraulic retention times. In pilot scale operations carried out by Jesis and Owen [57], they reported that when the volatile solid concentrations were between 30,000 to 40,000 mgL^{-1} during denitrification operation, 99% of influent nitrates could be removed under hydraulic retention times as low as 6 min. In a related study, Rabah and Dahab [56] during their evaluation of the use of fluidized-bed biofilm reactors for nitrate removal concluded that the FBBR system is capable of handling an exceptionally high nitrate nitrogen concentration of 1000 mg(N)L^{-1} with very high removal efficiency, up to 99.8%. The authors noted that higher denitrification rates can be achieved at relatively low superficial velocities because it is possible to maintain high biomass concentration at lower velocities. However, there is a minimum practical velocity below which agglomeration of media would occur and the process may fail [56]. The efficiency of the FBBR can be up to 10 times greater than that of the activated sludge system and typically occupies 10% of the space required by stirred tank reactors of similar capacities [56]. Higher biomass concentration in the FBBRs (40, 000 mgL^{-1}) compared to 3000 mgL^{-1} in the activated sludge have been shown to be the reason for the greater efficiency [58].

Anaerobic treatment of wastewater in fluidized bed reactors is another area that has been studied extensively [59]. In this article, Iza [59] presented theoretical basis for design and operation of a fluidized bed reactor for anaerobic treatment of wastewater. The anaerobic fluidized bed technology offers a number of advantages for treating wastewater including high concentration of biomass attached to the dense support that makes it possible to operate them at high dilution rate without cell washout. In these reactors no plugging, gas hold-up or channeling occurs.

Prior to the development of UASB, interest in anaerobic treatment systems in wastewater treatment was scarce [11]. Interestingly, the development of UASB saw a significant increase in anaerobic removal of various chemicals from the wastewater using these reactors. The examples include anaerobic removal of pentachlorophenol [60], nitrogen removal [61], dechlorination using Dehalospirillum multivorns [62], anaerobic treatment of municipal solid leachate [63], and starch degradation [64]. More studies on this subject have been reviewed in Veeresh et al., [42] (anaerobic treatment of phenol and cresols in UASB reactors). The success of the UASB concept relies on the establishment of a dense sludge bed in the bottom of the reactor which is usually a result of microbial growth and incoming sludge.

Seghezzo et al., [11] reported that in a pilot plant UASB reactor, internal mixing was not optimal for treating sewage (4–20°C) which produced dead space and hence reduced process efficiency. In order to improve the process efficiency, an adequate influent distribution was sought. The use of effluent recirculation in combination with a taller reactor (a larger height to diameter ratio) resulted in the expanded granular sludge bed (EGSB). Usually expended bed reactors, as opposed to EGSB, have biofilm that is adsorbed onto support particles. An example of expanded bed reactor is that of Tsuno et al., [43] who degraded pentachlorophenol (PCP) in a biological expanded-bed reactor anaerobically. In this reactor the granular activated carbon was used as a support.

Generally, total nitrogen removal from domestic or industrial wastewater streams is achieved in two steps: microbial nitrification of ammonium (aerobic process) followed by denitrification (anaerobic process) or reduction of formed nitrate to nitrogen. This conventional method employs a sequence of aerobic and anoxic processes in order to provide the two different environmental conditions [65]. However, studies have shown that these two important steps can occur simultaneously in one reactor in a process called simultaneous nitrification and denitrification (SND). In SND process, nitrification is restricted to the outer oxic zone of formed microbial flocs, whereas denitrification occurs predominantly in the inner anoxic zones [65]. To test the hypothesis that SND is a physical phenomenon, Pochana and Keller [66] carried out experiments to determine the effect of floc size on SND. Typical floc sizes as measured in their experiments were 50—110 µm, which is large. Such large floc sizes could create an anoxic zone inside the flocs leading to denitrification. Pochana and Keller [66] concluded that a substantial anoxic mass fraction exists in the center of the biomass floc resulting from an oxygen diffusion limitation into the floc. The rates of nitrogen removal (coupled nitrification-denitrification process) or productivities/specific productivities are shown in Table 2. The sequencing batch and single activated sludge flocs reactors require some O_2 to effect nitrification (NO_3^-/NO_2^- generation), which is the precursor for denitrification process.

Table 2. Productivities of different reactors employed for nitrification and denitrification/during the pretreatment of domestic or industrial wastewater streams.

Reactor Type	Removal rates (Productivities or specific productivities)		Reference
	Nitrification	**Denitrification**	
Activated sludge flocs (Single reactor)	24 µmol N g MLSS^{-1} h^{-1}	6 µmol N g MLSS^{-1} h^{-1}	[65]
Sequencing batch	19 mg NH$_3$ L^{-1}h^{-1}	13.5 mg N L^{-1}h^{-1}	[66]
Chemostat (Continuous)	5.6 µmol NH$_3$ h^{-1} mg protein^{-1}	5.6 µmol N h^{-1} mg protein^{-1}	[67]
Chemostat (Continuous)	2.6 µmol NH$_3$ h^{-1} mg protein^{-1}	4.1 µmol N h^{-1} mg protein^{-1}	[68]
Continuous	250 µmol NH$_3$ L^{-1}h^{-1}	400 µmol N L^{-1}h^{-1}	[69]

MLSS = Mixed liquor suspended solid

In contrast to previous view that denitrification occurs under anaerobic conditions [67] it (denitrification) has been shown to occur under aerobic conditions with a wide range of bacteria [68,70]. Robertson and Kuenen [71] observed that under fully aerobic conditions, Thiosphaera pantotropha carries out the following reactions sequentially and simultaneously, in the presence of a suitable electron donor such as acetate.

$$NH_4^+ \longrightarrow NH_2OH \longrightarrow NO_2^-$$
$$NO_2^- \longrightarrow N_2O \longrightarrow N_2$$
$$O_2 \longrightarrow H_2O$$

This implies that the organism can convert ammonia into nitrogen gas without intermediary accumulation of nitrite. Investigating the reason why this organism denitrifies under aerobic conditions, Robertson and Kuenen [72] demonstrated that denitrifying enzymes were present even when the organism was growing aerobically without nitrate.

The last decade has witnessed an increased interest in membrane bioreactors for wastewater treatment [73,74]. The membrane-aerated biofilm reactor (MABR), whereby the biomass is immobilized on membranes through which oxygen is supplied seems to be the most promising design. Results from studies with MABRs have been reported for the degradation of phenol [40], chlorophenols [41], xylene [75], and ammonia [76]. Stripping losses of volatile organic compounds are minimized, and the oxygen partial pressure in the gas compartment allows easy control of oxygen penetration into the biofilm; the dissolved oxygen gradient across the membrane and the biofilm offers an ideal environment for aerobic strains, and foaming due to surfactants can be prevented [74]. For high strength wastewaters, the possibility of enhanced oxygen penetration depths makes MABRs an attractive option for pollutant biodegradation. However, Casey et al. [74] reported that an excessive growth of biofilm is frequently observed. Therefore, biofilm growth should be controlled when operating this reactor. Although a significant amount of work has been performed on the use of biofilm reactors in wastewater treatment, it is beyond the scope of this article to discuss this work in greater detail.

Biofilms for Gas and Odor Treatment

Traditionally, industrial waste gases have been treated by physico-chemical methods known as adsorption, scrubbing, condensation, and oxidation processes [77]. Biological waste gas treatment is an attractive and environmentally friendly alternative to physico-chemical methods. Industrial waste gases can serve as energy or carbon sources for microbial metabolism. In addition, inorganic waste gases (H_2S, NH_3) may be treated directly by employing autotrophic microorganisms which have the ability to utilize CO_2 as a carbon source for anabolism [78]. Koe and Yang [79] during their evaluation on how to drastically reduce or eliminate the impact of air polluting emissions from wastewater treatment plant suggested that open sources of odorous emissions such as inlet works, primary sedimentation units, aeration tanks, final clarifiers, sludge processing units, and wastewater channels should be covered up and the odorous air be treated before discharging to the ambient atmosphere.

The biofilter, trickling biofilter, and bioscrubber are three major bioreactor designs frequently employed for the treatment of waste gas [78]. A biofilter consists of a filter-bed composed of a carrier (sawdust, compost, dry wastewater sludge, etc.) for the active microorganisms and as nutrient source [77]. Biofilters operate by facilitating the transfer of odorous gas from waste air blown through the biofilters into biofilms around particles of biofilter medium in which bacteria, fungi and other microorganisms are immobilized. On the other hand, waste gas treatment in trickling biofilters involves use of a biological filter continuously fed with a liquid medium and packed with a synthetic carrier on which biofilms grow [77]. Trickling biofiltration has been used, especially outside the United States, for removal of odorous waste gases such as H_2S [80]. Several species of microorganisms can oxidize hydrogen sulfide to form odorless sulfuric acid. Thiobacillus thiooxidans is capable of oxidizing H_2S at low pH [81]. For effective H_2S odor control, an ideal habitat for the growth of sulfide-oxidizing bacteria should be created and competing microbes which normally predominate in aerobic treatment processes should be excluded. De Beer et al., [82] demonstrated that the channels surrounding the cell clusters could increase the supply of oxygen and other nutrients to cells within the biofilm, thus relating structure to function [14]. The biofilm structure appears to be largely determined by the production of slime-like matrix of extracellular polymeric substances, which provide the structural support to the biofilm [14]. The structure of biofilms is largely determined by a number of biological factors such as microorganism growth rate, motility, cell signaling, and the production of extracellular polymeric substances. The physical growth environment may also play a significant role in the determination of the biofilm structure [14], and hence the efficiency. However, excessive biofilm development can lead to clogging of the filter-bed of the reactor [78].

Biomass growth and biofilm development can be limited by reducing nutrient supply although this may decrease reactor performance since higher biomass growth shows higher substrate consumption rates [83]. Therefore, it is important to find a balance between excessive biomass growth to prevent biofilter clogging and the odorous gases removal efficiency. Furthermore, waste gases that are characterized by high concentrations of water-soluble pollutants can be treated with bioscrubber. The bioscrubber consists of two reactors. The first reactor is an absorption column where pollutants are absorbed in a liquid phase. The liquid phase goes to the second reactor, which consists of a filter with an activated carbon medium that supports microbial growth. The high bioactivity in the bioscrubber enhances conversion of waste gases into nonhazardous and less odorous compounds. The effluent leaving the bioscrubber can be re-circulated to the absorption column; this technology allows for good gas cleaning when the gaseous pollutants are highly water soluble [77].

Ottengraf [78] reported that the rate of mass transfer of a given compound to be removed or deodorized is determined by the product of the overall mass transfer coefficient, the total contact area in the column, and the average driving force. Therefore, the absorption of a compound will be higher if its concentration in the wastewater is low and its solubility in water is high [78]. The control of operating parameters to the microorganisms in these bioreactors can sometimes be challenging.

Production of Industrial Chemicals in Biofilm Reactors

Biofilms and Biofilm Reactors in Ethanol Production

Bland et al. [84] produced ethanol in an attached film expanded bed bioreactor of Zymomonas mobilis. The cells of Z. mobilis were adsorbed onto vermiculite and the culture formed an active biofilm. Based on the total volume of the reactor, a productivity of 105 $gL^{-1}h^{-1}$ was obtained at a dilution rate of 3.6 h^{-1}. Usually, in a control batch or free cell reactor a productivity of <4 $gL^{-1}h^{-1}$ is achieved. The increased/enhanced productivity reported here is due to the formation of active biofilm onto the adsorbent.

Adsorbed cells of Saccharomyces cerevisiae were used in a packed bed continuous bioreactor to produce ethanol from molasses [4]. The cells were immobilized onto a support of natural origin, possibly sugarcane bagasse. It has been reported that the cells were immobilized by natural mode, which is likely to be adsorption. The amount of cells that was adsorbed onto this support was 0.13 gg^{-1} support. In this biofilm reactor, the authors reported a productivity of 28.6 $gL^{-1}h^{-1}$ as compared to 3.35 $gL^{-1}h^{-1}$ in a free cell continuous process. The dilution rates in the biofilm reactor and free cell continuous system were 0.47 h^{-1} and 0.65 h^{-1}, respectively. Although immobilized cell reactors (such as this biofilm reactor) are typically operated at higher dilution rates than the free cell continuous reactors, it is not clear why the authors used a lower dilution rate in the biofilm reactor. Since carbon utilization for newly growing cells was reduced, product yield was improved as compared to a batch reactor.

Since ion exchange resins have charge on them, bacterial cells can be adsorbed onto the resins thus forming biofilm layers. This concept was employed by Krug and Daugulis [85] to produce ethanol in high productivity reactors using Z. mobilis. To find a suitable adsorbent, 10 ion exchange resins, activated carbon, and ceramic chips were examined. A cationic macroreticular resin was shown to be the most efficient adsorbent to immobilize cells of Zymomonas mobilis. The immobilized cells were used in a continuous column and 100 gL^{-1} glucose was fed to the reactor. As a result of formation of biofilm, the reactor productivity was measured at 135.8 $gL^{-1}h^{-1}$ (void volume based productivity, P_{dv} = 377.4 $gL^{-1}h^{-1}$). The reactor stopped working due to excessive cell growth and plugging after a period of 200 h of operation.

Other reports on ethanol production in biofilm reactors are those of Kunduru and Pometto [86] and Demirici et al. [8]. Kunduru and Pometto [86] studied ethanol production in continuous reactors using biofilm supports of polypropylene or plastic composite. Employing a culture of Z. mobilis and a bacterial support of polypropylene, a staggeringly high productivity of 536 $gL^{-1}h^{-1}$ was obtained at a dilution rate of 15.36 h^{-1}. In a control free cell fermentation, a productivity of 5 $gL^{-1}h^{-1}$ was obtained at a dilution rate of 0.5 h^{-1}. The biofilm reactor was fed from the top, thus collecting product at the bottom of the reactor.

Kunduru and Pometto [86] used another biofilm reactor of S. cerevisiae adsorbed onto a plastic composite support and reported a productivity of 76 $gL^{-1}h^{-1}$ at a dilution rate of 2.88 h^{-1}. The reactor productivity in a control reactor was 5 $gL^{-1}h^{-1}$ at a dilution rate of 0.5 h^{-1}. Unlike the above biofilm reactor, S. cerevisiae biofilm reactor was fed at the bottom, and the product was obtained from the top. It is suggested that for a proper comparison both the reactors should have been fed in the same direction.

In order to enhance biofilm formation, Demirici et al. [8] developed a new support material for the growth of S. cerevisiae. A mixture of ground soybean hulls (or oat hulls), complex nutrients, and polypropylene was extruded at high temperature into disks and rings. It is likely that heat sensitive nutrients were inactivated during extrusion. Also, polypropylene film may have covered the nutrients, thus making them unavailable to the culture for cell growth. Since no data have been provided on the time period of formation of biofilm or thickness of biofilm, it is difficult to compare this support with other supports.

In a more recent study, Qureshi et al. [87] produced ethanol in a biofilm reactor of genetically engineered Escherichia coli from xylose. The biofilm was formed on clay brick particles, and the reactor was operated continuously for 103 days. The reactor was operated at various flow rates, and reactor productivity was found to be improved compared to a free cell batch process. Table 3 compares ethanol productivities obtained in biofilm reactors of various cultures.

Table 3. A comparison of production of ethanol in adsorbed cell biofilm reactors

System/Support	Reactor Type	Culture	Productivity [gL^{-1}h^{-1}]	Reference
Biofilm Reactors				
Resin	Packed bed	Z. mobilis	135.8 (P_{dT}), 377.4 (P_{dv})	[85]
Vermiculite	Packed bed[d]	Z. mobilis	105.0 (P_{dT}), 210 (P_{dv})	[84]
Sugarcane bagasse[a]	Packed bed	S. cerevisiae	28.6[b]	[4]
Polypropylene	Packed bed[e]	Z. mobilis	536[c]	[86]
Plastic composite	Packed bed	S. cerevisiae	76[c]	[86]
Cell Recycle	CSTR	Z. mobilis	200[b]	[88]
Batch/Continuous suspended cell (Control)				
Continuous	CSTR	Z. mobilis	5.0[b]	[86]
		S. cerevisiae	5.0[b]	[86]
Continuous	CSTR	S. cerevisiae	3.35[b]	[4]

a: The support was reported as an adsorbent of natural origin (perhaps sugarcane bagasse)
b: Not reported whether based on total reactor volume or void volume
c: Not reported (possibly based on void volume)
d: Cone shaped
e: Trickling packed bed
P_{dT} – Productivity based on total reactor volume
P_{dv} – Productivity based on reactor void volume

Biofilms and Biofilm Reactors for Butanol Production

Butanol is an important industrial chemical that can be produced from a number of carbohydrates using a number of microbial cultures. Butanol can be used as a fuel and has higher/greater energy content than ethanol. Production of butanol has been investigated in batch, fed-batch, free cell continuous, immobilized cell continuous, and cell recycle continuous reactors [1]. Continuous immobilized cell and cell recycle reactors offer higher productivities than batch and free cell continuous reactors. In addition to achieving a high productivity, a major advantage of immobilized cell technology is that there is no cell washout at high dilution rates.

Adsorption is a technique which does not require any chemicals for cell immobilization and can be easily performed inside the reactor. In order to immobilize cells of Clostridium acetobutylicum, the reactor is packed with an adsorption support followed by inoculation with the culture. The adsorption process varies from 2–3 days to weeks depending upon the culture, support, and the reactor. The culture forms cell layers (biofilm) on the support [5,6,22].

An early report of adsorption of cells of C. acetobutylicum for the production of butanol was that of Forberg and Haggstrom [5]. These authors used beechwood shavings to adsorb cells. The reactor was fed continuously with a glucose solution (and nutrient dosing). Over a period of time, an active biofilm was formed on the wood shavings, and a reactor productivity as high as 1.53 gL^{-1}h^{-1} was observed (compared to <0.1–0.35 gL^{-1}h^{-1} in control batch fermentation). This work was followed by experiments examining the production of butanol in adsorbed cell biofilm reactor of C. acetobutylicum from whey permeate [6]. It should be noted that biofilm formation on this support was quick, and a reactor productivity of

4.5 $gL^{-1}h^{-1}$ was observed, which was superior to any previously reported butanol production system. Following these reports, Welsh et al. [89] investigated the use of a number of adsorption supports for butanol production by C. acetobutylicum in batch and continuous systems. The adsorbents used were coke, kaolinite, and Gel White (a montmorillonite clay). Coke was reported to be superior to other supports for adsorption. A maximum concentration of acetone butanol ethanol (ABE) in the effluent of the reactor was reported to be 12 gL^{-1} at a dilution rate of 0.1 h^{-1}, thus resulting in a productivity of 1.2 $gL^{-1}h^{-1}$.

Following above reports, an intensive study was performed on the adsorption of C. acetobutylicum on a number of supports and biofilm formation (Table 4, 5). It has been observed that C. acetobutylicum and C. beijerinckii form visual biofilm layers in 2–4 days (in packed bed reactors), and reactors become productive after 4th day of continuous operation. The techniques of adsorption and reactor operation have been reported previously [6,38,89]. It has been observed that not all the supports are suitable for adsorption (Table 4). It has also been observed that during biofilm formation onto bonechar, the culture produces higher concentration of polysaccharide between day 2 and 4. During this period, up to 2.04 gL^{-1} polysaccharide production was observed as opposed to 0.95 gL^{-1} during day 5–30 (Fig. 2). As described in the previous section, the cultures that were used for adsorption for butanol production have flagella, which perhaps help bring the cells closer to the surface of support. In addition, charge on the support and cell is likely to aid in initial adsorption or bringing the cells closer to the support surface.

Table 4. Biofilm formation characteristics of Clostridium acetobutylicum/C. beijerinckii onto various supports

Support	Characteristics
Bonechar	*C. acetobutylicum* culture - Adsorption is quick - Biomass layers (biofilms) become visible in 3–4 days time - Between day 2 and 4, the culture produces polysaccharide in high concentrations (2.04 gL^{-1} broth as compared to 0.95 gL^{-1} broth from day 5 to 30) - Once initial layers appear, biomass accumulation is quick - Desorption does not occur at high dilution rates - < than 25% cells were desorbed when adsorbed cell particles were agitated at 200–300 rpm (in shake flasks on shaker) at pH 2.7 for 18–24 h at 30°C - During initial stages (2–4 days) the culture produced high concentrations of acids (~6–9 gL^{-1}) followed by becoming solventogenic - During solventogenic stages fluctuations in solvent concentrations were less
Glass beads	*C. acetobutylicum* culture - Biomass accumulation takes much longer than bone char - During initial stages (2–4 days) higher amount of polysaccharide production does not occur - Cells do not stick to the support as firmly as onto bonechar - Reactor produces <20% solvents as compared to bonechar adsorbed cells - Reactors are not stable as solvent concentration fluctuates - Cells can easily be washed off
Glass wool, Polypropylene tow, and stainless steel wire balls	*C. acetobutylicum* culture - <20% biomass accumulated than in bonechar packed reactor - Cells do not stick to the support firmly and can be desorbed easily - Reactors are not stable and poor solventogenesis occurred
Clay brick (Ref. 38)	*C. beijerinckii* culture - Cells stick firmly as in case of bonechar and reactors were solventogenic

Table 5. Production of solvents in packed bed biofilm reactors of C. acetobutylicum/C. beijerinckii

Culture/support	Maximum Solvent [gL⁻¹]	Maximum productivity [gL⁻¹h⁻¹]	Accumulated biomass [gL⁻¹ reactor vol]	Biomass accumulation [gg⁻¹ support]
C. acetobutylicum				
Bonechar	9.3 (0.30)	6.50 (1.5)	74.0	0.087
Glass beads	3.0 (0.31)	0.93 (0.31)	65.0	0.044
Glass wool	3.0 (0.10)	0.30 (0.10)	3.1	0.050
Polypropylene tow	2.3 (0.25)	0.58 (0.25)	0.8	-
Stainless steel wire balls	2.0 (0.07)	0.15 (0.07)	1.0	-
C. beijerinckii				
Clay brick [Ref 38]	7.9 (2.00)	15.8 (2.00)	73.7	0.093

Numbers in bracket are dilution rates (h⁻¹) at which solvent and productivity were obtained
- Values not calculated

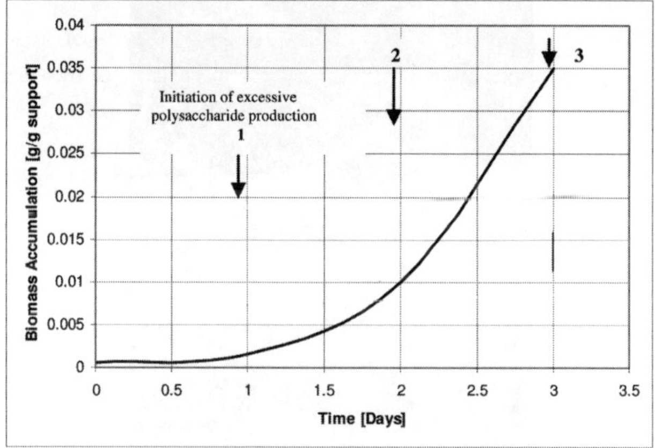

Figure 2. Production of polysaccharide and accumulation of cell mass during the initial 3 days of adsorption of cells of C. acetobutylicum onto bonechar for the production of butanol from whey permeate in a packed bed reactor. 1. Initiation of excessive production of polysaccharide; 2. Maximum growth and attachment starts; 3. Biofilms become visible and polysaccharide production continues.

Some supports accumulated more cell concentration (C. acetobutylicum) and were more solventogenic than the others (Table 5). At this stage, we are not aware what makes some supports better than others for biofilm formation and cell accumulation. From some supports it was easier to wash away the cells while from others such as bonechar and clay brick it was more difficult (Table 4). During our studies on butanol production, it was observed that approximately 0.9–1.0 gL⁻¹ cells were present in the effluent [7,90] of the reactor. We have demonstrated that the cells that are present in the effluent of the reactor are those that grew on the surface of the support, rather than those that grew in liquid medium inside the reactor [90], suggesting that a tremendous amount of activity occurs on the surface of the biofilm in C. beijerinckii/C. acetobutylicum cultures. The thickness of biofilm that is formed in C. acetobutylicum or C. beijerinckii cultures can range from few cell layers to as many as 35 or more. Figure 3 shows adsorbed cells and biofilm formed by C. acetobutylicum onto bonechar. Similar observations on biofilm formation were observed for C. beijerinckii [38].

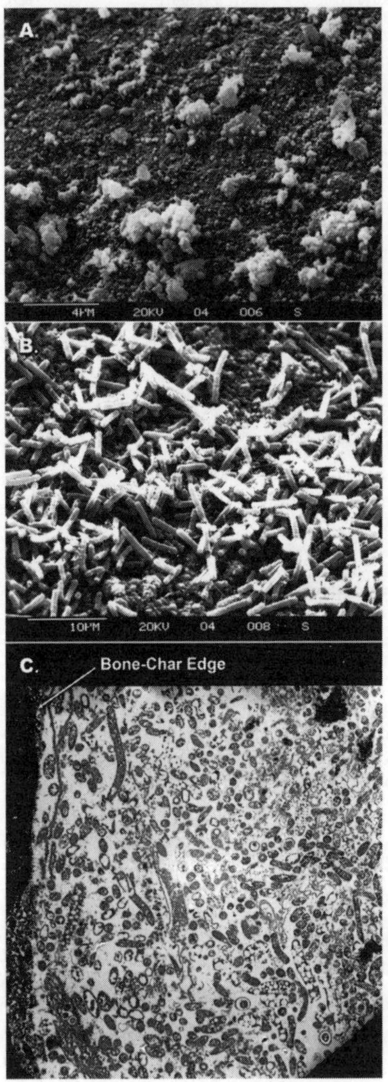

Figure 3. Scanning electron micrograph of adsorbed cells of C. acetobutylicum P262 onto bonechar. a) bonechar (magnification 5500); b) adsorbed cells onto bonechar (magnification 2200); c) transmission electron micrograph of adsorbed cells (magnification 2300). Similar figures (3b, c) with different magnification were published previously in the following article: Qureshi N, Paterson AHJ, Maddox IS: Model for continuous production of solvents from whey permeate in a packed bed reactor using cells of Clostridium acetobutylicum immobilized by adsorption onto bonechar. Appl Microbiol Biotechnol 1988, 29:323–328. Figure 3 is reprinted with permission from Springer, Germany (see above article).

Intensive research has been done on butanol production in various types of reactor systems [1,39,91,92]. The biofilm reactor systems that have been used for butanol production include vertical packed bed reactor (PBR), horizontal PBR, compartmentalized reactor, double series reactors, and FBR. The PBRs and FBRs are different in the sense that FBR is started with support <10% of its volume while packed beds are filled up to 90% of their volume. In packed bed reactors, as cell growth occurs, they are often blocked due to excessive cell growth while in FBRs this does not occur. In FBRs, the bed is fluidized either by recycling fermentation broth, using anaerobic gases (N_2 or CO_2 & H_2 in case butanol fermentation) or air (for other aerobic systems). Cell growth occurs all around the support particles and over a period of time the volume of biofilm particles becomes many fold greater than the support particle (Fig. 3, 4). It should be noted that in FBRs cell growth occurs on the particles in spite of broth's high flow rates [44]. In this fluidized bed reactor liquid flow velocity of the order 40–60 ms-1 was maintained. The reader is advised that despite such a high flow velocity, the culture maintains its growth as a biofilm. We have not calculated the shear rate on the biofilm particles. The reactor was used for the production of butanol from whey permeate in continuous operation for >4 months (unpublished results—Qureshi & Maddox). Newly adsorbed C. acetobutylicum cells onto bonechar grow in an exponential manner and accumulation of biomass continues with time. Figure 5 shows a picture of a fluidized bed reactor used for the production of butanol from whey permeate.

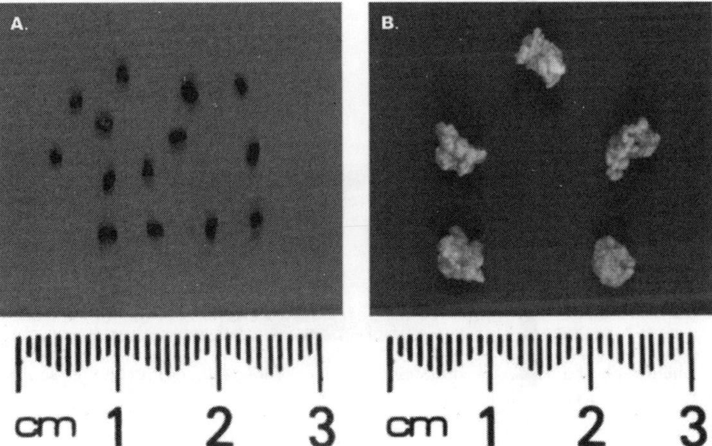

Figure 4. Photographs of biofilm particles of C. acetobutylicum P262 used in a fluidized bed reactor for the production of butanol from whey permeate. A) bonechar particles; B) biofilm particles after growth (bonechar particles are covered with biofilm layers).

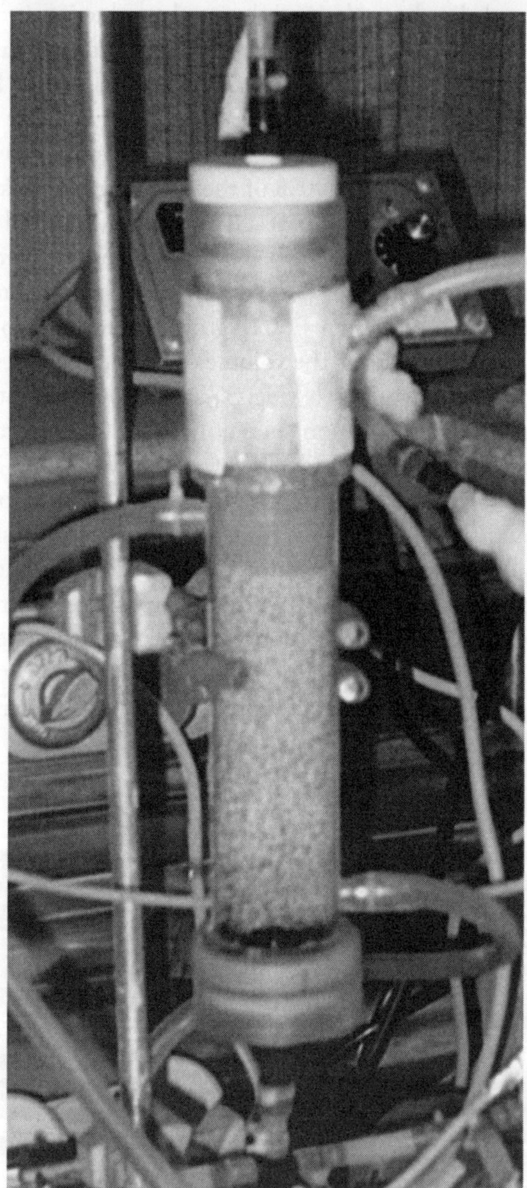

Figure 5. A photograph of a fluidized bed bioreactor (inside volume 450 cm³) used to produce butanol from whey permeate using C. acetobutylicum P262.

Among the various types of reactors used for butanol production, adsorbed cell biofilm reactors (cells adsorbed onto bonechar and clay brick) offered the highest reactor productivities. The reactor productivities that have been achieved in these reactors ranged from 6.5 [39] to 15.8 [38] gL^{-1}h^{-1} (as compared to 0.10–0.38 gL^{-1}h^{-1} in batch reactors). Membrane cell reactors also offer high productivities (6.5 gL^{-1}h^{-1}) [93,94]; however, biofilm reactors were superior to these reactors (Table 5). Of the various supports tested, bonechar and clay brick were found to be most suitable, and strong biofilms were formed on these supports. C. acetobutylicum was adsorbed onto bonechar while C. beijerinckii was adsorbed onto clay brick. Attempts were made to desorb the adsorbed cells of C. acetobutylicum. In order to achieve this, 100 g bonechar with adsorbed cells (25 days old reactor) was transferred to a 500 mL conical flask. The pH of the solution/reaction mixture was adjusted to 2.7, and the mixture was placed on a rotary shaker at 250 rpm for 18 to 24 h. After this period <30% cells were desorbed from the bonechar.

Biofilms in 2,3-Butanediol Production

In an attempt to improve reactor productivity in 2,3-butanediol fermentation, Maddox et al. [23] immobilized cells of Klebsiella pneumoniae on to bonechar. The cells of K. pneumoniae were adsorbed in a similar manner as C. acetobutylicum [6]. During the 2,3-butanediol fermentation, a productivity of 11.7 gL^{-1}h^{-1} was obtained, which was the highest reported productivity. Prior to this work Shazer and Speckman [95] reported a productivity of 1.04 gL^{-1}h^{-1} in 2,3-butanediol fermentation using Bacillus polymyxa in a membrane cell reactor. This work clearly demonstrated that bonechar adsorbed cells of K. pneumoniae result in superior productivities. Table 6 compares reactor productivity achieved in biofilm reactor as compared to various other reactor types. It should be noted that although high reactor productivity was obtained in the adsorbed cell reactor, cells did not adsorb on to bonechar as strongly as C. acetobutylicum. Rather, cells were entrapped in between bonechar particles. However, it is anticipated that there were a significant amount of cells sitting on the surface of bonechar as bonechar surface area was large. At the end of fermentation, it was observed that unlike cells of C. acetobutylicum, K. pneumoniae cells were washed away easily. It is not known whether K. pneumoniae cells produce polysaccharide which adds/facilitates adsorption of cells to the surface of bonechar or other surfaces. Even though K. pneumoniae cells do not form firm layers of cells, these reactors are still highly productive.

Table 6. A comparison of 2,3-butanediol productivity in a packed bed biofilm reactor with productivities in other reactor types

Reactor Type	Culture	Substrate	Productivity [gL^{-1}h^{-1}]	Reference
Biofilm, continuous	K. pneumoniae	Whey permeate	11.70	[23]
Batch (control)	A. aerogenes	Glucose	1.10	[96]
Continuous reactor (free cells)	K. pneumoniae	Glucose	4.25	[97]
Immobilized cell continuous	K. pneumoniae	Whey permeate	2.30	[98]
Cell recycle, continuous	B. polymyxa	Whey permeate	1.04	[95]
Cell recycle, continuous	K. pneumoniae	Glucose	9.84	[99]
Cell recycle, continuous	E. aerogenes	Glucose	5.40	[100]

Production of Other Chemicals in Biofilm Reactors

Other examples of production of industrial chemicals produced in biofilm reactors include acetic acid or vinegar, lactic acid, succinic acid, and fumaric acid. Acetic acid production in trickling bed biofilm reactors is a mature technology and is exercised at the commercial level [13]. In addition to trickling bed biofilm reactor, a submerged process was also developed in late 1940s. The acetic acid is produced by one of the bacteria grouped in the two genera, Gluconobacter and Acetobacter. The species that are used commercially include Acetobacter aceti, A. pasteurianus, and Gluconobacter oxydans. In the trickling bed biofilm reactor (volume 60 m^3), beechwood shavings are packed and the starting material (alcohol solution) is sprayed over the surface. To this solution, initially, nutrients and bacteria are added for growth of biofilm on the beechwood shavings. The liquid trickles to the bottom of the reactor containing acetic acid. In order to increase concentration of acetic acid, the liquid is cooled and pumped back to the top of the reactor. Of the alcohol added, approximately 90% is converted to acetic acid during the trickling process. Approximately 120 gL^{-1} acetic acid is obtained in 72 h, thus resulting in a productivity of 1.67 gL^{-1}h^{-1}.

Production of lactic acid in biofilm reactors is another example of industrial chemical production in such reactors. Demirci et al. [101] evaluated a number of supports for biofilm formation using lactic acid producing cultures. It has been reported that the best biofilms were obtained with Pseudomonas fragi, Streptomyces viridosporus, and Thermoactinomyces vulgaris when used in combination with polypropylene composite chips. The polypropylene composite chips contained polypropylene and 25% (w/w) agricultural material. The mixture of these components was extruded through an extruder to form chips of desired dimensions. Following this, a number of reports appeared from the same group on synthesizing, evaluating, and using various supports for biofilm formation and lactic acid production [102-105]. In one of the reports [104], lactic acid was produced in repeated batch cultures in a biofilm reactor. The reactor productivity was improved from 2.78 to 4.26 gL$^{-1h^{-1}}$. A maximum lactic acid concentration of 60 gL^{-1} was produced in biofilm reactors where plastic composite support was used for adsorption.

In a study on the production of lactic acid by adsorbed cells of Rhizopus oryzae, the culture was immobilized on a fibrous-bed and used in a bioreactor [37]. The fibrous bed was a sheet of 100% cotton cloth onto which the culture was adsorbed. In this reactor (fed-batch), a productivity of 2.5 gL^{-1}h^{-1} was obtained with a high yield of 90% and a high product concentration of 127 gL^{-1}. Glucose was used as a substrate. When glucose was replaced with cornstarch, yield improved to 100% and productivity decreased to 1.65 $gL^{-1}h^{-1}$. Using starch as a substrate, a product concentration of 126 gL^{-1} was achieved.

Other reports on using cell support for cell growth and lactic acid production are those of Park et al. [106] and Sun et al. [107]. Park et al. [106] used 3 gL^{-1} mineral support (Aid-Plus; ML-50D, Mizusawa Chemical Co., Niigata, Japan) and 5 ppm polyethylene oxide to flocculate the culture and change mycelial morphology from a large pellet to mycelial flocs. Sun et al. [107] immobilized cells of R. oryzae in polyurethane foam cubes. There are other reports on the use of immobilized cell technology to produce lactic acid, however, they have not been mentioned either due to space limitation or studies are not directly related to biofilm formation.

Biofilm reactors have also been used successfully for the production of fumaric acid [108] and mineral ore treatment [13]. In an interesting study, Cao et al. [108] used plastic discs to adsorb cells of R. oryzae to produce fumaric acid from glucose. The use of the biofilm reactor resulted in an increase in reactor productivity from 0.9 $gL^{-1}h^{-1}$ in a free cell stirred-tank reactor to 4.25 $gL^{-1}h^{-1}$ in the biofilm reactor. In the latter reactor, fumaric acid concentration up to 85 gL^{-1} was obtained from 100 gL^{-1} glucose. The fermentation time was shorter and took 20 h as compared to 72 h in the free cell reactor.

Succinic acid is a chemical that has been produced in biofilm reactors. The industrial potential for succinic acid fermentation was recognized as early as the late 1970s [109]. Succinic acid ($HOOCCH_2CH_2COOH$) is a dicarboxylic acid, which can be used as a feedstock chemical for the production of high value products such as 1,4-butanediol, tetrahydrofuran, adipic acid, γ-butyrolactone, and n-methylpyrrolidone [109] for applications in agriculture, food, medicine, plastics, cosmetics, and textiles. In a recent study on succinic acid production using Actinobacillus succinogenes, Urbance et al. [110] employed the customized plastic composite support (PCS) [111] and 20 other different PCS blends with and without mineral salt additions and evaluated 20 simulated repeated-batch fermentations using $MgCO_3$ for pH control and CO_2 supply. The customized plastic composite support (PCS) blends were screened for biofilm formation and succinic acid production. Succinic acid concentrations, percentage yield of succinic acid, and biofilm formation for each PCS blend were determined and no correlation between biofilm formation and succinic acid production was observed. However, the customized PCS blend for A. succinogenes in succinic acid production demonstrated 70% yields for succinic acid compared to 64% yield for suspended cell bioreactor [110]. Table 7 shows production of various chemicals in biofilm reactors.

Table 7. Production of various other chemicals in biofilm reactors

Product/Reactor Type	Adsorption support	Productivity [gL⁻¹h⁻¹]	Reference
Acetic acid			
Trickling bed biofilm reactor	Beechwood shavings	1.67 (120)	[13]
Lactic acid			
Agitating continuous reactor	Fibrous bed (cloth)	2.5 (126)	[37]
Fumaric acid			
Rotary continuous reactor	Plastic discs	4.25 (85)	[108]
Strirred-tank (control)	None	0.91	[108]
Succinic acid			
Repeated batch fermentations	Plastic discs	-	[110]

Numbers in bracket – product concentration in gL⁻¹

Enhanced Rates of Production of Chemicals in Biofilm Reactors

Length of Operation of Biofilm Reactors

Packed bed reactors often block due to excessive cell growth. It should be noted that reactor blockage depends on a number of factors including cell growth rate, packing density of the support, and supply of nutrients. This type of reactor has been operated ranging from 2 weeks to 3 months. Tyagi and Ghose [4] used a packed bed biofilm reactor of S. cerevisiae for a period of 35 days, while Qureshi et al. [87] used a packed bed biofilm reactor of E. coli for a period of 103 days for ethanol production in continuous operation. However, it was observed that packed bed biofilm reactors of C. acetobutylicum/C. beijerinckii blocked sooner than 103 days due to enhanced cell growth of these cultures. In order to prolong life of the reactor, feed media deficient in nutrients should be attempted as used by Qureshi & Maddox [6] and Qureshi et al. [90]. It has been observed that this type of reactor blocks at the bottom where fresh feed allows excessive cell growth. In the upper part of the reactor, minimal growth occurs due to product inhibition as in case of butanol and ethanol production. In such cases, inverting the reactor can prolong life of reactor. In addition to the reactor blockage due to excessive growth, influent to the reactor plays an important role in prolonging life of the reactor. Reactor feed may contain suspended and particulate solids, in particular with wastewater influents, which may block the reactor. It is suggested that such influents be filtered or centrifuged to remove suspended and particulate solids to prolong reactor's life.

Fluidized bed reactors do not block due to excessive growth and they can be operated for a long period of time (>4 months). It is also viewed that UASB and EGSB can be operated for long periods. Table 8 shows length of operation of different reactors for the production of various chemicals, their productivities and dilution rates. Biofilm reactors are highly productive as compared to other reactor systems. The reader is advised to refer to the production of various chemicals in biofilm reactor systems (in this article) to be able to compare their production rates with the other non-biofilm reactor systems.

Table 8. Length of operation of various biofilm reactors used for the production of different chemicals

Chemical Produced	Reactor Type	Length of operation [Days]	Dilution rate [h⁻¹] (Productivity [gL⁻¹h⁻¹])	Reference
Butanol	Packed bed	61 days	0.30–1.00 (0.98–4.10)	[6]
	Fluidized bed	>4 months	0.33–1.37 (1.65–5.10)	Unpublished data[1]
Lactic acid	Various reactors	Reviewed in ref 79 (Table 1)	- -	[37]
Ethanol	Packed bed	35 days	0.12–0.48 (7.80–28.60)	[4]
	Packed bed	103 days	0.04–0.12 (1.10–2.58)	[87]
	Packed bed	60 days	0.50–5.76 (5.00–74.88)	[86]

[1] Qureshi & Maddox
- Not reported

Barriers in Biofilm Reactors

In adsorbed cell biofilm reactors of C. acetobutylicum, it was identified that there were four different cell types: growing cells, butanol producing cells, dead cells, and inactive cells (non-growing, nutrient requiring) [7]. Cells that were involved in butanol production were only a fraction of the total cells. For example, the concentration of cells in the reactor was approximately 74 gL⁻¹, while the butanol producing cell mass was <10% of the total cells. The amount of dead cells or spores occupied most of the space in the reactor. It is viewed that if sporulation is blocked, the reactor productivity could be increased by many fold. This would improve the process economics of butanol production in biofilm reactors. At this stage we are not aware if this is applicable to the other organisms such as ethanol, 2,3-butanediol, succinic, acetic (vinegar), lactic and fumaric acid producers. It is suggested that this be investigated for the cultures that produce these chemicals. In UASB internal mixing is not optimal which reduces efficiency of the reactor [11]. This produces dead space in the reactor. For that reason expanded granular sludge bed (EGSB) are investigated [11].

Diffusion Limitations

Usually biofilms contain multiple layers of cells. The thickness of the biofilm may vary from a few to many μm. An increase in the biofilm particle diameter affects hydrodynamic conditions in the reactor including fluidization characteristics etc [59]. In order to measure the thickness of biofilm in C. acetobutylicum culture (PBR), an electron transmission micrograph was taken of a particle and it was identified that the biofilm was made up of >30 cell layers (Fig. 3c). In order for the cells to be active and be taking part in the reaction, nutrients and substrate must diffuse/penetrate to the inner layers of cells. However, it is likely that the nutrients and substrate are used up by the outer cell layers before they reach the innermost cell layers. If this is true, the innermost layers would neither survive nor take part in the reaction. Another example where the thickness of cell layers is an important consideration is bioparticles in a fluidized bed reactor. In these reactors the size of the bioparticles is much bigger than the bioparticle in PBR and cell layers are >>30. Accumulation of so many cell layers adds to the diffusion resistance to the substrate and nutrients. In order to keep the diffusion resistance to a minimum possible level, the size of the bioparticle should be kept to a minimum level while still keeping productivity of the reactor high. This should increase the rate of reaction and benefit the process economics.

In aerobic biofilm processes, such as oxidative degradation of toxic chemicals and production of acetic acid in trickling bed biofilm reactors, a constant supply of oxygen is essential. The oxygen should be dissolved in the liquid and be transported to the innermost layers. The penetration depth of oxygen should be 100% of the biofilm thickness. If the bioparticle size is large, then the inner layers would be starved of oxygen and the cells would die thus decreasing the conversion efficiency of the process. Supply of oxygen rather than air to the reactor would improve diffusion of oxygen to the inner layers; however, it would add to the cost of the process. Hence, size of the biofilms should also be kept to low to keep the reactor productive. In aerobic wastewater biofilm reactors oxygen is an important substrate/nutrient [12]. For anaerobic systems oxygen is toxic.

In addition to the above limitations, an additional limitation comes from the toxicity of product/s itself. Many of the fermentation products are toxic to the cells that produce them. Examples of such toxic products are those that have been described in the earlier section of this article. Butanol is toxic to the cells of C. acetobutylicum/C. beijerinckii and at higher concentrations it kills the cells. In the biofilm layers, the diffused substrate is converted to the products such as butanol. It is not known how quickly the produced butanol diffuses out of the cell layers. It is conceivable that accumulated butanol or other chemicals kill the cells before it is diffused out. It is also likely that a combination of nutrient deficiency and toxicity affects the cells more adversely.

Industrial/Pilot-Plant Level Biofilm Reactors

Wastewater Treatment

Biofilm reactors have successfully been used in wastewater treatment [9-12,43,59,61]. In these industrial biofilm reactors cell mass concentration as high as 30–40 gL-1 could be maintained [12,58]. As a result of superior efficiency, biofilm reactors are being used throughout the world with a number of full scale application for industrial and wastewater treatment. Examples of these reactors operating in The Netherlands and Brazil are shown in Fig. 6.

Figure 6. Full scale biofilm reactors: (a) biothane biofilm airlift suspension and expanded granular sludge blanket (Biobed) reactors at Gist Brocades, Delft (The Netherlands); (b) Pagues CIRCOX (foreground; 140 m³) and internal circulation (background; 385 m³) reactors at a brewery in Brazil. Reprinted from "Nicolella C, van Loosdrecht MCM, Heijnen SJ: Particle-based biofilm reactor technology. Trends in Biotechnology 2000, 18: 312–320, with permission from Elsevier, United Kingdom.

Acetic Acid/Vinegar Production

Commercial production of acetic acid or vinegar using biofilm reactors has been exercised for many years. Production of these chemicals has been reported by Crueger & Crueger [13]. Large biofilm fermentors of size up to 60,000 L have been used. Often beechwood shavings are used as a support for biofilm formation. For this system, trickling bed reactors have been used with an exit product concentration up to 120 gL⁻¹ and a productivity of 1.67 gL-1h-1. A description of the process has been given in previous sections.

Butanol Production

Butanol production in biofilm reactors has been practiced in numerous types of reactors at laboratory scale [6,38,39,44] with superior productivity to batch, fed-batch, and free cell continuous fermentations. Two of the most prominently used reactors are packed bed and fluidized bed reactors. In these reactors, productivities of the order of 4.5–15.8 $gL^{-1}h^{-1}$ have been achieved as compared to productivities of 0.10–0.38 $gL^{-1}h^{-1}$ in batch reactors. Given the scenario of increasing petroleum prices, it is suggested that fluidized bed reactors be scaled up to pilot plant level in view to further commercialize this fermentation.

Other Processes

Production of other industrial chemicals such as lactic acid and 2,3-butanediol should be exercised at pilot plant level. Nicolella et al. [12] reported that biofilm reactors are in operation at industrial scale throughout the world. Use of biofilm reactors is anticipated to be economical for the production of these industrial chemicals.

Future Directions & Conclusions

A comparison of biofilm reactors with other reactor systems suggests that biofilm reactors are simple and offer higher productivities than other reactor systems. In biofilm reactors, cells can be adsorbed within the reactor without the use of any chemicals, and the reactors can be operated for long period of times. This would help in reducing the process cost. These reactors are already in use for wastewater treatment and acetic acid/vinegar production by fermentation. It is clear that their use at bench scale has been consistently increasing for the production of various other chemicals. As productivities in these simple biofilm reactors are high, their full potential should be employed for biotechnological/biological conversion processes. For further reading on biofilms and their formation, the reader is referred to the comprehensive articles published by Costerton et al., [112] and O'Toole et al., [113].

Authors' Contributions

NQ would like not to mention the contributions made by individual authors. However, it is stated that all the authors made significant contributions to deserve to be contributing authors of this comprehensive article on "Biofilm Reactors."

Note

**Mention of trade names of commercial products in this article is solely for the purpose of providing scientific information and does not imply recommendation or endorsement by the United States Department of Agriculture.

Acknowledgements

N. Qureshi would like to thank Michael A Cotta (United States Department of Agriculture, National Center for Agricultural Utilization Research [USDA/NCAUR], Peoria, IL) for his encouragement during the preparation of this article and reading the manuscript critically. NQ is grateful to Professor Eberhard Morgenroth (Department of Civil and Environmental Engineering, University of Illinois, Urbana, IL) for his help with UASB reactors. Help from Christopher Skory (USDA/NCAUR) is also acknowledged. NQ would like to thank Holly Brining and Mark Maroon for their help during the preparation of this manuscript and Don Fraser for photographic/scanning work. NQ & ISM are grateful to Mr. Doug Hopcroft (AgResearch Ltd., Palmerston North, New Zealand; formerly known as Department of Scientific and Industrial Research [D.S.I.R; Biotechnology Division]) for performing the electron microscopy.

References

1. Maddox IS: The acetone butanol ethanol fermentation: recent progress in technology. Biotechnol Genetic Eng Reviews 1989, 7:190–220.

2. Qureshi N, Blaschek HP: Evaluation of recent advances in butanol fermentation, upstream, and downstream processing. Bioproc Biosys Eng 2001, 24:219–226.

3. Mehaia MA, Cheryan M: Ethanol production in a hollow fiber bioreactor using Saccharomyces cerevisiae. Appl Microbiol Biotechnol 1984, 20:100–104.

4. Tyagi RD, Ghose TK: Studies on immobilized Saccharomyces cerevisiae. I. Analysis of continuous rapid ethanol fermentation in immobilized cell reactor. Biotechnol Bioeng 1982, 24:781–795.

5. Forberg C, Haggstrom L: Control of cell adhesion and activity during continuous production of acetone and butanol with adsorbed cells. Enz Microbial Technol 1985, 7:230–234.

6. Qureshi N, Maddox IS: Continuous solvent production from whey permeate using cells of Clostridium acetobutylicum immobilized by adsorption onto bonechar. Enz Microbial Technol 1987, 9:668–671.

7. Qureshi N, Paterson AHJ, Maddox IS: Model for continuous production of solvents from whey permeate in a packed bed reactor using cells of Clostridium acetobutylicum immobilized by adsorption onto bonechar. Appl Microbiol Biotechnol 1988, 29:323–328.

8. Demirci A, Pometto AL III, Ho KLG: Ethanol production by Saccharomyces cerevisiae in biofilm reactors. J Ind Microbiol Biotechnol 1997, 19:299–304.

9. Lettinga G, van Nelsen AFM, Hobma SW, de Zeeuw W, Klapwijk A: Use of the upflow sludge blanket (USB) reactor concept for biological wastewater treatment, especially for anaerobic treatment. Biotechnol Bioeng 1980, 22:699–734.

10. Lettinga G, Roersma R, Grin P: Anaerobic treatment of raw domestic sewage at ambient temperature using a granular bed UASB reactor. Biotechnol Bioeng 1983, 25:1701–1723.

11. Seghezzo L, Zeeman G, van Lier JB, Hamelers HVM, Lettinga G: A review: The anaerobic treatment of sewage in UASB and EGSB reactors. Bioresource Technol 1998, 65:175–190.

12. Nicolella C, van Loosdrecht MCM, Heijnen SJ: Particle-based biofilm reactor technology. Trends in Biotechnol (TIBTECH) 2000, 18:312–320.

13. Crueger W, Crueger C: Organic acids. In "Biotechnology" A textbook of industrial microbiology. Sinauer Associates, Inc., Sunderland. MA; 1989.

14. Stoodley P, Sauer K, Davies DG, Costerton JW: Biofilms as complex differentiated communities. Annu Rev Microbiol 2002, 56:187–209.

15. Costerton JW, Stewart PS, Greenburg EP: Bacterial biofilms: a common cause of persistent infections. Science 1999, 284(5418):1318–1322.

16. Mortensen KP, Conley SN: Film fill fouling in counterflow cooling towers: mechanisms and design. CTI J 1994, 15:10–25.

17. McDonogh R, Schaule G, Flemming HC: The permeability of biofouling layers on membranes. J Membr Sci 1994, 87:199–217.

18. Carpentier B, Cerf O: Biofilm and their consequences with particular reference to hygiene in the food industry. J Appl Bacteriol 1993, 75:499–511.

19. Taras M, Hakansson K, Guieysse B: Continuous acetonitrile degradation in packed-bed bioreactor. Appl Microbiol Biotechnol 2005, 66:567–574.

20. Hall ER: Biofilm reactors in anaerobic wastewater treatment. Biotech Adv 1987, 5:257–269.

21. Meyer A, Wallis FM: Development of microbial biofilms on various surfaces for the treatment of heavy metal containing effluents. Biotechnol Tech 1997, 11(12):859–863.

22. Qureshi N, Maddox IS: Novel bioreactors for the ABE fermentation using cells of Clostridium acetobutylicum immobilized by adsorption onto bonechar. In Fermentation technologies: Industrial applications. Edited by: Yu PL. London: Elsevier Appl Sci Publ; 1990.

23. Maddox IS, Qureshi N, McQueen J: Continuous production of 2,3-butanediol from whey permeate using cells of Klebsiella pneumoniae immobilized on to bonechar. New Zealand J Dairy Sci Technol 1988, 23:127–132.

24. Sponza DT: Anaerobic granule formation and tetrachloroethylene (TCE) removal in an upflow anaerobic sludge blanket (UASB) reactor. Enz Microbial Technol 2001, 29:417–427.

25. Annachhatre AP, Bhamidimarri SMR: Microbial attachment and growth in fixed-film reactors: process startup considerations. Biotech Adv 1992, 10:69–91.

26. Watnick P, Kolter R: Biofilm, City of microbes. J Bacteriol 2000, 182(10):2675–2679.

27. Davies DG, Chakrabarty AM, Geesey GG: Exopolysaccharide production in biofilms: substratum activation of alginate gene expression by Pseudomonas aeruginosa. Appl Environ Microbiol 1993, 59:1181–1186.

28. Donlan RM, Costerton JW: Biofilms: survival mechanisms of clinically relevant microorganisms. Clin Microbiol Rev 2002, 15(2):167–193.

29. Costerton JW, Lewandowski Z, Caldwell DE, Korber DR, Lappin-Scott HM: Microbial biofilms. Annu Rev Microbiol 1995, 49:711–745.

30. Allison DG, Ruiz B, SanJose C, Jaspe C, Gilbert P: Extracellular products as mediators of the formation and detachment of Pseudomonas fluorescens biofilms. FEMS Microbial Lett 1998, 167:179–184.

31. Characklis WG, McFeters GA, Marshall KC: Physiological ecology in biofilm systems. In Biofilms. Volume 37. Edited by: Characklis WG, Marshall KC. New York: John Wiley and Sons; 1990:67–72.

32. Qureshi N, Karcher P, Cotta M, Blaschek HP: High-productivity continuous biofilm reactor for butanol production. Appl Biochem Biotechnol 2004, 113–116:713–721.

33. Cowan MM, Warren TM, Fletcher M: Mixed species colonization of solid surfaces in laboratory biofilms. Biofouling 1991, 3:23–34.

34. Bücks J, Mozes N, Wandrey C, Rouxhet PG: Cell adsorption control by culture conditions. Appl Microbiol Biotechnol 1988, 29:119–128.

35. Donlan RM: Biofilms: microbial life on surfaces. Emerg Inf Dis 1992, 8(9):881–890.

36. Huang WC, Ramey DE, Yang ST: Continuous production of butanol by Clostridium acetobutylicum immobilized in a fibrous bed bioreactor. Appl Biochem Biotechnol 2004, 113:887–898.

37. Tay A, Yang ST: Production of L-(+)-lactic acid from glucose and starch by immobilized cells of Rhizopus oryzae in a rotating fibrous bed bioreactor. Biotechnol Bioeng 2002, 80:1–12.

38. Qureshi N, Schripsema J, Lienhardt J, Blaschek HP: Continuous solvent production by Clostridium beijerinckii BA101 immobilized by adsorption onto brick. World J Microbiol Biotechnol 2000, 16:377–382.

39. Qureshi N, Maddox IS: Reactor design for the ABE fermentation using cells of Clostridium acetobutylicum immobilized by adsorption onto bonechar. Bioprocess Eng 1988, 3:69–72.

40. Woolard CR, Irvine RL: Biological treatment of hypersaline wastewater by a biofilm of halophilic bacteria. Water Environ Res 1994, 66:230–235.

41. Wobus A, Ulrich S, Roske I: Degradation of chlorophenols by biofilms on semi-permeable membranes in two types of fixed bed reactors. Water Sci Technol 1995, 32:205–212.

42. Veeresh GS, Kumar P, Mehrotra I: Treatment of phenol and cresols in upflow anaerobic sludge blanket (UASB) process: A review. Water Research 2005, 39:154–170.

43. Tsuno H, Kawamura M, Somiya I: Anaerobic degradation of pentachlorophenol (PCP) in biological expanded-bed reactor. Water Sci Technol 1996, 34(5–6):335–344.

44. Qureshi N, Maddox IS: Integration of continuous production and recovery of solvents from whey permeate: use of immobilized cells of Clostridium acetobutylicum in fluidized bed bioreactor coupled with gas stripping. Bioprocess Engineering 1991, 6:63–69.

45. Lazarova V, Manem J: Innovative biofilm treatment technologies for water and wastewater treatment. In Biofilm II: Process analysis and applications. Edited by: Bryers JD. New York: Wiley-Liss Press; 2000:159–206.

46. Rodgers M, Zhan XM: Moving-medium biofilm reactors. Rev Environ Sci Bio/Technol 2003, 2:213–224.

47. Kargi F, Dincer AR: Salt inhibition effects in biological treatment of saline wastewater in RBC. J Environ Eng 1999, 125:966–971.

48. Kargi F, Eker S: Rotating-perforated-tubes biofilm reactor for high-strength wastewater treatment. J Environ Eng 2001, 127:959–963.

49. Gönec E, Harremoes P: Nitrification in rotating disc systems. Water Res 1985, 19:119–127.

50. Kargi F, Uygur A: Effect of liquid phase aeration on performance of rotating biodisc contactor treating saline wastewater. Environ Technol 1997, 18:623–630.

51. Lapara TM, Knopka A, Nakatsu CH, Alleman JE: Thermophilic aerobic treatment of a synthetic wastewater in a membrane-coupled bioreactor. J Ind Microbiol Biotechnol 2001, 26:203–209.

52. Vogelaar JCT, Bouwhuis E, Klapwijk A, Spanjers H, van Lier JB: Mesophilic and thermophilic activated sludge post-treatment of paper mill process water. Water Res 2002, 36:1869–1879.

53. Tripathi CS, Allen DG: Comparison of mesophilic and thermophilic aerobic biological treatment in sequencing batch reactors treating bleached Kraft pulp mill effluent. Water Res 1999, 33:836–846.

54. Quesnel D, Nakhla G: Utilization of an activated sludge for the improvement of an existing thermophilic wastewater treatment system. Environ Eng 2005, 131:570–578.

55. Jeris JS, Owen RW, Hickey R, Flood F: Biological fluidized-bed treatment for BOD and nitrogen removal. J Water Pollut Control Fed 1977, 49:816–831.

56. Rabah FKJ, Dahab MF: Nitrate removal characteristics of high performance fluidized-bed biofilm reactors. Water Res 2004, 38:3719–3728.

57. Jesis JS, Owen RW: Biological fluid-bed treatment for BOD and nitrogen removal. J Water Pollut Control Fed 1977, 49:816–821.

58. Shieh WK, Keenan JD: Fluidized bed biofilm reactor for wastewater treatment. Adv Biochem Eng Biotechnol 1986, 33:131–169.

59. Iza J: Fluidized bed reactors for anaerobic wastewater treatment. Water Sci Technol 1991, 24(8):109–132.

60. Ye FX, Shen DS, Feng XS: Anaerobic granule development for removal of pentachlorophenol in an upflow anaerobic sludge blanket (UASB) reactor. Process Biochem 2004, 39:1249–1256.

61. Schmidt JE, Batstone DJ, Angelidaki I: Improved nitrogen removal in upflow anaerobic sludge blanket (UASB) reactors by incorporation of Annammox bacteria into the granular sludge. Water Sci Technol 2004, 49(11–12):69–76.

62. Horber C, Christiansen N, Arvin E, Ahring BK: Improved dechlorinating performance of upflow anaerobic sludge blanket reactors by incorporation of Dehalospirillum multivorans into granular sludge. Appl Env Microbiol 1998, 64(5):1860–1863.

63. Iza J, Keenan PJ, Switzenbaum MS: Anaerobic treatment of municipal solid waste landfill leachate: Operation of a pilot scale hybrid UASB/AF reactor. Water Sci Technol 1992, 25(7):255–264.

64. Fang HHP, Hui HH: Effect of heavy metals on the methanogenic activity of starch-degrading granules. Biotechnol Lett 1994, 16(10):1091–1096.

65. Satoh H, Nakamura Y, Ono H, Okabe S: Effect of oxygen concentration on nitrification and denitrification in single activated sludge flocs. Biotechnol Bioeng 2003, 83:604–607.

66. Pochana K, Keller J: Study of factors affecting simultaneous nitrification and denitrification (SND). Water Sci Technol 1999, 39:61–68.

67. Robertson LA, van Niel EDJ, Torremans RAM, Kuenen JG: Simultaneous nitrification and denitrification in aerobic chemostat cultures of Thiosphaera pantotropha. Appl Environ Microbiol 1988, 54:2812–2818.

68. Robertson LA, Cornelisse R, De Vos P, Hadioetomo R, Kuenen JG: Aerobic denitrification in various heterotrophic nitrifiers. Antonie van Leeuwenhoek (Historical Archive) 1989, 56:289–299.

69. Jensen K, Sloth NP, Risgaard-Petersen N, Rysgaard S, Revsbech NP: Estimation of nitrification and denitrification from microprofiles of oxygen and nitrate in model sediment systems. Appl Environ Microbiol 1994, 60:2094–2100.

70. van Niel EDJ, Braber KJ, Robertson LA, Kuenen JG: Heterotrophic nitrification and aerobic denitrification in Alcaligenes faecalis strain TUD. Antonie van Leeuwenhoek (Historical Archive) 1992, 62:231–237.

71. Robertson LA, Kuenen JG: Thiosphaera pantotropha gen nov sp nov, a facultative anaerobic, facultative autotrophic sulphur bacterium. J Gen Microbiol 1983, 129:2847–2855.

72. Robertson LA, Kuenen JG: Aerobic denitrification: a controversy revived. Archive Microbiol 1984, 139:351–354.

73. Brindle K, Stephenson T: The application of membrane biological reactors for the treatment of wastewaters. Biotechnol Bioeng 1996, 49:601–610.

74. Casey E, Glennon B, Hamer G: Oxygen mass transfer characteristics in a membrane-aerated biofilm reactor. Biotechnol Bioeng 1999, 62:183–192.

75. Debus O, Wanner O: Degradation of xylene by a biofilm growing on a gas permeable membrane. Water Sci Technol 1992, 26:607–616.

76. Brindle K, Stephenson T: Nitrification in a bubbleless oxygen mass transfer membrane bioreactor. Water Sci Technol 1996, 34:261–267.

77. Kennes C, Thalasso F: Waste Gas Biotreatment Technology. J Chem Technol Biotechnol 1998, 72:303–319.

78. Ottengraf SPP: Biological systems for waste gas elimination. Trends Biotechnol 1987, 5:132–136.

79. Koe LCC, Yang F: A bioscrubber for hydrogen sulfide removal. Water Sci Technol 2000, 41:141–145.

80. Ottengraf S, Van Den Oever A: Kinetics of organic compound removal from waste gases with a biological filter. Biotechnol Bioeng 1983, 25:3089–3102.

81. Shinabe K, Oketani S, Ochi T, Matsumura M: Characteristics of hydrogen sulfide removal by Thiobacillus thiooxidans KS1 isolated from a carrier-packed biological deodorization system. J Ferment Bioeng 1995, 80:592–598.

82. De Beer D, Stoodley P, Roe F, Lewandowski Z: Effects of biofilm structures on oxygen distribution and mass transport. Biotechnol Bioeng 1994, 43:1131–1138.

83. Morgenroth E, Schroeder ED, Chang DPY, Scow KM: Nutrient limitation in a compost biofilter degrading hexane. J Air Waste Managem Assoc 1996, 46:300–308.

84. Bland RR, Chen HC, Jewell WJ, Bellamy WD, Zall RR: Continuous high rate production of ethanol by Zymomonas mobilis in an attached film expanded bed fermentor. Biotechnol Lett 1982, 4:323–328.

85. Krug TA, Daugulis AJ: Ethanol production using Zymomonas mobilis immobilized on an ion exchange resin. Biotechnol Lett 1983, 5:159–164.

86. Kunduru MR, Pometto AL III: Continuous ethanol production by Zymomonas mobilis and Saccharomyces cerevisiae in biofilm reactors. J Ind Microbiol Biotechnol 1996, 16:249–256.

87. Qureshi N, Brining H, Iten L, Dien B, Nichols N, Saha B, Cotta MA: Adsorbed cell dynamic biofilm reactor for ethanol production from xylose and corn fiber hydrolysate. The 36th Great Lakes Regional Meeting of the American Chemical Society, Peoria, IL, October 17–20, 2004

88. Rogers PL, Lee KJ, Skotnicki ML, Tribe DE: Ethanol production by Zymomonas mobilis. Adv Biochemical Eng 1982, 23:37–84.

89. Welsh FW, Williams RE, Veliky IA: Solid carriers for a Clostridium acetobutylicum that produced acetone and butanol. Enz Microbial Technol 1987, 9:500–502.

90. Qureshi N, Lai LL, Blaschek HP: Scale-up of a high productivity continuous biofilm reactor to produce butanol by adsorbed cells of Clostridium beijerinckii BA101. Transactions Institution Chemical Engineers (Trans IChemE) (Chemical Engineering Research and Design) 2004, 82(C2):164–173.

91. Qureshi N, Blaschek HP: Butanol production from agricultural biomass. In Advances in Food Biotechnology. Edited by: Shetty K, Pometto A, Paliyath G. New York, NY: CRC Press; 2005:in press.

92. Ezeji TC, Qureshi N, Blaschek HP: Industrially relevant fermentations. In Handbook on Clostridia. Edited by: Durrie P. New York, NY: CRC Press; 2005.

93. Pierrot P, Fick M, Engasser JM: Continuous acetone butanol fermentation with high productivity by cell ultrafiltration and recycling. Biotechnol Lett 1986, 8:253–256.

94. Afschar AS, Biebl H, Schaller K, Schugerl K: Production of acetone and butanol by Clostridium acetobutylicum in continuous culture with cell recycle. Appl Microbiol Biotechnol 1985, 22:394–398.

95. Shazer WH, Speckman RA: Continuous fermentation of whey permeate by Bacillus polymyxa. J Dairy Sci 1984, 67(Suppl 1, D5G):1216–1218.

96. Syblayrolles JM, Goma G: Butanediol production by Aerobactor aerogenes NRRL B199: effect of initial substrate concentration and aeration agitation. Biotechnol Bioeng 1984, 26:148–155.

97. Ramachandran KB, Goma G: Effect of oxygen supply and dilution rate on the production of 2,3-butanediol in continuous bioreactor by Klebsiella pneumoniae. Enz Microbial Technol 1987, 9:107–111.

98. Lee HK, Maddox IS: Continuous production of 2,3-butanediol from whey permeate using Klebsiella pneumoniae immobilized in calcium alginate. Enz Microbial Technol 1986, 8:409–411.

99. Ramachandran KB, Goma G: 2,3-butanediol production from glucose by Klebsiella pneumoniae in a cell recycle system. J Biotechnol 1988, 9:39–46.

100. Zeng AP, Biebl H, Deckwer WD: Production of 2,3-butanediol in a membrane bioreactor with cell recycle. Appl Microbiol Biotechnol 1991, 34:463–468.

101. Demirci A, Pometto AL III, Johnson KE: Evaluation of biofilm reactor solid support for mixed-culture lactic acid production. Appl Microbiol Biotechnol 1993, 38:728–733.

102. Demirci A, Pometto AL III, Johnson KE: Lactic acid production in a mixed-culture biofilm reactor. Appl Environ Microbiol 1993, 59:203–207.

103. Ho KLG, Pometto AL III, Hinz PN, Demirci A: Nutrient leaching and end product accumulation in plastic composite supports for L-(+)-lactic acid biofilm fermentation. Appl Environ Microbiol 1997, 63:2524–2532.

104. Ho KLG, Pometto AL III, Hinz PN: Optimization of L-(+)-lactic acid production by ring and disc plastic composite supports through repeated-batch biofilm fermentation. Appl Environ Microbiol 1997, 63:2533–2542.

105. Ho KLG, Pometto AL III, Hinz PN, Dickson JS, Demirci A: Ingredient selection for plastic composite supports for L-(+)-lactic acid biofilm fermentation by Lactobacillus casei subsp. rhmnosus. Appl Environ Microbiol 1997, 63:2516–2513.

106. Park EY, Kosakai Y, Okabe M: Efficient production of L-(+)-lactic acid using mycelial cotton-like flocs of Rhizopus oryzae in an air-lift bioreactor. Biotechnol Prog 1998, 14:699–704.

107. Sun Y, Li YL, Bai S: Modeling of continuous L(+)-lactic acid production with immobilized R. oryzae in an airlift bioreactor. Biochem Eng J 1999, 3:87–90.

108. Cao N, Du J, Gong CS, Tsao GT: Simultaneous production and recovery of fumaric acid from immobilized Rhizopus oryzae with a rotary biofilm contactor and an adsorption column. Appl Env Microbiol 1996, 62:2926–2931.

109. Zeikus JG, Jain MK, Elankovan P: Biotechnology of succinic acid production and markets for industrial products. Appl Microbiol Biotechnol 1999, 51:545–552.

110. Urbance SE, Pometto AL III, DiSpirito AA, Demirci A: Medium evaluation and plastic composite support ingredient selection for biofilm formation and succinic acid production by Actinobacillus succinogenes. Food Biotechnol 2003, 17:53–65.

111. Cotton JC, Pometto AL III, Gvozdenovic-Jeremic J: Continuous lactic acid fermentation using a plastic composite support biofilm reactor. Appl Microbiol Biotechnol 2001, 57:626–630.

112. Costerton JW, Cheng KJ, Geesey GG, Ladd TI, Nickel JC, Dasgupta M, Marrie TJ: Bacterial biofilms in nature and disease. Annu Rev Microbiol 1987, 41:435–464.

113. O'Toole G, Kaplan HB, Kolter R: Biofilm formation as microbial development. Annu Rev Microbiol 2000, 54:49–79.

Techno-Economic Analysis for the Conversion of Lignocellulosic Biomass to Gasoline via the Methanol-to-Gasoline (MTG) Process

S. B. Jones, Y. Zhu

SUMMARY

Biomass is a renewable energy resource that can be converted into liquid fuel suitable for transportation applications. As a widely available biomass form, lignocellulosic biomass can have a major impact on domestic transportation fuel supplies and thus help meet the Energy Independence and Security Act renewable energy goals (EISA 2007).

With gasification technology, biomass can be converted to gasoline via methanol synthesis and methanol-to-gasoline (MTG) technologies. Producing a gasoline product that is infrastructure ready has much potential. Although the MTG technology has been commercially demonstrated with natural gas conversion, combining MTG with biomass gasification has not been shown. Therefore, a techno-economic evaluation for a biomass MTG process based on currently available technology was developed to provide information about benefits and risks of this technology. The economic assumptions used in this report are consistent with previous U.S. Department of Energy Office of Biomass Programs techno-economic assessments.

The feedstock is assumed to be wood chips at 2000 metric ton/day (dry basis). Two kinds of gasification technologies were evaluated: an indirectly-heated gasifier and a directly-heated oxygen-blown gasifier. The gasoline selling prices (2008 USD) excluding taxes were estimated to be $3.20/gallon and $3.68/ gallon for indirectly-heated gasified and directly-heated. This suggests that a process based on existing technology is economic only when crude prices are above $100/bbl. However, improvements in syngas cleanup combined with consolidated gasoline synthesis can potentially reduce the capital cost. In addition, improved synthesis catalysts and reactor design may allow increased yield. This is shown in the figure below (the ranges for the base case and the improved case reflect differences in gasifier types).

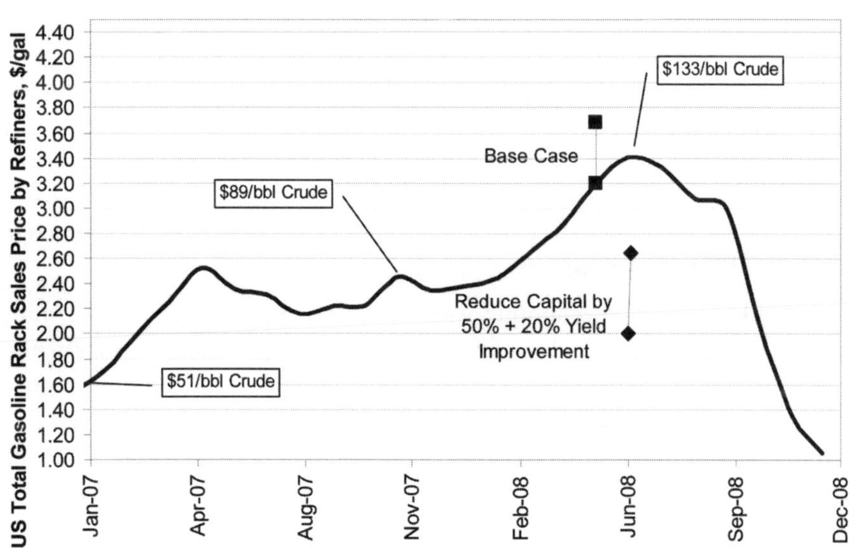

Figure 1

Introduction

Biomass is an important domestic resource that has the potential to make a significant impact on domestic fuel supplies and thus help meet the Energy Independence and Security Act renewable energy goals (EISA 2007). This study is part of an ongoing effort within the Department of Energy to meet the renewable energy goals for liquid transportation fuels. In particular, this report assesses the potential for gasoline production from biomass via the methanol-to-gasoline process.

Biomass conversion to synthesis gas by gasification has been demonstrated. The syngas, rich in CO and H_2, can be further used to produce methanol, ethanol or other chemicals and liquid fuels. Among these products, methanol has been widely used as a solvent and also as the raw material for many chemicals and liquid fuels. In the 1970s, a methanol-to-gasoline (MTG) process was developed by Mobil in response to the Arab oil embargo (Chang, 1992). This technology was based on the direct conversion of methanol to hydrocarbons catalyzed by zeolite ZSM-5 (Olson, et al. 1981). A commercially operated gas-to-gasoline plant using Mobil's MTG technology was built in New Zealand in the 1980s. In this plant, methane was converted into syngas by steam reforming, and then syngas was used to make methanol. The methanol was then converted to gasoline by a fixed-bed MTG process. Another type of MTG process is the fluid-bed type, which has not yet been commercially demonstrated (Chang, 1992). Currently, a coal-based MTG demonstration plant is under construction in China for start-up in late 2008/mid-2009 using MTG technology from ExxonMobil (Tabak, et al., 2008) and a commercial scale plant is being planned. ExxonMobil has also announced plans for two coal-to-gasoline plants in the United Stated using MTG technology (Ondry 2008, DKRW 2008). Although natural gas has been used and coal will be used for gasoline production, biomass feedstock has not yet been commercially used for MTG production.

A number of techno-economic assessments, using process design and simulation models, have been conducted for biomass gasification to fuels and chemicals such as methanol (Hamelinck and Faaij, 2001), Fischer-Tropsch liquid transportation fuels (Hamelinck, et al. 2003), hydrogen (Hamelinck and Faaij, 2001; Spath, et al. 2005), and ethanol (Aden, et al. 2005; Phillips, et al. 2007). Limited research and analysis work has been conducted for a biomass conversion to gasoline via MTG. In this study, process and economic models were developed for a biomass gasification based MTG system using existing technology. In the following sections, the design basis, simulation methods, and results are described. The sensitivity analysis results will also be discussed.

Process Design Basis and Modeling Approach

A simplified block diagram for the Biomass MTG system is shown in Figure 2-1. In this system, wood chips are converted to synthesis gas in a gasifier. Raw syngas is sent to a tar reformer, a particulate scrubber, and finally a sulfur removal unit. A steam reformer is used to convert CH_4 to H_2 and CO and to adjust the H_2/CO ratio to that required by methanol synthesis. Excess CO_2 is removed in an amine unit. The syngas is then compressed and sent to the methanol synthesis section to produce feed for the MTG process. Part of the purge gas from methanol synthesis is used to produce hydrogen by a pressure swing adsorption (PSA) unit; the remainder purge gas is used as fuel. Raw methanol is converted to hydrocarbons and water in the MTG reactors. The raw gasoline product stream is separated to produce fuel gas, liquefied petroleum gas (LPG), light gasoline, and heavy gasoline. The heavy gasoline is further treated with hydrogen from the PSA to meet the final gasoline specifications. Steam generated in the processes is collected and sent to the steam cycle for power generation. Some steam is used in steam reforming and other processes.

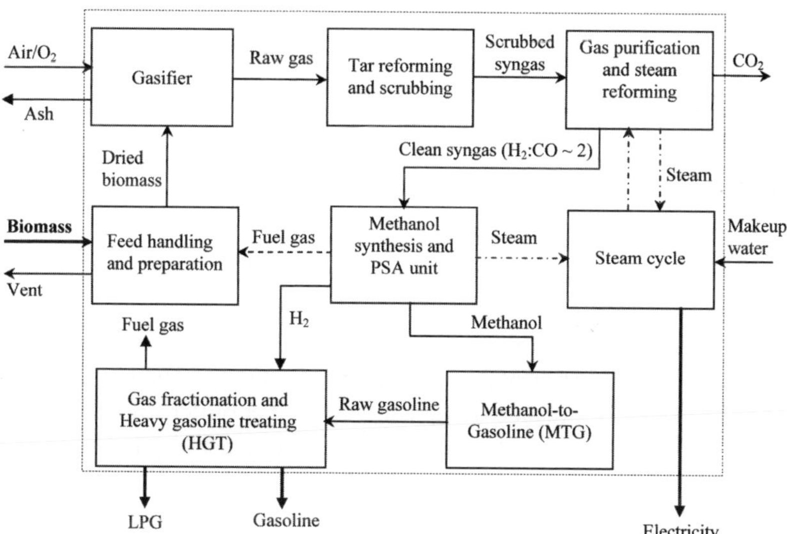

Figure 2-1 Block Diagram of Biomass-to-Gasoline via MTG System

Process Design Basis

The biomass-to-gasoline via MTG processing consists of eight main sections. Each section is briefly described in the follow paragraphs

Feed Handling and Preparation

The feedstock is assumed to be wood chips at a moisture content of 50 wt%. The wood feed rate is 2000 dry metric ton per day. The wet wood chips enter rotary dryers and are dried to a moisture content of 12 wt% (Spath, et al. 2005). For the indirectly-heated gasifier based system, the hot flue gas from the char combustor is used for biomass drying. For the directly-heated gasifier, the char combustor flue gas is insufficient for feed drying. The dryer used for this gasifier is assumed to have an auxiliary burner to provide additional heat. Fuel for the burner comes from the methanol and MTG syntheses purge gas. The dried biomass is then conveyed to the gasifier.

Gasification

Two types of gasifiers are considered in this study: a low pressure indirectly-heated gasifier and a pressurized oxygen-blown, directly-heated gasifier. The indirectly-heated gasifier consists of two vessels: a gasifier and a combustor. Dried wood is fed into a low-pressure indirectly heated entrained flow gasifier. Steam extracted from the steam cycle is sent to the gasifier at a flow rate of 0.4 lb of steam/lb of bone dry wood to fluidize the bed and to supply a portion of the heat required for the gasifier. The gasifier is assumed to be operated at 870°C (1598°F) and 23 psia. The gasifier is mainly heated by circulating olivine particles between the gasifier and the separate combustor. Olivine is synthetic sand that serves as the heat carrier in the gasifier. Char formed in the gasifier is carried out of the gasifier along with the olivine, separated in a series of cyclones and sent to the fluidized bed combustor, where air is used to burn the char, thereby reheating the olivine.

The oxygen-blown, directly-heated gasifier is a pressurized fluidized bed gasifier that also uses steam to fluidize the gasifer bed and to provide a portion of the heat. A pressurized cryogenic air separation unit provides purified oxygen at 99.5% for the gasifier at 350 psia and 15.6°C (60°F). The mass flow rate of oxygen is varied to achieve an 870°C (1600°F) gasifier outlet stream temperature. The dried wood is fed using a lock hopper feeder system and pressurized by compressed CO_2 recovered from the syngas purification process. The CO_2 gas used in the lock hopper is fed at a flow rate of 0.03 lb of CO°/lb dried wood and compressed to 330 psi. A small amount of MgO is also added to the gasifier to react with potassium in the ash to prevent agglomeration in the gasifier bed.

The indirectly-heated gasifier is modeled using the correlations reported in Spath, et al. (2005) which in turn is based on data from a Battelle-Columbus Laboratory (BCL) process development unit gasifier. The performance of the oxygen-blown, directly-heated pressurized gasifier was predicted using correlations derived from an experimental Institute of Gas Technology (IGT) gasifier (Evans, 1988).

Tar Reforming and Gas Scrubbing

During gasification, a relatively small fraction of biomass is converted into tars consisting mostly of aromatic and poly-aromatic hydrocarbons. The nitrogen in the biomass is converted to ammonia. The raw syngas from the cyclone in the gasifier section is sent to a catalytic tar cracker, which is assumed to be a bubbling fluidized bed reactor. A portion of the syngas tar, methane, and other light hydrocarbons in the raw gas are converted to CO and H_2, and some of the ammonia is converted to N_2 and H_2. The conversion percentage for each compound is reported in Spath, et al., (2005). The gas enters the tar reformer at the gasifier outlet temperature and exits the reformer at 750°C (1,383°F). The syngas is further cooled to 149°C (300°F) and sent to a wet scrubber to remove other impurities, such as particulates, residual ammonia and residual tars.

Gas Purification and Steam Reforming

The scrubbed syngas is compressed to 450 psia. A liquid phase oxidation (LO-CAT) process followed by a ZnO bed is used to remove sulfur that would otherwise poison downstream catalysts. The LO-CAT process is assumed to remove the sulfur to a concentration of 10 ppm H_2S, and then the ZnO bed polishes the syngas to less than 1 ppmv (Spath, et al. 2005).

Syngas leaving the ZnO bed is sent to a steam reformer to convert the remaining methane and light hydrocarbons to additional syngas and to adjust the H_2:CO ratio via the water-gas shift reaction. The main steam reformer reactions are:

$$C_nH_m + nH_2O \leftrightarrow (n + m/2)H_2 + nCO \tag{1}$$

$$CO + H_2O \leftrightarrow CO_2 + H_2 \tag{2}$$

Before the syngas is sent to the steam reformer, it is mixed with high temperature steam from the steam cycle and the steam generated by the methanol synthesis reactor. Reactions take place between 800 and 900°C (1472 and 1652°F). The steam reformer is fired with off-gas from the methanol synthesis section purge and the MTG syntheses section. The H_2: CO ratio is adjusted to ~ 2, as required by methanol synthesis reaction. The converted syngas passes through several heat exchangers to recover heat by generating saturate high pressure steam and superheated high pressure steam. The cooled syngas from the reforming process is further cooled by air cooling and cooling water. Excess CO_2 is removed in an amine unit. The clean syngas is then compressed to 1450 psia.

Methanol Synthesis and PSA Unit

The compressed clean syngas is sent to a low-pressure methanol synthesis process. The principle reactions are:

$$CO + 2H_2 \leftrightarrow CH_3OH \tag{3}$$

$$CO_2 + 3H_2 \leftrightarrow CH_3OH + H_2O \tag{4}$$

The first reaction is the primary methanol synthesis reaction, and the second one represents a small fraction of carbon dioxide in the feed that acts as a promoter for the primary reaction.

A simplified process flow diagram of methanol synthesis is shown in Figure 2-2.

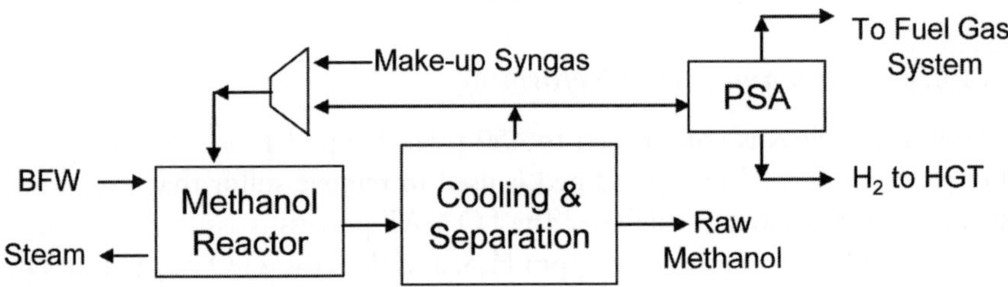

Figure 2-2 Block Diagram of Methanol Synthesis

The syngas from the steam reformer is compressed and preheated before entering the methanol reactor. Methanol synthesis temperatures and pressures typically range between 230 to 270°C (446 to 578°F) and 735 to 1470 psia, respectively (Cheng, 1994, Fiedler, et al., 2000). The methanol reactor is assumed to be isothermal, low-pressure (850 psia), gas-phase shell and tube type with ZnO/CuO catalyst in the tubes. The methanol synthesis reaction is highly exothermic and reaction heat is removed by generating medium pressure steam on the shell side of the reactor. The hot reactor product vapor is cooled by recycled compressed syngas and further cooled by air and then cooling water. The product stream is sent to a flash tank where liquid raw methanol is separated from the non-condensable gases. Approximately ninety-five percent of the vapor phase is compressed and recycled to the methanol reactor. The liquid product is further reduced in pressure to produce raw methanol at about 95 wt% purity. The raw methanol is sent to the MTG process. No distillation equipment is provided to produce high purity methanol.

Part of the purged gas from the methanol process is fed to a PSA unit where high purity (99.9%) hydrogen is produced for use in the heavy gasoline treating (HGT) unit. This unit hydrotreats the heavy gasoline fraction from the MTG process and is described more fully in the next sections.

Methanol-to-Gasoline (MTG) Process

Methanol is converted to gasoline by the Mobil MTG process. The principle reactions are (Chang, 1992, Cheng, 1994):

$$\frac{n}{2}\left[2CH_3OH \Leftrightarrow CH_3OCH_3 + H_2O\right] \xrightarrow{\;-nH_2O\;} C_nH_{2n} \rightarrow n\left[CH_2\right] \tag{5}$$

$\left[CH_2\right]$ = average formula for the paraffin-aromatic mixture

First methanol is partially dehydrated to dimethyl ether (DME) using a methanol dehydration catalyst. This reaction is followed by olefin formation and finally aromatic/paraffin formation over a zeolite catalyst (ZSM-5). The final hydrocarbon product is predominantly in the gasoline boiling range, with some LPG and fuel gas generated as well.

Mobil has developed two reactor configurations: an adiabatic fixed-bed and a circulating fluidized-bed process (Kaneko, et al. 2002). ZSM-5 undergoes deactivation with time. In fixed bed reactors, the catalyst ages in a "band-aging" mechanism that travels down the length of the reactor. Methanol break-through occurs when the age-band reaches the bottom of the reactor and indicates the need for catalyst regeneration. Regeneration is accomplished by burning off organic deposits with air (Cheng, 1994).

Circulating fluidized beds have the potential to simplify heat and catalyst activity management by providing continuous catalyst regeneration.

The fixed bed concept was demonstrated on a commercial scale in New Zealand using natural gas as the feed stock. This plant produced 14,500 bpd of gasoline and used one DME reactor and four MTG reactors operating in parallel, with a fifth MTG reactor off-line for regeneration. Multiple reactors minimize bed pressure drop and allow a staggered operating sequence to prevent significant methanol breaking through to the product (Cheng, 1994). The fluidized bed configuration has been demonstrated on a pilot scale only.

The simulation of the MTG process for this work is based on a study by the W.R. Grace Co. (1982 a,b,c) using the fixed bed process. A simplified flow diagram is shown in Figure 2-3.

The MTG process is highly exothermic and heat management is performed in two steps. First, raw methanol is sent to a single fixed bed dehydration reactor to convert some of the methanol to dimethyl ether (DME) and water. Preconversion to DME reduces the exotherm in the next set of reactors. The DME reactor effluent is combined with recycle gas and sent to two conversion reactors operating in parallel. A third reactor is offline for catalyst regeneration. The mixture is converted to hydrocarbons and water over a zeolite catalyst in the conversion reactors. The reaction is highly exothermic and is controlled by using a large re-cycle gas-to-DME reactor effluent ratio. Approximately ninety-nine percent of the non-condensable gas is combined with the effluent from the dehydration unit and sent to the conversion unit to achieve a recycle ratio of 9 moles of recycle gas per mole of methanol feed (Cheng, 1994). The remaining gas is combined with the hydrocarbon liquid and sent to the gas fractionation unit.

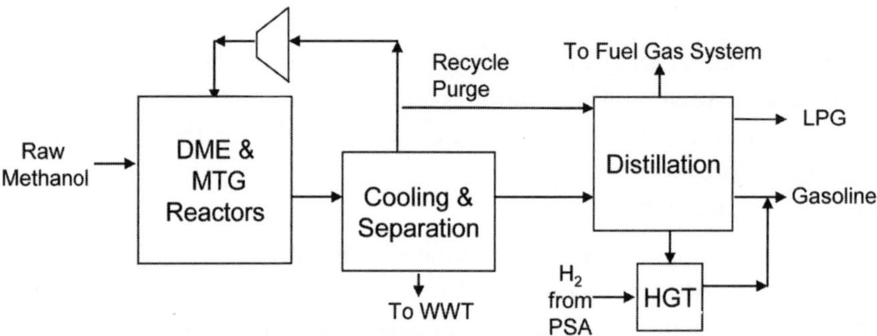

Figure 2-3 Block Diagram of Methanol-to-Gasoline Synthesis

The hydrocarbon product typically yields about 85 wt% high-quality gasoline. Almost no hydrocarbons with carbon numbers greater than ten are found in the product because of the shape selective nature of the conversion catalyst (Chang, 1992). The process operates at essentially 100% conversion of methanol until the catalyst is deactivated by carbon formation. The deactivated catalyst is regenerated by burning off the coke from the catalyst using air combustion.

Gas Fractionation and HGT

The hydrocarbons from the MTG process are separated in a gas fractionation unit to produce fuel gas, liquefied petroleum gas (LPG), light gasoline and a heavy gasoline feed stream. The gas fractionation unit consists of three fractionation towers, the de-ethanizer, the de-butanizer, and the gasoline splitter.

The product stream from the MTG process gas-liquid separator is first sent to a de-ethanizer tower to remove ethane and lighter hydrocarbons for fuel gas from the product. The de-ethanizer bottoms are then fractionated in the debutanizer to produce the feed to the gasoline splitter. The de-butanizer produces a light over head consisting of C4 and lower hydrocarbons, which is suitable for sale as lique-fied petroleum gas (LPG). Most of the de-butanizer bottoms (a stream containing C5+ hydrocarbon) are sent to the gasoline splitter. A small portion of the de-butanizer column bottoms is recycled to the de-ethanizer column to act as a lean oil solvent. This assists in the separation of the light gases from the liquids in the de-ethanizer. In the gasoline splitter, light gasoline containing hydrocarbons in the C5 to C9 range are separated in the overhead flow. The light gasoline is cooled and then sent to gasoline storage as gasoline product. The bottom flow from the gasoline splitter is heavy gasoline with hydrocarbons numbering C9 and above. This heavy fraction contains durene, a low melting point component, which if not removed, can cause freeze point problems in the gasoline. The heavy gasoline is treated in the Heavy Gasoline Treater (HGT).

The HGT unit removes durene via hydrogenation. The hydrogen produced from the PSA unit is combined with the heavy gasoline feed from the gasoline splitter. The mixture is sent to the catalytic hydrotreating reactor. The effluent is cooled by process streams and air before it is sent to the gas-liquid separator. Most of the vapor is compressed and recycled to the hydrotreater, with a small purge going to fuel gas system. The liquid flow from the separator is stabilized in the product stripper. The stablilized heavy gasoline is combined with the lighter gasoline cut from the de-butanizer, cooled and sent to product storage.

In the indirectly-heated gasifier case, some of the LPG is used as fuel for the steam reformer. The remaining LPG is available for sale as a by-product. In the directly-heated gasifier case, all of the LPG is used as fuel for the steam reformer and the wood dryer.

Power Generation

Saturated steam is generated by cooling process streams in the gasifier, steam reformer and MTG synthesis areas. It is then superheated and sent to a steam turbine to generate power for the entire plant. Excess power is sold. The steam cycle also provides medium and low pressure process steam for use in the system.

Analysis Approach

The process simulation for the gasifier based biomass-to-gasoline via MTG system was developed in CHEMCAD 6.1. The capital and operating costs were assembled in an EXCEL spreadsheet using information from the CHEMDAD simulation. A discounted cash flow method was used to estimate the product selling price.

Simulation and Economic Assumptions

This section describes the performance and cost simulation assumptions.

Simulation Assumptions

Table 3-1 shows the main assumptions for the indirectly-heated and directly-heated gasifier based cases.

Table 3-1 Operating Conditions Used in the Simulations

Gasification		
Gasifier Type	Indirectly-Heated Gasifier	Directly-Heated Gasifier
Gasifier Pressure, psia	23	330
Gasifier Temperature, °C (°F)	870 (1598)	871 (1600)
Biomass Feed, metric ton/d, dry basis	2000	2000
Oxidant	Air	Oxygen
Dried Biomass, moisture fraction %	12	12
Tar Reformer, °C (°F)/P, psia	717 (1323) / 20	717 (1323) / 327
Steam Reforming		
Reformer outlet temperature, °C (°F)	900 (1652)	900 (1652)
Reformer outlet pressure, psia	423	430
Reformed gas H_2/CO	2.1	2.1
Methanol Synthesis and Purification		
Reactor outlet temperature, °C (°F)	260 (500)	260 (500)
Reactor outlet pressure, psia	840	840
kg/h methanol/L catalyst	0.9	0.9
Methanol-to-Gasoline		
Dehydration unit (DME reactor)		
Outlet temperature, °C (°F)	400 (752)	400 (752)
Outlet pressure, psia	385	385
kg/h hydrocarbon/L catalyst	0.96	0.96
Conversion unit (MTG reactor)		
Outlet temperature, °C (°F)	390 (733)	390 (733)
Outlet pressure, psia	300	300
Recycle ratio: moles recycle gas /mole methanol	9	9
kg/h hydrocarbon/L catalyst	0.25	0.25
Heavy Gasoline Treating		
Reactor outlet temperature, °C (°F)	282 (540)	282 (540)
Reactor outlet pressure, psia	470	470
H_2 partial pressure at reactor inlet, psia	423	423

Economic Assumptions

Most of the base equipment costs for gas purification and conditioning and steam cycle and power generation sections of the plant were obtained from Spath, et al. (2005). The estimation of the equipment cost for the two types of gasifiers is based on Hamelinck and Faaij (2002). Capital costs for methanol synthesis and methanol-to-gasoline synthesis were estimated using ASPEN ICARUS 2006.5 and the W.R. Grace Report (Grace, 1982 a,b,c).

All capital costs are reported in 2008 dollars. Equipment cost escalation is calculated by using the Chemical Engineering Plant Cost Index (CEPCI). The total capital investment is factored from installed equipment costs as shown in Table 3-2. Table 3-3 lists the assumptions used to estimate the production costs.

Table 3-2 Project Investment Factors

	% of TPEC
Total Purchased Equipment Cost (TPEC)	100%
Purchased Equipment Installation	39%
Instrumentation and Controls	26%
Piping	31%
Electrical Systems	10%
Buildings (including services)	29%
Yard Improvements	12%
Total Installed Cost (TIC)	247%
Indirect Costs	
Engineering	32%
Construction	34%
Legal and Contractors Fees	23%
Project Contingency	37%
Total Indirect	126%
Total Project Investment	373%

Table 3-3 Operating Cost Assumptions

	Value used in model, 2008 basis	Units or Basis	Reference
Raw Materials			
Hybrid poplar chips	60	$/dry short ton	Aden, 2008
Olivine makeup	258	$/short ton	Phillips, *et al.* 2007*
Ash disposal	44	$/short ton	Phillips, *et al.* 2007*
Tar cracker catalyst	8.06	$/lb	Phillips, *et al.*, 2007*
Reformer catalyst	25.4	$/lb	SRI PEP 2007*
Methanol catalyst	9.69	$/lb catalyst	SRI PEP 2007
MTG zeolite catalyst	60	$/lb catalyst	SRI PEP 2007
DME catalyst	11.6	$/lb catalyst	SRI PEP 2007
Utilities			
Waste water treatment	2.4	$/100 ft^3	Phillips, *et al.* 2007
Cooling tower makeup	245	¢/1000 gal	Phillips, *et al.* 2007
Electricity	6.68	¢/kWh	EIA, 2008b
Economic Assumptions			
Stream Factor	90%		estimated
MACRS Depreciation, yrs	7		Phillips, *et al.* 2007
Plant life, yrs	20		Phillips, *et al.* 2007
Construction Period	2.5 years		Phillips, *et al.* 2007
1st 6 months expenditure	8%		
Next 12 months expenditure	60%		
Last 12 months expenditure	32%		
Start-up time	6 months		Phillips, *et al.* 2007
Revenues	50%		
Variable Costs	75%		
Fixed Costs	100%		
Working Capital	5% of Total Capital Investment		Phillips, *et al.* 2007
Land	6% of Total Purchased Equipment Cost (taken as 1st year construction expense)		Phillips, *et al.* 2007
Internal Rate of Return	10%		Phillips, *et al.* 2007
* Reference value escalated to 2008 dollars using the producer price index			

Results and Analysis

This section describes the main performance and cost simulation results for the biomass-to-gasoline via MTG systems. The results of the sensitivity analysis are also discussed.

Performance Results and Discussion

Table 4-1 shows the main performance results for the biomass-to-gasoline via MTG systems.

Table 4-1 Performance Analysis Results

Case	Indirectly-Heated Gasifier	Directly-Heated Gasifier
Gasifier pressure, psi	23	330
Gasifier temperature, °C (°F)	870 (1598)	871 (1600)
H_2:CO ratio in scrubbed syngas, molar	0.9	1.4
Char production, lb/hr	33970	7510
Feed		
Wood chips, metric ton/d, dry basis	2000	2000
Natural Gas, lb/hr	--	--
Feed Preparation		
Directly-heated dryer	Heat from char burner exhaust only	Heat from char burner exhaust plus auxiliary fuel gas burner
Products		
Gasoline, mmgal/y	40	40
LPG, mmgal/y	7.2	--
Methanol Synthesis		
Recycle ratio: moles recycle gas /moles makeup gas	5.6	2.0
Power Consumption, MW		
Air separation unit	--	-8.6
Lock hopper gas compressor	--	-0.2
Dryer air blower	--	-0.1
Char burner air compressor	-4.8	-0.5
Syngas compressor	-16.0	-1.9
Reformer air compressor	-1.1	-1.8
Reformer flue gas blower	-0.5	-0.6
Clean syngas compressor	-3.8	-4.6
Methanol syn. recycle compressor	-2.7	-2.4
MTG recycle & hydrogen compressor	-3.0	-3.7
Steam turbines	24.4	56.0
Net power, MW	**-7.5**	**32**
Carbon Efficiency, %	**32%**	**29%**
Thermal Efficiency, %, higher heating value (HHV) basis	**47%**	**41%**

In general, one would expect that the directly-heated gasifier would produce more final product than an indirectly-heated gasifier due to the former's lower char production and thus higher syngas rate. However, the biomass dryer for the directly-heated gasifier needs an external burner to supply the dryer heat load. Fuel to the auxiliary burner is supplied by using all of the LPG plus a higher purge rate in the methanol synthesis step. The higher methanol off-gas purge rate reduces the yield of methanol and hence reduces the final gasoline yield. This effect could be mitigated by 1) using natural gas in the dryer burner, or 2) by using an indirectly-heated dryer using steam as the heat source. Analysis of these options is beyond the scope of this work. Lack of LPG product in the directly-heated gasifier case reduces the carbon and the thermal efficiency as compared to the indirectly-heated gasifier case. However, the larger volumes of gas through the syngas generation and cleanup steps greatly increase the amount of steam recuperated, and thus results in a net export of power.

Cost Results and Discussion

Table 4-2 shows the capital and operating cost breakdown for each section of the plant.

Table 4-2 Capital and Operating Cost Summary

Case	Indirectly-Heated Gasifier		Directly-Heated Gasifier	
CAPITAL COSTS	Million $	% of Total	Million $	% of Total
Air separation unit	--	--	$10	8%
Feed prep and drying	$11	11%	$12	9%
Gasification with tar reforming, heat recovery, scrubbing	$17	16%	$41	31%
Syngas cleanup & compression	$33	32%	$27	20%
Methanol synthesis	$10	10%	$8	6%
MTG, gas fractionation, and HGT	$24	23%	$24	16%
Steam system and power generation	$7	7%	$11	8%
Remainder off-site battery limits (OSBL)	$2	2%	$2	1%
Total Purchased Equipment Cost (TPEC)	$104	100%	$136	100%
Total Installed Cost (TIC), mm$	$258		$335	
Total Indirect Cost, mm$	$132		$272	
Total Project Investment, mm$	$383		$499	
OPERATING COSTS	$/gal	% of total	$/gal	% of total
Biomass	1.07	34%	1.07	29%
Natural Gas	0.00	0%	0.00	0%
Catalysts & Chemicals	0.29	9%	0.38	10%
Waste Disposal	0.04	1%	0.05	1%
Electricity and other utilities	0.13	4%	-0.38	-10%
LPG credit	-0.30	-9%	0.00	0%
Fixed Costs	0.50	16%	0.61	17%
Capital Depreciation	0.47	15%	0.62	17%
Average Income Tax	0.33	10%	0.43	12%
Average Return on Investment	0.65	20%	0.91	25%
Estimated Selling Price at 10% ROI, $/gal gasoline	3.20	100%	3.68	100%

Methanol and MTG synthesis, gas fractionation, and HGT represents about one-third of the total equipment cost. The total capital cost of the directly-heated gasifier based system is about 30% higher than that of the indirectly-heated gasifier based system. The directly-heated gasifier is pressurized and requires an Air Separation Unit (ASU) whereas the indirectly-heated gasifier is at atmospheric pressure.

For both gasifier cases, approximately 30% of the production cost is feedstock related. The indirectly-heated gasifier based system has by-products credit of LPG and no export electricity. The directly-heated gasifier has credits from electricity production, and no LPG credit. Therefore, the selling price of gasoline for the indirectly-heated gasifier based system is lower than that of the directly-heated gasifier based system.

As shown in Figure 4-1, the gasoline refiners rack sales prices from January 2007 to December 2008 ranged from a low in December 2008 of $1.05/gallon to a high of $3.41/gallon in June 2008. The June 2008 gasoline price corresponds to the EIA reported peak of over $130/barrel (bbl) crude oil. Superimposed on this plot is the estimated selling price for the indirectly-heated gasifier case (horizontal line at $3.20/gallon). The estimated gasoline selling prices for the two gasifier models are at or above the high end of the rack market price as shown in Figure 4-1. Thus, the base case gasification and MTG technology investigated in this study appears to be economically viable only at crude prices above $100/bbl. To give this another perspective, if taxes are assumed to be an additional 25% on top of the rack price, then the pump price for the base case indirectly-heated gasifier is over $4.25/gallon.

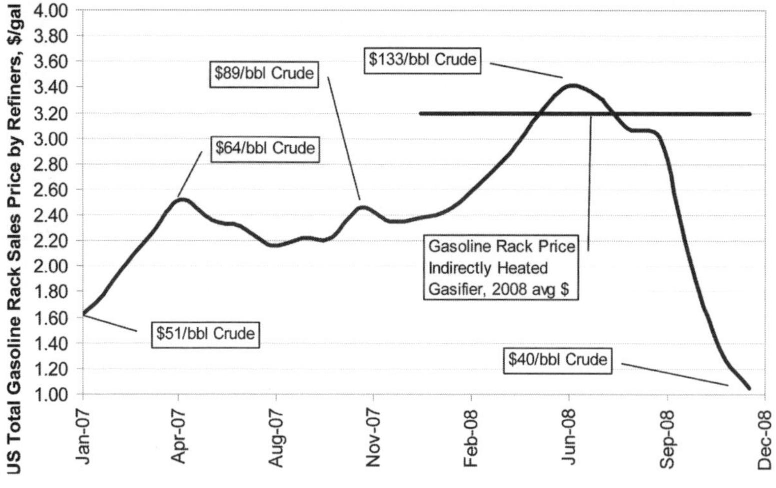

Figure 4-1 US Refiners Gasoline Rack Sales Price and World Average Crude Prices (EIA, 2009)

Sensitivity Analysis

A sensitivity analysis was conducted to investigate the effects of different cost assumptions. The feedstock cost, total capital cost and yield were varied to determine their economic impact. The feedstock cost was chosen because it represents between 30% of the total production cost as shown in Table 4-2. The capital cost was chosen because it has potential for cost reductions.

Feedstock Cost

Figure 4-2 shows a range of feedstock prices superimposed on the rack gasoline sale price. For each price range, the upper point corresponds to the directly heated gasifier. Very low feedstock costs (below $20/dry ton) are required to approach the medium gasoline rack price for this time frame. As little or no feedstock is available at this price, this effect alone suggests that producing gasoline with current technology will never be viable unless crude prices are extremely high.

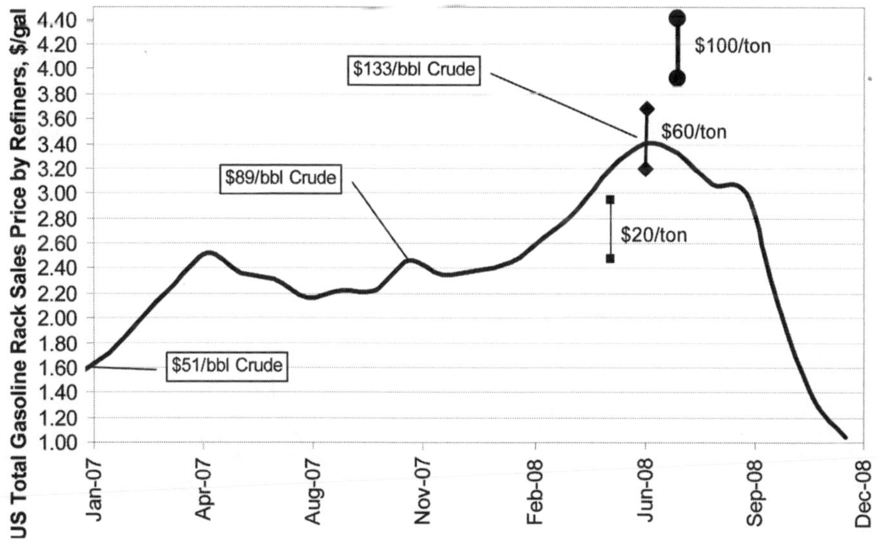

Figure 4-2 Effect of Feedstock Cost on Gasoline Price

Sensitivity to Capital Cost and Yield

There is scope for reducing the capital costs for the syngas cleanup and for the entire MTG synthesis process. The base case assumes that the gas cleanup step includes a tar reformer and a separate steam reformer. Additionally, the MTG process uses six fixed bed reactors:

- one for syngas to methanol synthesis,
- one for methanol to dimethyl ether (DME) conversion,
- three for methanol/DME conversion to gasoline and
- one for hydrogenating the heavy gasoline fraction to remove durene.

Research into improved syngas cleanup is ongoing (Phillips, 2007). Pilot scale tests have shown that a continuous fluidized bed reactor can achieve nearly the same conversion to gasoline as the multi-fixed bed process (Cheng, 1994). MTG reactor design that avoid large recycle streams for temperature control, such as fluid bed MTG or shell and tube reactors, can reduce capital costs in the synthesis area. Additionally, combined synthesis steps, such as once through methanol/DME production could potentially reduce capital costs and product losses. This sensitivity explores the effect of capital cost reductions.

Figure 4-3 shows the producers selling price for three capital costs levels superimposed on the historic rack price data. The 100% bars are the base case, with the indirectly-heated gasifier the lower point and the directly-heated gasifier the upper. Reducing the capital to 75% of the base case brings the estimated rack price down by $0.40/gallon. This corresponds to a crude price just above $100/bbl. Further reducing the capital cost for the MTG synthesis step to 50% of the base case reduces the gasoline producers selling price by nearly $1/gallon, but high crude costs are still needed. It appears that capital costs reduction alone may be insufficient to make this process viable.

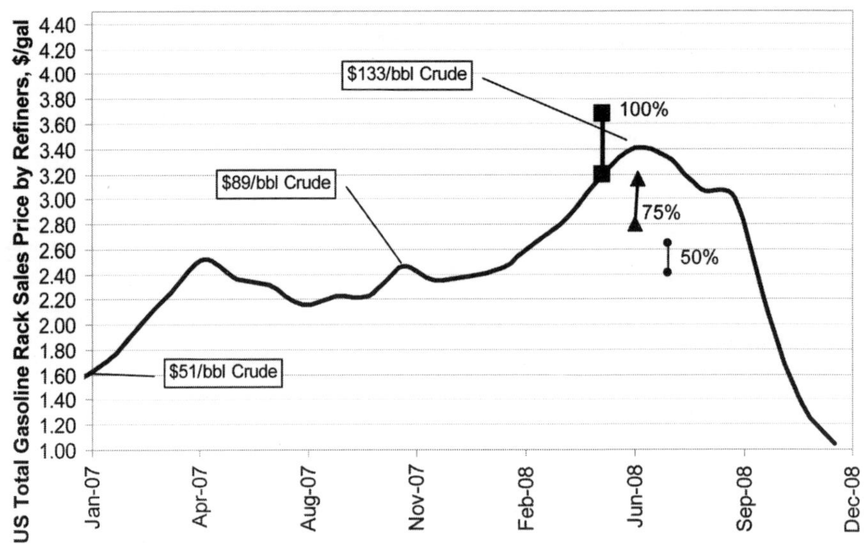

Figure 4-3 Effect of Capital Cost on Gasoline Price

Figure 4-4 shows the effect of reduced capital and improved yield. If all of the potential capital improvements are realized, and yield losses in the synthesis step are minimized, then the methanol-to-gasoline process estimated rack selling price can be lowered to approximately $2.00/gallon and thus becomes economically attractive.

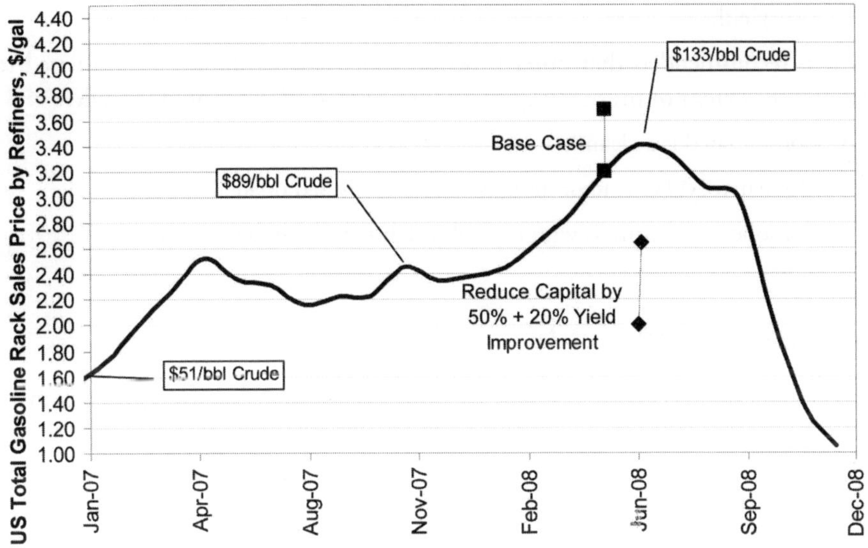

Figure 4-4 Effect of Capital Cost Reduction and Yield Improvement

Conclusions and Recommendations

A techno-economic analysis (TEA) of a conventional biomass to gasoline process was conducted, based on existing technology. This process produces an infrastructure ready fuel, with a vast available market. The simulation results showed that the estimated gasoline selling prices for the indirectly-and directly-heated gasifier based systems result in rack selling prices higher than the average of the gasoline refiners rack price in the past year. The specific conclusions are:

1. The conventional MTG process is not commercially competitive at this scale unless crude prices are well above $100/bbl.

2. The main capital costs areas are 1) gas cleanup steps (tar cracking, sulfur and other impurities removal and methane reduction by steam reforming) and 2) the MTG process, including methanol production, gas fractionation and HGT.

3. Sensitivity analysis indicated that:

    a. Very low feedstock costs are required to make the fixed-bed MTG process economic.

    b. Reducing the capital costs through consolidated gas cleanup and improved synthesis steps may significantly reduce the costs

    c. Economic viability will likely require reduced capital and improved yield.

It is recommended that other processing alternatives should be simulated to identify the extent of process improvements. Such options should include:

- Consolidated gas cleanup,

- Alternative MTG reactor designs such as fluid-bed or shell & tube,

- Consolidated synthesis with once through DME/Methanol reactors.

Acknowledgements

The authors thank DOE's biomass program for funding this work and acknowledge the modeling work performed by the National Renewable Energy Laboratory (NREL), which is publicly available and serves as the basis for gasification and syngas conditioning portion of the models.

References

1. Aden, A. 2008. Biochemical Production of Ethanol from corn Stover: 2007 State of Technology Model. NREL/TP-510-43205. National Renewable Energy Laboratory, Golden, CO. May 2008. http://www.nrel.gov/docs/fy08osti/43205.pdf

2. Aden, A., P. Spath, and A. Atherton. 2005. The Potential of Thermochemical Ethanol Via Mixed Alcohols Production. Milestone Completion Report, FY05-684. National Renewable Energy Laboratory, Golden, CO. October 2005.

3. Chang, C.D., 1992. The New Zealand Gas-to-Gasoline Plant: An Engineering Tour de force. Catalysis Today, 13(1992):103-111.

4. Cheng, W., Kung, H. 1994. Methanol Production and Use. Marcel-Dekker. 1994.

5. DKRW website 2008. http://www.dkrwenergy.com/fw/main/Home-140.html.

6. Evans, R.J., R.A. Knight, M. Onischak, S.P. Babu, 1998. Development of Biomass Gasification to Produce Substitute Fuels. Institute of Gas Technology for Pacific Northwest Laboratory, PNL-6518, Richland, WA.

7. EIA, 2008a. Natural Gas Monthly. Energy Information Administration, Office of Oil and Gas, U.S. Department of energy, Washington, DC. http://www.eia.doe.gov/oil_gas/natural_gas/data_publications/natural_gas_monthly/ngm.html

8. EIA, 2008b. Electric Power Monthly—August 2008. DOE/EIA-0226. Energy Information Administration, Office of Coal, Nuclear, Electric and Alternative Fuels, U.S. Department of Energy, Washington, DC. http://www.eia.doe.gov/cneaf/electricity/epm/epm.pdf

9. EIA, 2009. Topics for Petroleum Prices. Energy Information Administration, Office of Oil and Gas, U.S. Department of energy, Washington, DC. http://tonto.eia.doe.gov/dnav/pet/pet_pri_top.asp

10. EISA 2007. "U.S. Energy Independence and Security Act of 2007," Public Law Number 110-140, signed 19 December, 2007; Title II. 2007. frwebgate.access.gpo.gov/cgi-bin/getdoc.cgi?dbname=110_cong_bills&docid=f:h6enr.txt.pdf

11. Fiedler, E., G. Grossmann, D. B. Kersebohm, G. Weiss, and C. Witte. 2000. Methanol. Ullmann's Encyclopedia of Industrial Chemistry, Electronic Release, 7th ed., Wiley-VCH, Weinheim, 2007.

12. Grace & Co. 1982a. Preliminary Design and Assessment of a 50,000 BPD Coal-to-Gasoline-to-Methanol Plant. Process Engineering and Mechanical Design Reports Volume II of V. DOE/ET/14759-T1-Vol.2-14B, Prepared by W.R. Grace &Co. for the U.S. Department of Energy, August 1982

13. Grace & Co. 1982b. Preliminary Design and Assessment of a 12,500 BPD Coal-to-Gasoline-to-Methanol Plant. Process Engineering and Mechanical Design Reports Volume II of IV. DOE/ET/14759-T3-Vol.2-14B, Prepared by W.R. Grace &Co. for the U.S. Department of Energy, August 1982

14. Grace & Co. 1982c. Preliminary Design and Assessment of a 12,500 BPD Coal-to-Gasoline-to-Methanol Plant. Capital and Operating Costs Estimates Volume I of IV. DOE/ET/14759-T3-Vol.1-18B, Prepared by W.R. Grace &Co. for the U.S. Department of Energy, August 1982

15. Hamelinck, C.N., A.P.C. Faaij, 2001. Future prospects for production of methanol and hydrogen from biomass, NWS-E-2001-49, ISBN 90-73958-84-9, September 2001.

16. Hamelinck, C.N. and A.P.C. Faaij, 2002. Future Prospects for Production of Methanol and Hydrogen from Biomass. Journal of Power Sources, 111 (1):1-22. 18 September 2002.

17. Hamelinck, C.N., A.P.C. Faaij, H. den Uil, and H. Boerrigter. 2003. Production of FT Transportation Fuels from Biomass; Technical Options, Process Analysis and Optimization and Development Potential. NWS 90-393-3342-4. Utrecht University. The Netherlands, March 2003.

18. Kaneko, T., F. Derbyshire, E. Makino, D. Gray, and M. Tamura, 2002. Coal Liquefaction. Ullmann's Encyclopedia of Industrial Chemistry, Electronic Release, 7th ed., Wiley-VCH, Weinheim, 2007.

19. Ondry, G., 2008. Coal-to-Gasoline process will make U.S. debut. Chemical Engineering, 115(September 2008):16.

20. Phillips, S., A. Aden, J. Jechura, and D. Dayton, 2007. Thermochemical Ethanol via Indirect Gasification and Mixed Alcohol Synthesis of Lignocellulosic Biomass. NREL/TP-510-41168. April 2007.

21. Schuster, et al. 1985. Continuous Preparation of Ethanol. U.S. Patent 4,517,391. May 14, 1985.

22. Spath, P., A. Aden, T. Eggerman, M. Ringer, B. Wallace, J. Jechura, 2005. Biomass Hydrogen Production Detailed Design and Economics Utiltizing the Battelle Columbus Laboratory Indirectly Heated Gasifier. NREL/TP-510-37408. National Renewable Energy Laboratory, Golden, CO.

23. SRI PEP 2003 Yearbook International. SRI Consulting, United States. Menlo Park, CA. 2003.

24. SRI PEP 2007 Yearbook International. SRI Consulting, United States. Menlo Park, CA. 2007.

25. Tabak, S., X. Zhao, A. Brandl, and M. Heinritz-Adrian, 2008. An Alternative Route for Coal to Liquid Fuel - ExxonMobil Methanol-to-gasoline (MTG) Process. First World Coal-to-Liquids Conference, April 3-4, 2008, Paris, France.

Aluminum Hydride: A Reversible Material for Hydrogen Storage

Ragaiy Zidan, Brenda L. Garcia-Diaz, Christopher S. Fewox,
Andrew Harter, Ashley C. Stowe and Joshua R. Gray

ABSTRACT

Hydrogen storage is one of the challenges to be overcome for implementing the ever sought hydrogen economy. Here we report a novel cycle to reversibly form high density hydrogen storage materials such as aluminium hydride. Aluminium hydride (AlH_3, alane) has a hydrogen storage capacity of 10.1 wt% H_2, 149 kg H_2/m^3 volumetric density and can be discharged at low temperatures (< 100ºC).[1,2] However, alane has been precluded from use in hydrogen storage systems because of the lack of practical regeneration methods. The direct hydrogenation of aluminium to form AlH_3 requires over 105 bars of hydrogen pressure at room temperature and there are no cost effective synthetic means. Here we show an unprecedented reversible cycle to form alane electrochemically, using alkali metal alanates (e.g. $NaAlH_4$, $LiAlH_4$) in aprotic *solvents. To complete the cycle, the starting alanates can be regenerated by direct hydrogenation of the dehydrided alane and the alkali hydride being the other compound formed in the electrochemical cell. The process of forming NaAlH₄ from NaH and Al is well established in both solid state and solution reactions.[3-7] The use of adducting Lewis bases is an essential part of this cycle, in the isolation of alane from the mixtures of the electrochemical cell. Alane is isolated as the triethylamine (TEA) adduct and converted to pure, unsolvated alane by heating under vacuum.*

Discovering efficient and economic methods for storing hydrogen is critical to realizing the hydrogen economy. The US Department of Energy (DOE) is supporting research to demonstrate viable materials for on-board hydrogen storage. The DOE goals are focused on achieving a storage system of 6 mass% H_2 and 45 kg H_2/m^3 by 2010 and developing a system reaching 9 mass% H_2 and 81 kg H_2/m^3 by 2015.[8]

Researchers worldwide have identified a large number of compounds with high hydrogen capacity that can fulfill these gravimetric and volumetric requirements. Unfortunately, the majority of these compounds fail to fulfill the thermodynamic and kinetic requirements for on-board storage systems. Alane has the gravimetric (10.1 mass% H_2) and the volumetric (149 kg H_2/m^3) density needed to meet the 2010 DOE goals. In addition, rapid hydrogen release from alane can be achieved using only the waste heat from a fuel cell or a hydrogen internal combustion engine.[9] The main drawback to using alane in hydrogen storage applications is unfavorable hydriding thermodynamics. The direct hydrogenation of aluminium to alane requires over 10^5 bars of hydrogen pressure at room temperature as shown in equation (1). The impracticality of high hydriding pressures has precluded alane from being considered as a reversible hydrogen storage material.

$$Al + \frac{3}{2}H_2 \xrightarrow{10^5\,bar\,25°C} AlH_3 \tag{1}$$

The typical formation route for alane is through the chemical reaction of an alanate compound (e.g. -LiAlH₄) with aluminium chloride in diethyl ether:

$$3LiAlH_4 + AlCl_3 \xrightarrow{ether} 4AlH_3 - Et_2O + 3LiCl \tag{2}$$

This reaction yields dissolved alane etherate, $AlH_3 \bullet Et_2O$, and precipitates lithium chloride. Alane can then be separated from the ether by heating en vacuo.[2, 10-12] The synthesis of AlH_3 by this methods also results in the formation of alkali halide salts such as LiCl. The formation of these salts becomes a thermodynamic sink because of their stability. For a cyclic process, lithium metal must be recovered from lithium chloride by electrolysis of a LiCl/KCl melt at 600°C and requires at least −429 kJ/mol of energy equivalent to the heat of formation and heat of fusion of LiCl.[13] In addition, -117 kJ/mol is required to regenerate LiAlH$_4$. The large amount of energy required to regenerate AlH$_3$ from spent aluminium and the alkali halide makes this chemical synthesis route economically impractical for a reversible AlH$_3$ storage system.

For these reasons, our research focused on developing a novel cycle to cost effectively regenerate AlH$_3$ under practical conditions. The proposed cycle uses electrolysis and catalytic hydrogenation of products to avoid both the high hydriding pressure for aluminium and the formation of stable by-products such as LiCl. The cycle, shown in Figure 1, utilizes electrochemical potential to drive the formation of alane and alkali hydride from an ionic alanate salt (e.g. sodium alanate, NaAlH$_4$). The starting alanate is regenerated by reacting spent aluminium with the byproduct alkaline hydride in the presence of a titanium catalyst under moderate hydrogen pressure. In comparison to chemical methods, using the electrochemical cycle should, in principle, cost the heat of formation of NaAlH$_4$, which is -115 kJ/mol.

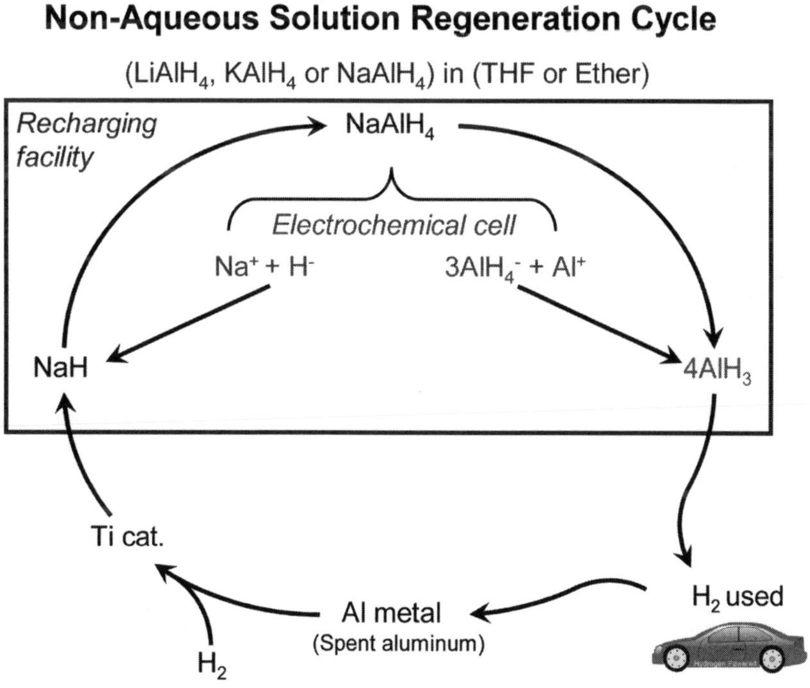

Non-Aqueous Solution Regeneration Cycle

(LiAlH$_4$, KAlH$_4$ or NaAlH$_4$) in (THF or Ether)

Figure 1. Proposed reversible fuel cycle for alane. All components of the electrochemical process can be recycled to continually afford a viable solid state storage material for the hydrogen economy.

The electrolysis reaction is carried out in an electrochemically stable, aprotic, polar solvent such as THF or ether. $NaAlH_4$ is dissolved in this solvent, forming the ionic solution (Na^+/AlH_4^-/THF) which is used as an electrolyte. Though not directed at the regeneration of alane, elaborate research and extensive studies on the electrochemical properties of this type of electrolyte has been reported.[14, 15] Although attempts in the past were made to synthesizes alane electrochemically,[16-18] none have shown isolated material or a characterized alane product.

Thermodynamic calculations were made to determine the reduction potentials for possible electrochemical reactions with $NaAlH^4$ in an aprotic solution (THF) with an aluminium electrode. From the half reaction potentials of all possible reactions in solution (shown in the supplementary information), the cell voltage for alane formation was calculated and a theoretical cyclic voltammagram was constructed (see Figure 2). The onset potential of the $NaAlH_4$ decomposition reaction to Al is shifted to approximately -2.05 V vs. SHE indicating greater than 250 mV of overpotential. A limiting current of 9.7 mA was reached near -1.5 V. AlH_3 formation is confirmed by observing the rapid precipitation of the insoluble $AlH_3 \bullet$ TEDA complex when triethylenediamine (TEDA) is added to the electrochemical cell. $AlH_3 \bullet$ TEDA was identified by Powder XRD. Two separate reaction mechanisms can produce alane at the aluminium electrode. One mechanism is the oxidation of the alanate ion to produce alane, an electron, and hydrogen as shown in equation (3):

$$AlH_4^- \rightarrow AlH_3(THF) + \frac{1}{2}H_2 \uparrow + e^- \tag{3}$$

The other possible electrochemical reaction is the reaction of the alanate ion with the aluminium anode to form alane. In this reaction, no hydrogen is evolved and the reaction will consume the Al electrode as in equation (4).

$$3AlH_4^- + Al \rightarrow 4AlH_3(THF) + 3e^- \tag{4}$$

Experimental observations confirm that under certain conditions the anode is consumed as shown in Figure 3 as discussed below.

Once the correct operating voltage for the formation of alane was calculated, constant voltage experiments were performed. During these experiments, the current was steady and increased slightly with time. The electrochemical production of alane is not slowed by the formation of AlH_3. In contrast to previous reports, no visible signs of alane formation are observed and the alane produced by our method is completely dissolved in solution as a THF adduct.[17] During electrolysis, dendritic material was deposited on the platinum counter electrode. This material was collected and determined to be Na_3AlH_6 from XRD data.

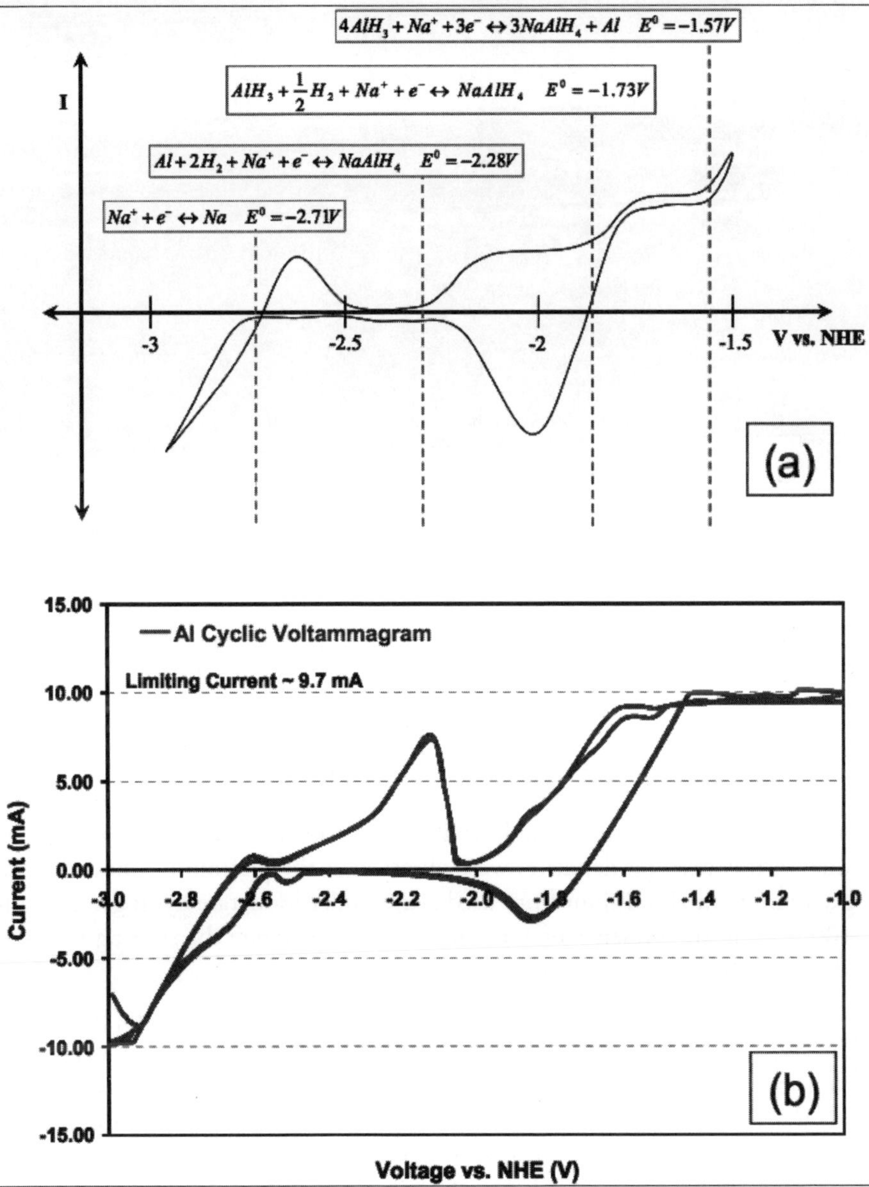

Figure 2. Experimental and hypothetical cyclic voltammagrams for the electrochemical formation of alane. (a) A hypothetical cyclic voltammagram was formulated from the equilibrium potential data for possible reactions and the anticipated state of each species generated. (b) Bulk electrolysis experiment at an aluminum wire electrode for a cell containing a 1.0 M solution of NaAlH4 in THF at 25oC.

Figure 3. The aluminum electrode is consumed during the reaction at +1.5 V vs. SHE.

Experiments were conducted to determine the feasibility of plating sodium at the platinum cathode to complete the cycle. The platinum cathode and aluminium anode potentials were -2.89 V and -1.31 V respectively. Plating of Na metal was observed at the cathode while alane was produced at the aluminium anode. In this case, no dendrites were observed at the platinum cathode. Reacting sodium with aluminium from used alane under moderate hydrogen pressure (~100 bars) in the presence of a Ti catalyst will regenerate the starting material, $NaAlH_4$, leading to a reversible cycle.

In addition to determining the electrochemical processes for producing AlH_3, recovering AlH_3 from the solution is a major step of this cycle. The separation of alane from $AlH_3 \bullet Et_2O$ is well established and affords pure AlH_3.[10-12] However, separation of the $AlH_3 \bullet THF$ adduct is more complicated because it decomposes when heated under vacuum. Therefore, adducts such as triethylamine (TEA) were added to the reaction product to stabilize the alane during purification. Adduct free alane is recovered by heating the neat liquid $AlH_3 \bullet TEA$ en vacuo.

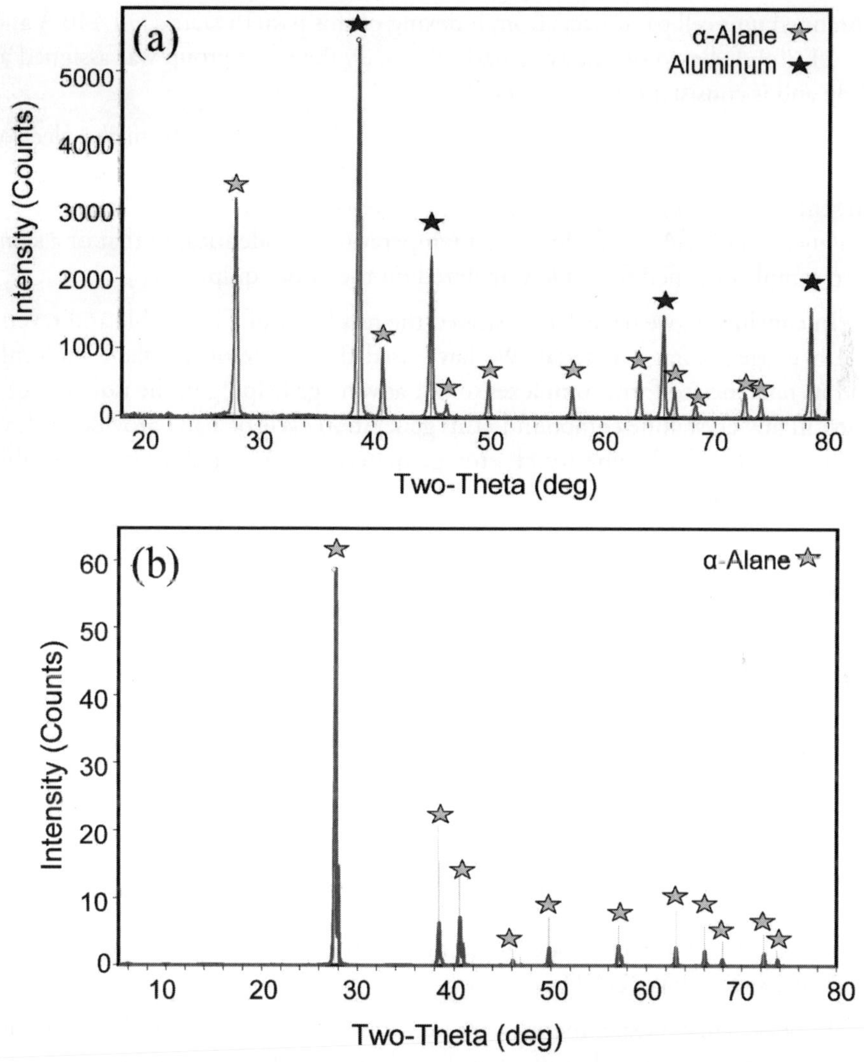

Figure 4. XRD patterns for products recovered from an electrochemical cell. (a) Alane separated from reaction mixture as the THF adducts. When heated under vacuum to remove THF, the solid partially decomposes, losing hydrogen and affording aluminum. (b) Alane is separated using triethylamine to stabilize the adduct, prevent polymerization, and increase yield.

Alane recovered from the electrochemical cells was characterized by powder X-ray diffraction, Raman spectroscopy, and thermal gravimetric analysis (TGA). Powder X-ray diffraction patterns data for two different separation methods are shown in Figure 4. Heating the $AlH_3 \bullet$ THF product after removing left over starting materials (Figure 4a) showed the presence of a large amount of aluminum as well as α-alane. Separation using the TEA (Figure 4b) yields only α-alane. The unrefined unit cell parameters from indexing of this pattern were a = 4.446 Å and c = 11.809 Å. Based on the systematic absences, the space group was assigned as R-3c and is consistent with α-alane.[19]

Raman spectra and TGA data collected for AlH3 isolated from the electrochemical cell are available in the supplementary information. Raman modes present at 510, 715, 1050, and 1515 cm^{-1} are consistent with the literature for α-alane.[20] The TGA weight loss onset temperature was identical to that of a standard sample obtained from Dow analyzed on the same equipment.

In conclusion, we have demonstrated the feasibility of a recyclable and reversible hydrogen storage material. We have used the nature of the alane molecule and its tendency to form complexes to our advantage helping in the isolation of a pure, highly crystalline compound. This generation cycle of alane provides a clean facile route to a high capacity H_2 storage material while avoiding unrecoverable thermodynamic costs.

Methods

Electrochemical Preparation of AlH$_3$

Cell preparation and electrochemical experiments were performed in an argon environment using traditional air sensitive techniques. Electrochemical alane generation experiments were performed using an air-tight three electrode electrochemical cell. The working electrode was an aluminum sheet. Prior to experiments, the aluminum electrode was sanded and rinsed with acetone in an inert environment to remove any oxide layer on the electrode. A "leak-free" 3M KCl Ag/AgCl reference electrode (Warner Instruments) was used to measure the potential of the working electrode. A platinum foil counter electrode was used. The electrolyte was 1.0M $NaAlH_4$.

Electrochemical experiments were performed with a CH Instruments 660C potentiostat. After assembly of the cell and before cell characterization or bulk electrolysis, the working electrode was held at a constant potential of +1.3 V vs. SHE. Cyclic voltammograms were performed between -3.0 V and -1.0 V vs. SHE. Bulk electrolysis experiments were carried out at -1.5 V vs. SHE and experiments to investigate aluminium consumption were performed at +1.5 V vs. SHE.

Chemical Separation and Isolation of Alane

The product solution was decanted from the cell apparatus and filtered. An equal volume of pentane was added to this solution and stirred overnight. The resulting precipitate was filtered on a frit over celite. To the filtrate, 100 mL of a 4:1 mixture of triethylamine and pentane was added, stirred overnight and filtered over celite. The solvent was removed en vacuo leaving a viscous clear liquid. This liquid was heated at 65ºC in a sand bath for 2 hours to remove the remaining triethylamine. Approximately 15 mL of benzene was added to the flask and stirred. The solution was placed under vacuum and heated again at 65ºC. After another 2 hours the product rendered was a fine blue grey powder. This powder was rinsed with 3 times with 20 mL portions of THF and dried under vacuum.

Acknowledgements

We would like to thank DOE office of EERE (Energy Efficiency and Renewable Energy) Contract Number EB4202000. Our gratitude is extended to Dr. Robert Lascola for his expertise in characterizing materials using Raman spectroscopy. Special thanks are given to Joseph G. Wheeler for whose assistance and guidance we are extremely grateful.

Contribution of Authors

R.Z. was responsible for the idea of the reversible cycle of the regencration of alane using organic solvent and complex hydride salt and directing the project.

B.L.G-D. was responsible for the design of electrochemical cells and experiments as well as the analysis of experimental results. She was also responsible for the co-ordination of laboratory operations and interpretation of analytical experiments.

C.S.F. was responsible for chemical isolation and wet chemistry treatment of the electrochemical cells. He has also assisted in conducting and interpreting powder x-ray diffraction data.

R.Z., B.L.G-D and C.S.F. were responsible for writing the manuscript.

A. H. Started preliminary studies, conducted experiments, and characterized resultant products of the electrochemical experiments

A. S. helped in designing several parts of the apparatus, conducted experiments, and characterized resultant products of the electrochemical experiments.

J.R.G. contributed to experimental design and execution, as well as data analysis and interpretation.

References

1. Beattie, S., Humphries, T., Weaver, L. & McGrady, S. Watching the dehydrogenation of alane (AlH3) in a TEM. American Physical Society, 2008 APS March Meeting, New Orleans, Louisiana March 10–14, 2008, Abstract # L36.004.

2. Graetz, J. & Reilly, J. J. Decomposition kinetics of the AlH3 polymorphs. Journal of Physical Chemistry B, 109, 22181–22185 (2005).

3. Bogdanovic, B. & Schwickardi, M. Ti-doped alkali metal aluminium hydrides as potential novel reversible hydrogen storage materials. J. Alloys Compd., 253, 1-9 (1997).

4. Zidan, R. A., Takara, S., Hee, A. G. & Jensen, C. M. Hydrogen cycling behavior of zirconium and titanium-zirconium-doped sodium aluminum hydride. J. Alloys Compd., 285, 119–122 (1999).

5. Jensen, C. M., Zidan, R., Mariels, N., Hee, A. & Hagen, C. Advanced titanium doping of sodium aluminum hydride: segue to a practical hydrogen storage material. Int. J. Hydrogen Energy, 24, 461–465 (1999).

6. Bogdanovic, B., Brand, R. A., Marjanovic, A., Schwickardi, M. & Tolle, J. Metal-doped sodium aluminium hydrides as potential new hydrogen storage materials. J. Alloys Compd., 302, 36–58 (2000).

7. Sandrock, G., Gross, K. & Thomas, G. Effect of Ti-catalyst content on the reversible hydrogen storage properties of the sodium alanates. J. Alloys Compd., 339, 299–308 (2002).

8. Hydrogen, Fuel Cells & Infrastructure Technologies Program. Multi-Year Research, Development and Demonstration Plan: Planned Program Activities for 2005–2015. http://www1.eere.energy.gov/hydrogenandfuelcells/mypp/ (2007).

9. Sandrock, G. et al. Alkali metal hydride doping of alpha-AIH(3) for enhanced H-2 desorption kinetics. J. Alloys Compd., 421, 185–189 (2006).

10. Finholt, A. E., Bond, A. C. & Schlesinger, H. I. Lithium aluminum hydride, aluminum hydride and lithium gallium hydride, and some of their applications in organic and inorganic chemistry. J. Am. Chem. Soc., 69, 1199–1203 (1947).

11. Brower, F. M. et al. Preparation and properties of aluminum-hydride. J. Am. Chem. Soc., 98, 2450–2453 (1976).

12. Kato, T., Nakamori, Y., Orimo, S., Brown, C. & Jensen, C. M. Thermal properties of AlH3-etherate and its desolvation reaction into AlH3. J. Alloys Compd., 446, 276–279 (2007).

13. Wietelmann, U. & Bauer, R. J. in Ullmann's Encyclopedia of Industrial Chemistry 2005 (Wiley-VCH, Weinheim, 2005).

14. Senoh, H., Kiyobayashi, T., Kuriyama, N., Tatsumi, K. & Yasuda, K. Electrochemical reaction of lithium alanate dissolved in diethyl ether and tetrahydrofuran. J. Power Sources, 164, 94–99 (2007).

15. Senoh, H., Kiyobayashi, T. & Kuriyama, N. Hydrogen electrode reaction of lithium and sodium aluminum hydrides. Int. J. Hydrogen Energy, 33, 3178–3181 (2008).

16. Adhikari, S., Lee, J. J. & Hebert, K. R. Formation of aluminum hydride during alkaline dissolution of aluminum. J. Electrochem. Soc., 155, C16-C21 (2008).

17. Alpatova, N. M. Kessler, Y. M., Dymova, T. N. & Osipov, O. R. Physicochemical properties and structure of complex compounds of aluminum hydride. Russ. Chem. Rev., 37, 99–114 (1968).

18. Clasen, H., German P. 1141 623 (1962).

19. Turley, J. W. & Rinn, H. W. Crystal Structure of Aluminum Hydride. Inorganic Chemistry, 8, 18-& (1969).

20. Wong, C. P. & Miller, P. J. Vibrational spectroscopic studies of alane. J. Energetic Mat., 23, 169–181 (2005).

Solving the Structure of Metakaolin

Claire E. Whitea, John L. Provisa, Thomas Proffenb,
Daniel P. Rileyc and Jannie S. J. van Deventera

ABSTRACT

Metakaolin has been used extensively as a cement additive and paint extend-er, and recently as a geopolymer precursor. This disordered layered alumino-silicate is formed via the dehydroxylation of kaolinite. However, an accurate representation of its atomic structure has bever before been presented. Here, a novel synergy between total scattering and density functional modeling is pre-sented to solve the structure of metakaolin. The metastable structure is eluci-dated by iterating between least¬squares real-space refinement using neutron pair distribution function data, and geometry optimization using density functional modeling. The resulting structural representation is both energeti-cally feasible and in excellent agreement with experimental data. This accu-rate structure of metakaolin provides new insight into the local environment of the aluminum atoms, with evidence of the existence of tri¬coordinated alu-minum. By the availability of this detailed atomic description, there exists the opportunity to tailor chemical and mechanical processes involving metaka-olin at the atomic level to obtain optimal performance at the macro-scale.

Given that nanostructured, amorphous and disordered materials are ever gaining in technological importance, it is becoming essential to understand these materials at different length scales, including at the atomic level. By understanding local atomic environments, it will become possible to tailor the behavior of a material like metakaolin undergoing a chemical or mechanical process, leading to economic and efficiency savings in areas which are important to society.

Portland cement is produced through the calcination of limestone and currently accounts for around 8% of man-made CO_2 emissions. This chemistry intrinsically constrains the reduction of emissions achievable through improvements to the existing production route.[1] A major shift in cement chemistry is required in order to enable sustainable production of construction materials in a carbon-constrained world. Alkali-activated inorganic polymer cement, otherwise known as 'geopolymer', is increasingly being proven to be a suitable alternative material, and is now entering commercial-scale production in Australia and elsewhere in the world. The production of geopolymer cement emits 80% less CO_2 than Portland cement, indicating that it has the potential to playa major role in the 'greening' of the world's construction industry.[2] However, geopolymer cement reactions are not well understood (as is in fact the case with Portland cement) owing to the lack ofunderstanding of reactions between the disordered aluminosilicate precursors such as metakaolin and alkaline gel. Hence, the aim of this paper is to elucidate the atomic structure of metakaolin as a disordered layered aluminosilicate material. Knowledge of this atomic structure will enable macro-scale optimization of metakaolin synthesis and chemical reactions during geopolymer mix design.

There appears to be an almost complete deficit in the literature regarding the atomic structure of metakaolin. This material is synthesized via the calcination of kaolinite, with various industrially important uses apart from geopolymer cement: primarily as an intermediate phase in ceramic processing,[3] but also as a paint extender, cement additive[4] and in the manufacture of molecular sieves.[5] Given the extent to which metakaolin is used globally, it is surprising that not more has been done to elucidate the local atomic structure of this material.

Previous investigations into various aspects of metakaolin have made advances towards understanding this material. It is well known that conventional diffraction analysis shows a lack of long-range order, as evidenced by the absence of Bragg reflections. Various structural models have been proposed since the late 1950s,[6-8] however these are incompatible with the high degree of disorder which is observed experimentally. Nevertheless, there have been numerous investigations into the local environment of meta kaolin using techniques such as NMR.

Investigations involving NMR spectroscopy have reported various peak assignments and coordination distributions, especially in the case of ^{27}Al MAS NMR spectra. The reasons for this have been explained in part by Rocha, and are related to the relatively modest and differing MAS spinning rates and magnetic fields used by various authors.[9] This uncertainty in the local environment of aluminum will be resolved in this paper, which is pivotal to our understanding of geopolymer gel formation.

In order to obtain an accurate representation of the structure of metakaolin the techniques employed are the cutting edge of current day research. The current state of the art in experimental atomic structure solution of complex and disordered materials is the technique of total scattering,[10] which enables real-space structure refinements of disordered crystalline materials, nanoparticles and amorphous materials. In this study the technique of total scattering and utilization of the subsequent pair distribution function (PDF) is coupled with density functional modeling. Density functional theory (DFT) modeling provides a structure model which is energetically feasible but not necessarily in agreement with experimentally determined atom positions, while least-squares real-space refinement of the PDF will provide a set of atom positions which provide a PDF matching that is observed experimentally, but without considering any thermodynamic effects when selecting these positions. It is proposed that an iterative process involving these two techniques will be able to converge to a final structure which satisfies both the requirement to match experimental PDF and the requirement to be energetically feasible. Such results will enable modeling of the gel-phase conversion of disordered aluminosilicates and underpin the development of geopolymer cement as a viable Portland cement replacement.

The metakaolin sample studied was prepared from high-purity kaolinite via calcination at 750°C, and neutron diffraction data were collected up to high momentum transfer to produce a high quality PDF (see Methods). DFT modeling was performed on a supercell atomic structure (see Methods). To obtain the local structure of meta kaolin, an iterative process was used, alternating between real space least-squares refinements of neutron PDF data and DFT full-geometry optimizations, until convergence was reached.

The intennediate fits to the PDF data obtained during the iteration between PDF fitting and DFT calculation are shown in Fig. 1. The sidebar of this Figure also gives the total energy of the calculated structure at each step, which indicates whether or not each atomic arrangement is energetically feasible. It is apparent that the use of DFT alternating with neutron PDF data refmement moves the atomic structure from an energetically feasible but experimentally implausible configuration (Fig. 1a) to a configuration that is still energetically feasible but more experimentally plausible (Fig. 1c). The most noticeable discrepancy between the calculated and experimental PDFs in Fig. 1a is the large peak at ~5.5A. This feature correlates more closely with the experimental data in Fig. 1c, showing that the PDF refinement conducted in between the two rounds ofDFT full-geometry optimization greatly enhances the fit to the experimental data. However, the structure in Fig. 1c is once again in an energetically favorable configuration, which proves the need for both experimental and theoretical input in order to obtain a sensible result overall.

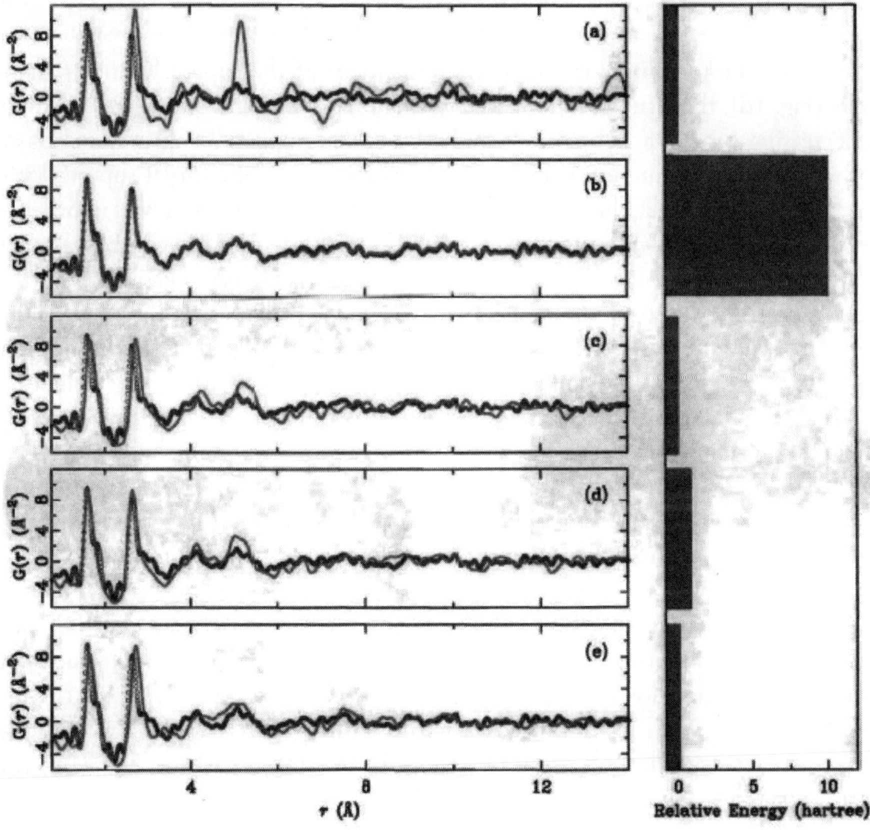

Figure 1. Pair distribution functions of metakaolin atomic structure, and associated total energies (sidebar; relative to lowest energy configuration at -38656.4 Ha). Blue circles represent the experimental PDF data; red lines are the PDFs calculated from the structures generated at various stages in the iterative process (a -e). The output structure of DFT becomes the input structure for PDF least-squares real space refinement, and so on. The total energy of each structure was calculated using DFT. (a) Initial DFT full geometry optimization. (b) Least-squares real space refinement of all atom positions. Note that the total energy of this atomic configuration implies that the structure is not energetically feasible. (c) DFT full-geometry optimization. (d) Least-squares real space refinement of model supercelliattice parameters. Note the small increase in total energy from previous iteration. (e) DFT full-geometry optimization, which produces a final metakaolin structure that is not only energetically feasible, but also maintains good agreement with the experimental PDF.

Although Fig. 1b shows the best agreement between the atomic structure (red) and experimental data (blue) ofany of the fits presented, the total energy of this structure is far from the minimum energy for metakaolin (as best as it can be defined for a metastable structure), as seen in the sidebar. Hence, this structure is not energetically feasible. This result is not surprising since all atoms were allowed to move freely from their original positions during the least-squares refinement, without any constraints on bond lengths or bond angles. However, subjecting this structure to DFT full-geometry optimization the structure attained the minimum energy (Fig. 1c), as well as showing better agreement with the neutron PDF data than the initial DFT -generated structure shown in Fig. 1a.

By then performing a least-squares refinement of the supercelliattice parameters (Fig. 1d), the structure underwent a small reduction in volume (8.4%), with a decreasing by 2.9%, b by 3.1 % and c by 2.4%. Supercell angles were not allowed to vary. A reduction similar to this is expected as the initial supercell was derived from twelve kaolinite unit cells, with water molecules removed in order to attain the correct stoichiometry of metakaolin (kaolinite unit cell contents change from $Al_4Si_4O_{18}H_8$ to $Al_4Si_4O_{14.5}H$). During the dehydroxylation ofkaolinite to form metakaolin, the removal of water molecules would be expected to cause the remaining structure to collapse slightly. Hence, this lattice parameter refinement is necessary to obtain the correct density. DFT was not used for supercelliattice parameter refinement since it is well known that DFT has difficulty in reproducing long-range dispersive interactions such as the long-range tails of van der Waals interactions (see for example ref.[11] for a discussion of this issue), aJthough significant progress in this area has been made recently and may open increased opportunities for future investigations to use this method.[12] Due to this difficulty, the unit cell would change size unrealistically if allowed to vary freely. Hence, here the change in supercell dimensions is obtained from the neutron PDF data by real-space least-squares fitting, holding all relative atom positions and supercell angles fixed while allowing supercelllattice parameters to vary, giving a structure as shown in Fig. Id. This optimized structure is then subjected to a final round of full-geometry optimization using DFT, to give the final structure (Fig. 2) with a fit to experimental PDF data as shown in Fig. 1e.

From Fig. 2a, it is evident that the resulting supercell atomic structure still retains 1:1 layering of Si and Al, although the layers are buckled. When viewed down the c-axis it is apparent that the buckling of the layers affects the positions of the atoms in all three dimensions, not just in the c-direction. This is visible in Fig. 2b, as the two pairs of layers in the supercell are no longer aligned Si with Si and Al with Al, which was the case in the parent kaolinite. The silica rings do not change significantly in connectivity during dehydroxylation, as all the -OH groups in kaolinite are bonded to the alumina layer, but shift and buckle to accommodate the large changes that occur throughout the alumina layers. This provides strong justification for the need to use a multi-layer supercell rather than a single Si-Allayer pair; this behavior would not be observable except with the use of a supercell, and has never before been demonstrated in any structural determination of metakaolin.

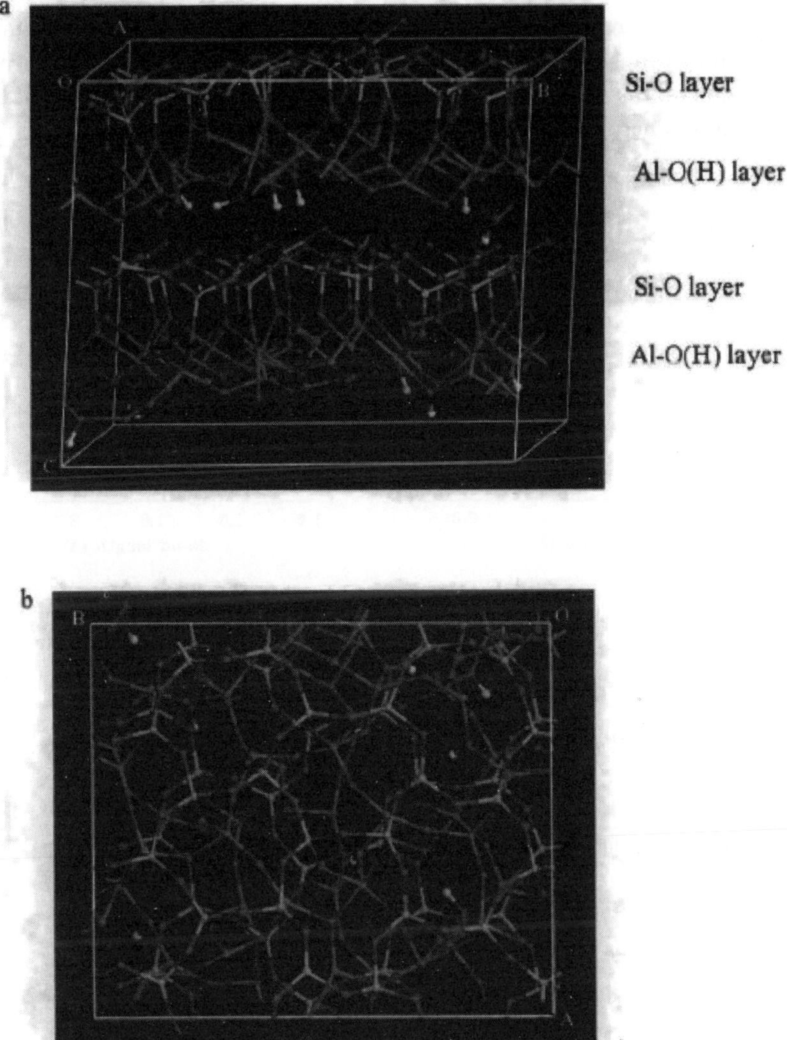

Figure 2. Atomic structure representation of meta kaolin. Silicon is yellow, aluminum is purple, oxygen is red, and hydrogen is white. (a) Side-on view showing the buckling of the layers, which is responsible for the lack of Bragg peaks in the X-ray and neutron diffraction patterns of metakaolin. (b) View along the c-axis, illustrating the subtle differences between the two silica layers in the supercell, and the more obvious differences between the alumina layers.

Metakaolin is a metastable structure, and the large supercell with a significant degree of disorder means that the free energy surface shows a very broad and flat 'valley' with very many configurations that are approximately equal in energy. The comparison with experimental PDF data provides a key means of selecting the appropriate configuration from among these possibilities. So, while it calU10t be stated that the structure presented here is a unique solution to the problem -certainly, a different structure would have been obtained if a different selection of hydroxyl groups were removed from the structure prior to the first full-geometry optimization -it must be considered to be in at least some way representative of the structural motifs present in metakaolin, and this is the first time such detailed information has been made available.

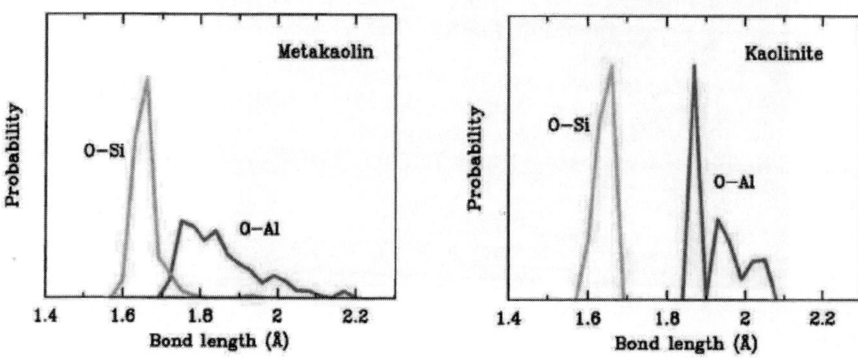

Figure 3. Histograms of bond lengths in metakaolin and kaolinite. The O-Si bond lengths remain mostly unchanged, whereas the O-AI bond length distribution broadens considerably. Kaolinite structure from ref.[13]

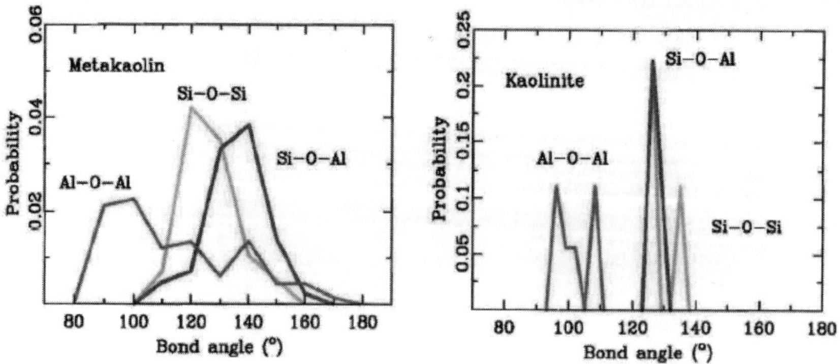

Figure 4. Histograms of bond angles in metakaolin and kaolinite. Bond angle distributions in metakaolin are broader than kaolinite, especially when comparing the angles containing aluminum. Kaolinite structure from ref.[13]

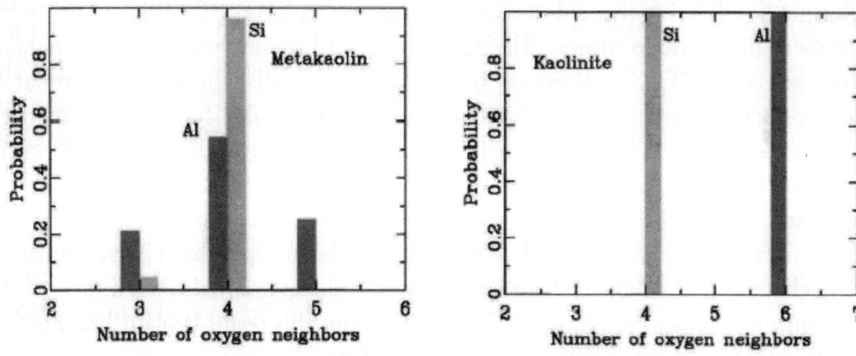

Figure 5. Histogram of the probability of finding N oxygen atoms around the silicon (yellow) and aluminum (purple) atoms. Cutoff radii were r_{Al-O} = 2.2 Å and r_{Si-O} = 2.0 Å. The histogram of meta kaolin' indicates the presence of both III-coordinated aluminum and silicon atoms, and the absence of VI-coordinated aluminum. Kaolinite structure from ref.[13]

In order to gain a better understanding of the resulting atomic structure of metakaolin, bond lengths (Fig. 3), bond angles (Fig. 4) and atom coordination numbers (Fig. 5) were determined. In each Figure the corresponding data for kaolinite are shown in order to aid understanding of the changes that occur due to dehydroxylation. Apart from the similarities with kaolinite, the bond lengths shown in Fig. 3 are also approximately similar to the values commonly observed in aluminosilicate glasses.[14-17] Bond length distributions in metakaolin are somewhat skewed (although still generally unimodal) due to the range of coordinations present. The mean O-Al bond length (1.9 ± 0.1 Å) is slightly longer (by ~0.1 Å) than those reported in the glass literature for charge-balanced tetrahedral O-Al; the absence of charge-balancing alkali or alkaline earth cations in metakaolin will be expected to alter the covalent/ionic nature of the bonds and therefore influence the bond length distribution. The fact that the O-Si bond length distribution is more similar to kaolinite than is the O-Al bond length distribution further highlights the fact that the silica layers in metakaolin undergo less alteration than the alumina layers due to dehydroxylation.

Fig. 4 illustrates that the metakaolin structure encompasses a large range of bond angles for each bond type, especially for angles containing aluminum atoms. Investigations into the bond angles of alumino silicate glasses by Benoit et al.[15] (Si-O-Si and Si-O-Al) exhibit very similar bond angle averages to the values presented in this investigation. There is little literature reporting Al(IV)-O-Al(IV) bond angles, as Lowenstein's "rule"[18] indicates that such bonds are strongly disfavored in aluminosilicate frameworks not containing explicit layers.

For analysis of coordination numbers, the defined distances used to calculate the probability are: r_{Al-O} = 2.2 Å, and r_{Si-O} = 2.0 Å, reflecting the possibility oflonger bonds being formed to aluminum atoms with more flexible coordination numbers than silicon atoms. The resulting coordination numbers for metakaolin are given in Fig. 5. These coordination number distributions are similar to those presented by Benoit et al.,[15] where ab-initio molecular dynamics simulations were performed on a Ca-aluminosilicate melt, with their results also showing the presence of three-coordinated Al and Si atoms. Of significant interest for the understanding of the metakaolin structure is the evidence that three-coordinated aluminum atoms exist, which has never before been reported. Given the structural changes that occur in the alumina layered during dehydroxylation, and the tendency for silica to retain its $SiO2$ configuration, this result is plausible. Further investigations are required to confirm this finding.

The coordination numbers for aluminum determined for this metakaolin atomic structure (Fig. 3) do in general agree with the NMR results of Rocha.[9] Variations occur in the fraction of aluminum which is 6-coordinated, which is not surprising given the range of calcination temperatures which can be used to make metakaolin. Here, metakaolin has been defined as kaolinite that has undergone 12.5% weight reduction through the removal of 7 in 8 hydrogen atoms, and 3.5 in 18 oxygen atoms. With only one in eight hydrogen atoms remaining, there does not appear to be scope for the presence of 6-coordinated Al within the metakaolin structure. It therefore seems likely that such sites are either located within residual kaolinite-like environments where dehydroxylation is locally incomplete, or in regions where the progression towards mullite (which also contains octahedral Al) is further advanced.

The structure of metakaolin presented here still retains the layers inherent in kaolinite even though large disturbances occur due to dehydroxylation. There were no specific constraints imposed requiring the retention of layering during the DFT and real-space refinements, and hence these results show that the structure does thermodynamically prefer to maintain the layers. It has also been shown by Rocha et al.[19] that metakaolin is able to partially rehydrate under hydrothermal conditions to reform kaolinite. Water molecules are therefore apparently able to infiltrate the residual layered structure and reverse to some extent the straining of the alumina layer. Further DFT modeling would be advantageous to simulate this rehydration on the atomic level, which would promote understanding of both dehydration and rehydration processes.

The availability of an accurate representation of the atomic structure of metakaolin enables further simulations to be performed investigating the dissolution of this structure and subsequent formation of other materials. Dissolution in various environments is of great importance, e.g. the production of zeolites from metakaolin, the addition of metakaolin to concrete, and the conversion of metakaolin and other disordered aluminosilicates into geopolymer concrete. Through understanding chemical processes at the atomic level, this gives rise to the opportunity to modify the local environment prevalent in the materials to achieve optimal performance at the macro-scale. Likewise, understanding the changes that occur upon heating of a layered aluminosilicate at the atomic level will lead to advances in the technology of ceramic production. This can be achieved through tailoring the heating rates, temperatures and the use of additives to obtain the optimal final structures (both crystallographic and in terms of macroscopic geometries) in firing of kaolinite-based ceramic bodies.

In terms of the understanding of the metakaolin structure, one important outcome of this study is evidence of the layers and their buckling as a result of dehydroxylation. It is this buckling of the layers which is responsible for the atomic disorder and subsequent absence of Bragg peaks in the diffraction pattern. As kaolinite undergoes dehydroxylation, water molecules are removed from the alumina layers, which results in strained aluminum sites. As a result, these aluminum atoms reorganize to relieve as much of the strain as possible. This in turn requires all atoms to shift in position, which leads to the buckling of both alumina and silica layers. The alumina layers undergo the greatest degree of reorganization, with the silica layers accommodating for these changes. This phenomenon has never before been explicitly shown at the atomic scale for metakaolin. Hence, this study of the atomic structure of metakao lin is an essential step forward in the structural analysis and understanding of complex disordered layered materials, which will underpin the analysis of their use in producing technologically important materials including ceramics and sustainable concretes.

In summary, we have determined an accurate representation of the atomic structure of metakaolin. By understanding the material at the local structure level there exists the potential to optimize the manner in which it is used at the industrial scale. The structure still retains the 1:1 layering inherent of the parent material kaolinite, although the layers are now buckled locally. The environment of aluminum is particularly important, as the strained nature of the alumina layers is responsible for the high reactivity of this material, especially in geopolymeric gels. It has been shown that the local structure ofmetakaolin contains a small percentage of three-coordinated aluminum, providing additional evidence that the alumina layers are strained. It has been shown that by using this iterative process of least-squares real-space refinement using neutron PDF data and energy minimization using DFT it is possible to attain a result for a complex material that is both energetically and experimentally favorable. The potential of this methodology is by no means limited to metakaolin, or even to disordered layered aluminosilicates. Application of a similar methodology to other technologically important "amorphous" materials can be expected to bring 'order to disorder' by enabling the analysis and manipulation of complex structures which have to date defied conventional analytical techniques.

Methods

Density functional modeling of metakaolin was carried out by performing full-geometry optimization using the code DMol[3] v4.2 (Accelrys, San Diego, CA).[20] Computations were run on 16 parallel processors, using the Australian Partnership for Advanced Computing (APAC) facility (now known as the National Computational Infrastructure National Facility), hosted by the Australian National University, Canberra. A previous investigation on kaolinite revealed that the appropriate exchange-correlation functional for aluminosilicates available in DMol[3] is the generalized gradient approximation BLYP,[13] hence this functional was used to model metakaolin. The numerical basis set used was double numerical (two atomic orbital for each occupied atomic orbital) plus a polarization d-function on all non-hydrogen atoms, and a polarizationp-function on all hydrogen atoms (DNP)[20] to account for hydrogen bonding. No pseudopotentials or effective core potentials were used. Convergence thresholds were set at 7.5×10^{-3} hartree for energy and 0.02 hartree/Å for maximum force. An SCF convergence of 10^{-5} hartree was used, along with $1 \times 1 \times 1$ k-point sampling and 0.05 hartree smearing to aid convergence. While such a degree of smearing is considered to be quite high, it was necessary here due to the strained state of the structure. The supercell used to model metakaolin was derived from a $3 \times 3 \times 2$ unit cell slab of kaolinite (at a temperature of 15 K), with cell parameters a = 15.4668Å, b = 17.8678Å, c = 14.7480Å, α = 91.926°, β = 105.046°, γ = 89.797°.[13] The initial structure used as input was obtained by systematically removing 84 of the 96 hydrog-n atoms and 42 of the 216 oxygen atoms from the kaolinite supercell. This removal amount is given by TGA results of the dehydroxylation of kaolinite. All inner hydrogen atoms were removed, with the remnant hydroxyls residing as the inner surface hydroxyls separating the aluminosilicate sheets.

Metakaolin was synthesized by placing high-purity kaolin (KGa-lb, Source Clay Repository, Columbia MO) into a furnace preheated to 750°C for 2 hours in air, and then letting the sample and furnace cool to room temperature. Time-of-flight neutron diffraction was carried out on the NPDF beamline at Lujan Center, Los Alamos National Laboratory.[21] The total scattering data were collected at 15K in a displex cryostat and a standard vanadium can. Standard data reduction was conducted using the PDFGetN software.[22] A sine damping function was applied to the total scattering function above Qdamp = 25Å^{-1}. A cubic spline background subtraction was perfonned, followed by a linear/parabolic subtraction to obtain the desired low-Q dependence. Extrapolation to 0Å^{-1} was carried out. A Fourier sine transfonn was applied to this scattering pattern (0 -30 Å^{-1}) to obtain the PDF.

Real space refinement was carried out starting from the DFT energy minimized metakaolin structure using the PDFfit2IPDFgui software,[23] using the experimental PDF as the observed pattern to which the structure was fitted. Atomic displacement ("thennal") parameters Ui were set to a nominal value of 0.0025 .

The first refinement involved refinement of structure scale function and all atom positions (x,y,z) within the structure. The resulting structure was used as the starting structure for another round of density functional full-geometry optimization. Then the structure once again was refined using PDFgui, this time only allowing the structure scale factor and supercelliattice parameters to change (excluding supercell angles). The output structure from PDFgui was then subjected to full-geometry optimization to obtain the metastable atomic structure of metakaolin.

Acknowledgements

This work was funded in part by the Australian Research Council (ARC) (including some funding via the Particulate Fluids Processing Centre, a Special Research Centre of the ARC), and in part by a studentship paid to Claire White by the Centre for Sustainable Resource Processing via the Geopolymer Alliance. The density functional modeling work was supported by an award under the Merit Allocation Scheme on the NCI National Facility at the ANU. Travel funding for the experimental work conducted at Los Alamos National Laboratory was provided through the ANSTO Access to Major Research Facilities Program. We thank Dr Hyunjeong Kim (LANL) for assistance on the NPDF beamline, Prof. Don Kearley (ANSTO) for useful discussions regarding density functional modeling, and Dr Kia Wallwork (Australian Synchrotron) for assistance in collecting the X-ray powder diffraction data. The PDF work was carried out on NPDF at the Lujan Center at Los Alamos Neutron Science Center, funded by DOE Office of Basic Energy Sciences. Los Alamos National Laboratory is operated by Los Alamos National Security LLC under DOE Contract DE-AC52-06NA25396. The upgrade ofNPDF has been funded by the NSF through grant DMR 00-76488. X-ray diffraction data were collected on the Powder Diffraction beamline (IOBM 1) at the Australian Synchrotron, Victoria, Australia. The views expressed herein are those of the authors and are not necessarily those of the owner or operator of the Australian Synchrotron.

References

1. Taylor, M.; Tam, c.; Gielen, D. "Energy efficiency and CO_2 emissions from the global cement industry," International Energy Agency (2006).

2. Duxson, P.; Provis, J. L.; Lukey, G. C.; van Deventer, J. S. J. The role of inorganic polymer technology in the development of 'green concrete.' Cement and Concrete Research 37,1590–1597 (2007).

3. Brindley, G. W.; Nakahira, M. The kaolinite-mullite reaction series. 1. A survey of outstanding problems. Journal of the American Ceramic Society 42, 311–314 (1959).

4. Sabir, B. B.; Wild, S.; Bai, J. Metakaolin and calcined clays as pozzolans for concrete: a review. Cement & Concrete Composites 23,441–454 (2001).

5. Murray, H. H. Clays, in Ullmann's Encyclopedia ofIndustrial Chemistry; John Wiley & Sons, Inc., Weinheim, Germany (2006).

6. Brindley, G. W.; Nakahira, M. The kaolinite-mullite reaction series. 2. Metakaolin. Journal ofthe American Ceramic Society 42, 314–318 (1959).

7. MacKenzie, K. J. D.; Brown, I. W. M.; Meinhold, R. H.; Bowden, M. E. Outstanding problems in the kaolinite-mullite reaction sequence investigated by Si-29 and AI-27 solid-state nuclear magnetic-resonance. 1. Metakaolinite. Journal of the American Ceramic SOCiety 68, 293–297 (1985).

8. Gualtieri, A.; Bellotto, M. Modelling the structure of the metastable phases in the reaction sequence kaolinite-mullite by X-ray scattering experiments. Physics and Chemistry ofMinerals 25, 442–452 (1998).

9. Rocha, J. Single-and triple-quantum AI-27 MAS NNIR study of the thennal transfonnation of kaolinite. Journal of Physical Chemistry B 103, 9801–9804 (1999).

10. Egami, T.; Billinge, S. J. L. Underneath the Bragg Peaks: Structural Analysis of Complex Materials; Pergamon (2003).

11. Grimme, S. Accurate description of van der Waals complexes by density functional theory including empirical corrections. Journal of Computational Chemistry 25, 1463–1473 (2004).

12. Ugliengo, P.; Zicovich-Wilson, C. M.; Tosoni, S.; Civalleri, B. Role of dispersive interactions in the layered materials: a periodic B3LYP and B3LYP-D* study of Mg(OH)2, Ca(OH)2 and kaolinite. Journal of Materials Chemistry 19, 2564–2572 (2009).

13. White, C. E.; Provis, J. L.; Riley, D. P.; Proffen, T.; van Deventer, J. S. J. What is the structure of kaolinite? Reconciling theory and experiment. J. Phys. Chem. B 113,6756–6765 (2009).

14. Petkov, V.; Billinge, S. J. L.; Shastri, S. D.; Himmel, B. Polyhedral units and network connectivity in calcium aluminosilicate glasses from high-energy X-ray diffraction Physical Review Letters 85, 3436–3439 (2000).

15. Benoit, M.; Ispas, S.; Tuckennan, M. E. Structural properties of molten silicates from ab initio molecular-dynamics simulations: Comparison between CaO¬Ah03-Si02 and Si02. Physical Review B 64, 224205 (2001).

16. Petkov, Y.; Gerber, T. ; Himmel, B. Atomic ordering in Caxl2AlxSil-x02 glasses (x = 0,0.34,0.5,0.68) by energy-dispersive x-ray diffraction. Physical Review B 58, 11982–11989 (1998).

17. Wu, Z. Y. et aI., Evidence for AlISi tetrahedral network is aluminosilicate glasses from Al K-edge x-ray-absorption spectroscopy. Physical Review B 60, 9216–9219 (1999).

18. Loewenstein, W. The distribution of aluminum III the tetrahedra of silicates and aluminates. American Mineralogist 39, 92–96 (1954).

19. Rocha, L.; Adams, L. M.; Klinowski, J. The rehydration of metakaolinite to kaolinite: Evidence from solid-state NMR and cognate techniques. Journal of Solid State Chemistry 89, 260–274 (1990).

20. DelIey, B. From molecules to solids with the DMo13 approach. Journal of Chemical Physics 113, 7756–7764 (2000).

21. Proffen, T. et ai., Building a high resolution total scattering powder diffracto-meter upgrade of NPD at MLNSC. Applied Physics A -Materials Science & Processing 74, S163-S165 (2002).

22. Peterson, P. F.; Gutmann, M.; Proffen, T.; Billinge, S. 1. L. PDFgetN: A user-friendly program to extract the total scattering structure function and the pair distribution function from neutron powder diffraction data. Journal of Applied Crystallography 33, 1192 (2000).

23. Farrow, C. L. et aI., PDFfit2 and PDFgui: computer programs for studying nanostructure in crystals. Journal of Physics-Condensed Matter 19, 335219 (2007).

An 8-Fold Parallel Reactor System for Combinatorial Catalysis Research

Norbert Stoll, Arne Allwardt, Uwe Dingerdissen
and Kerstin Thurow

ABSTRACT

Increasing economic globalization and mounting time and cost pressure on the development of new raw materials for the chemical industry as well as materials and environmental engineering constantly raise the demands on technologies to be used. Parallelization, miniaturization, and automation are the main concepts involved in increasing the rate of chemical and biological experimentation.

Introduction

The last few years have seen a dramatic increase in the importance of systems that allow reactions with reactant gases such as hydrogen, oxygen, and carbon monoxide at varying pressure levels. Accordingly, newly developed equipment boast increased reactions per time unit, reduction in reaction volume to be accommodated by evermore compact equipment designs, and increased levels of automation [1,2].

Current State of Technology

Currently, there are three conceptually distinct system approaches in reaction technology—single, parallel, and microreaction systems. Another distinguishing feature of each system is its area of application, which is limited to either heterogeneous or homogenous catalysis. Single-reaction systems only have one reaction vessel. Examples are HITEC-Zang LabKit, HELAuto-Lab, and reactors manufactured by Parr, Premex Reactor, or Buchi. These enable immediate influence on reaction parameters such as pressure, temperature, stirring intensity, and substance concentration. Controlled and directed changes in reaction parameters provide the possibility of reducing the number of experiments, with the elimination of unusable test results cutting research times.

The main advantage of parallel reaction systems is the large number of reaction vessels that allow simultaneous reactions at low pressure (Chemspeed ASW2000, HEL ChemScan lp, Argonaut Advantage Series 3400) or high Amtec SPR16, Argonaut Endeavor System, Symyx PPR, ChemspeedSLT100,HELChem-Scanhp). Low-pressure systems usually use a common gas supply, which means that individualpressure adjustmentin each reactionis notpossible, whereasmostreactors thatworkatpressures above10bar are equippedwith individualreactorvesselpressureregulators. They not only support the individual parameterization of each reaction vessel, but also provide substance supply under reaction pressure enabling a controlled reaction start. Highly parallelized high-pressure systems such as SymyxHPR,HEL CAT96, celisca HPMR 50–96, Premex Reactor 96x multire actor use a central gas supply, like the low-pressure parallel reactors.

The third group consists of micro reaction reactors such as Ehrfeld Mikrotechnik AG, IMM-Mainz GmbH, Syrris Africa. Their main advantage is the large reaction surface that decreases the time taken in the reaction compared to reactors operated in batch mode. Since microreactors use one stirrer, they operate in continuous mode. Their compact size allows for steep pressure and temperature gradients.

The low continuity in reactor volume becomes clear from parallel reaction systems that operate at reaction pressures above 10 bar; the devices on offer mostly have two-figure milliliter volumes, while the 2–3 ml volume range is only seldom covered (ChemspeedSLT100, HELCAT96).

Assignment Description

The most important catalytic processes in classical bulk chemistry include high-pressure processes with gases such as hydrogen and carbon monoxide. Up to now, these processes have only been sporadically used in materials development and modification. Performing hydration, carbonylation, and other related reactions under pressure (20–100 bar in all cases) together with the safety measures while providing conditions meeting equipment and safety standards requires thedevelopmentof suitable reaction systems [3].

Advances in automated reaction technology have shown a marked development in miniaturization in the reaction approach. One factor thathas made thisdevelopment possible is the refinement of analysis conditions. The most important reasons for this development are, firstly, the reduction in substance quantity required and therefore the experimental costs involved, and, secondly, the increasein thenumberof parallel reactions running relative to the working area available due to miniaturization. This leadst o an increased number of reactions in a given time, therefore increasing productivity. However, the more compact reaction systems often need to be adapted to conventional reaction systems with respect to area-to-volume ratios to make future upscaling to larger reaction systemspossibleata later stage. This especially applies to catalytically controlled reactions with gases [4].

Often, the most important factor in these heterogeneously or homogeneously catalysed reactions is the exclusion of moisture and/oroxygen. This makes planning for a protection gas system essential when constructing the corresponding equipment. Other parameters that should be covered in a corresponding reactor system are stability under pressures of up to 150 bar and constant operation at temperatures between –20 and +150°C. Additional requirements are chemical resistance and avoidance of contamination effects [5].

8-Fold Parallel Reactor Concept

The reactor system is based on an arrangement of up to three reactor arrays, each with eight individual auto claves based on a modular design [6–8]. The concept is aimed towards accommodating individual magnetic stirring systems for each reactor and individual pressure control at pressures up to150 bar. The system temperature is to be controlled for each array using a circulation thermostat.

The control concept is to consist of a hierarchically arranged distributed system with separated functionality and remote client connectivity. This will be based on a distributed arrangement of microcontroller components connected using a CAN field bus and a higher-level control computer (Figure 1).

Hardware Solution

General Description

The base modules are made of stainless steel in order to preserve their chemically inert qualities. Apart from the gas supply, they also have a pipette opening to enable the addition of air-sensitive and moisture-sensitive reagents under inert conditions. A third connection serves the reaction vessel also made of stainless steel, where the catalytic processes take place. The rotationally symmetric vessel has an innervolume of 3 ml, emphasizing the main aim of miniaturization in the construction design. The head of the vessel is shaped in such a way as to simplify sample processing by automated analysis.

The reagents are homogenized using magnetic stirrers driven by an integrated mobile stirrer unit installed underneath the reactor array. This is housed in a casing with a design based on a modification of a conventional aluminium heat sink. This ensures that there is sufficient heat dissipation from the coil-based magnetic stirrer drives despite their position near the temperature-controlled reactor.

The temperature is controlled by transporting the heating fluids through drilled openings in the reactor modules that form a closed circuit in combination with their respective end pieces. This ensures a very even temperature distribution. The circulation thermostat combined with DW Therm fluid provides a temperature control range of –40 to 200°C. The heating and cooling rates are 4.5°C and 2.5°C per minute, respectively, in temperatures ranging between 25°C and 100°C.

Each autoclave is connected to a supply system for inert and process gas, which also purges the system when connected to a vacuum supply.

The respective processes are automated using the integrated measuring and actuation devices. The actuation system is solely based on switch-overvalves.

These enable pressure control to an accuracy of under ±0.1 bar at target pressures of up to 150 bar. The 150 bar pressure limitation is due to the maximum operating pressure of the FAS Bacosol magnetic valves used. Setting the pressure in the eight reactors only takes five minutes. The control principle applied would still allow for the use of other components, thus allowing an extension of the pressure range.

Parker NOVA pneumatic membrane valves are used to seal off the reactor area from the supply system independently of pressure gradient.

The whole system is housed in a compact casing allowing simple integration into standardized laboratory extractor systems.

The results from the catalytic processes are returned to the control system to determine reactor pressure. This involves pressure sensors integrated into the supply system, thus providing continuous information on gas consumption.

Autoclave Array

The central unit of each array consists of a reactor block consisting of eight reactors made of V4A steel 1.4571 arranged inline in a block. The other parts of each reactor are connected up to the supply systemvia the gas supply. The interchangeable reaction vessels with a volume of 3 ml are connected into the reactor from underneath and are fixed into the reactor using a vessel nut. These are also made of stainless steel 1.4571. A Viton O-ring is used to seal off the reactor and the reactor vessel. This material ensures a broad range of chemical resistance properties. There are drilled openings in the reactor blocks for the silicon oil-based temperature control fluid for the circulation thermostat. The fluid reagents are dosed into the reaction vessels via an opening on the upper side of each individual reactor that can be sealed off using a screwed connection and ring gasket (Figure 2).

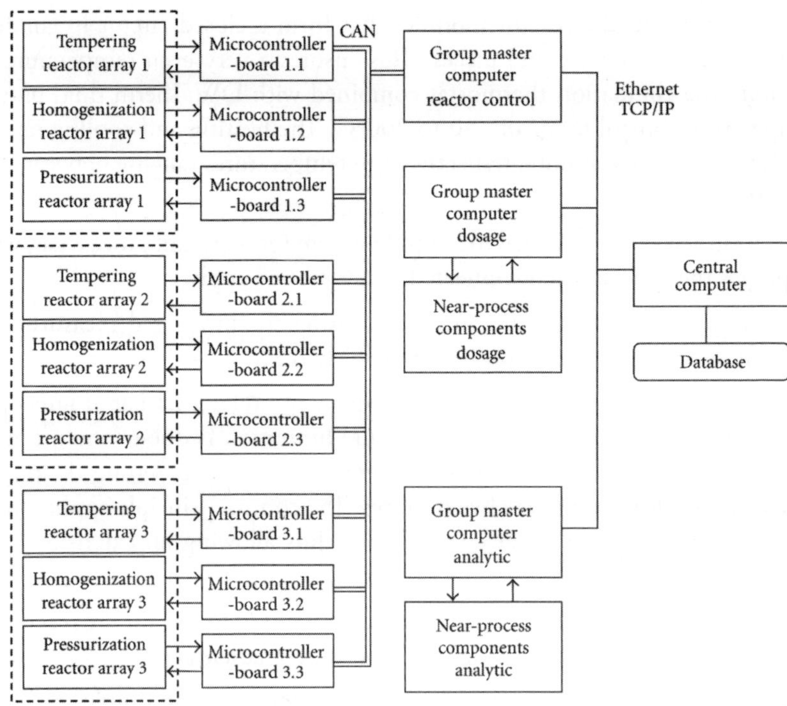

Figure 1. Control concept.

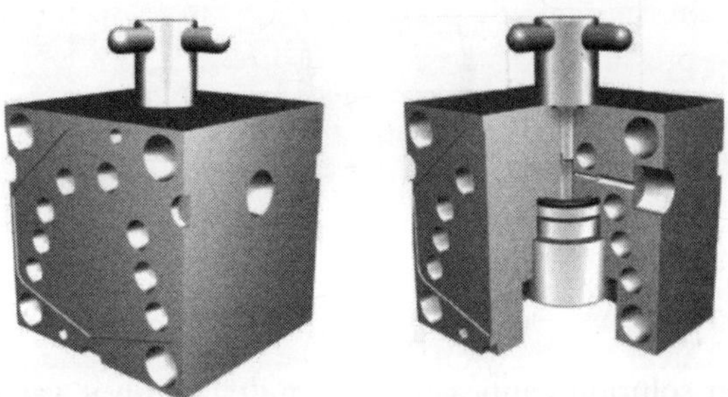

Figure 2. Individual reactor design.

Temperature Control

The reactor temperature is regulated by an external circulation thermostat that heats or cools the fluid in cavities constructed for the purpose, and pumps the fluid into the reactor block. The circulation thermostat used here is a Huber Kältemaschinen GmbH Unistat Tango device, which is designed as a closed system and ensures a temperature range between –40 and 200°C [9].

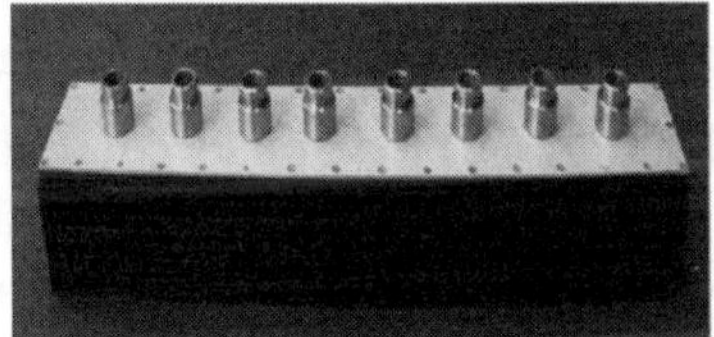

Figure 3. Stirrer unit.

The Unistat Tango has an RS232 interface and a controller unit connected to the local control system using a proprietary protocol [10].

Homogenization

Under the reactor a stirrer assembly is arranged in such a way thatit canbe moved towards therear, allowing there-action vessels to be mounted into the reactor block from underneath. The stirrer assembly ensures that the reagents are mixed effectively, which takes place under thecontrol of VARIOMAG MICRO stirrer drives, which drive the magnetic stirrers in the reaction vessels in a rotating field (Figure 3).

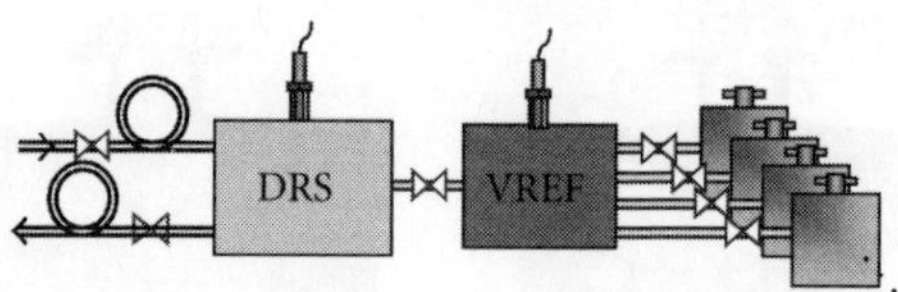

Figure 4. Basic reactor system design.

Each reaction solution canbeagitated simultaneouslyat varying speeds. The stirrers are equipped with a soft-start mechanisminordertooptimize start-up. Thestirrer unitis dimensioned in such a waythat the substances are mixed effectively even under extreme reaction conditions at temperatures of up to 200°C and pressuresofupto200 bar. The stirrerunit iscooledby compressed air to prevent over heating. This assembly is controlled via a CAN interface. Various LEDs show the state of the stirrer unit visually [11].

Gas Supply System

The mechanical design of the reactor ensures excellent operating safety at operating pressures above 150 bar. The centralized gas supply system uses conventional fluid automation systems (FAS) two-way valves, and can supplywith the reactor with reactive gas, inert gas, or a vacuum for purging (Figure 4).

The individual reactor modules are connected by separately controlled pneumatically actuated NOVA membrane valves with a reference volume. Information can be gained onthe inner pressure of the reactor, and therefore the stage of the reaction, from the sequential pressure equalization processes between the valve and each reactor.

The reference volume is connected to a pressure regulation line, which is also used in the gas supply for the system, for which it has in lets for process and inert gases as well as outlets for connection to a vacuum supply and air extraction system.

The detailed reactor structure results from integrating the various actuation and sensor components involved in the basic design (Figure 5).

The individual modules in each reactor block equipped with a temperature sensor (Pt100) are separated from the reference volume by membrane valves, which have an integrated high-accuracy PMP 4070 pressure sensor (0.04% of theendvalue). This sensorisintendedforrecordingpressure equalization processes. Pneumatic actuators are also used for blocking the reference volume to the pressure regulation line. The target pressures for each individual reactor are set in the pressure regulation line. The corresponding pressure values are measured using a PMP 1400 sensor. The temperature in the pressure regulation line is measured by a Pt100 sensor element.

The connected vacuum system can only be blocked using a conventional FAS Bacosol magnetic valve set in one operating correction due to the constant pressure gradient. A vacuum sensor (Norgren GAH 40 level controller) is used for recognizing potential supply outages in this area.

The pressure is controlled using magnetic valves selected for their dynamic characteristics. The valves are additionally connected using capillary tubing with an inner diameter of 0.5 mm to limit the flow. The other path installed into the outlet branch serves towards slow depressurization of the reaction area and is fitted with a capillary tube with a nominal diameter of 0.25 mm. Another magnetic valve prevents contamination of the pressure regulation line with the two capillary-combined output valves set against their operating direction.

Two more passive elements are integrated into the branches serving the gas supply. These consist of particle filters with a filter range of 5–9 µ and ball check valves with a trigger pressure of 0.14 bar. The valves are to protect the actuators from excess pressure acting against their operating direction.

Specialized PMP 1400 sensor sare integrated into the gas feed of the reactor to ensure real-time process monitoring. The sensors are positioned horizontally over the reference volume. The digital signal conversion electronics is housed in the terminal box already fitted.

The dimensions of the 8-fold parallel reactor are 57 × 65 × 30 cm. The other reactor arrays can be mechanically connected with additional connecting equipment.

Hardware Reactor Control

The parallelized reaction platform is automated using a distributed control system consisting of intelligent microcontroller units networked using a field bus protocol based on the CAN bus systemtoatop-level control computer.

Each of the decentralized controller units fulfils a complete control function. The functionality of the DCS is arranged into temperature control, homogenization, and pressure management. In the finished system, two of the micro-controller units were placed in a central electronic equipment rack that also accommodates the power supply units. Figure 6 shows the front and rear of the "19" rack unit. Each unit has at least one front-side status LED for information on operating status. Errors can also be shown visually on the controller units.

Figure 7 shows the structure of the system at process level.

Software Reactor Control

The system software consists of three separate function modules.

One of these modules is the control software in the electronic units, and governs the microcontrollers that control the hardware to ensure smooth operation in the subprocesses to be carried out. A part from controlling the actuators, the functions ofthis software module include reading sensor information and, if necessary, throwing a unit-specific error.

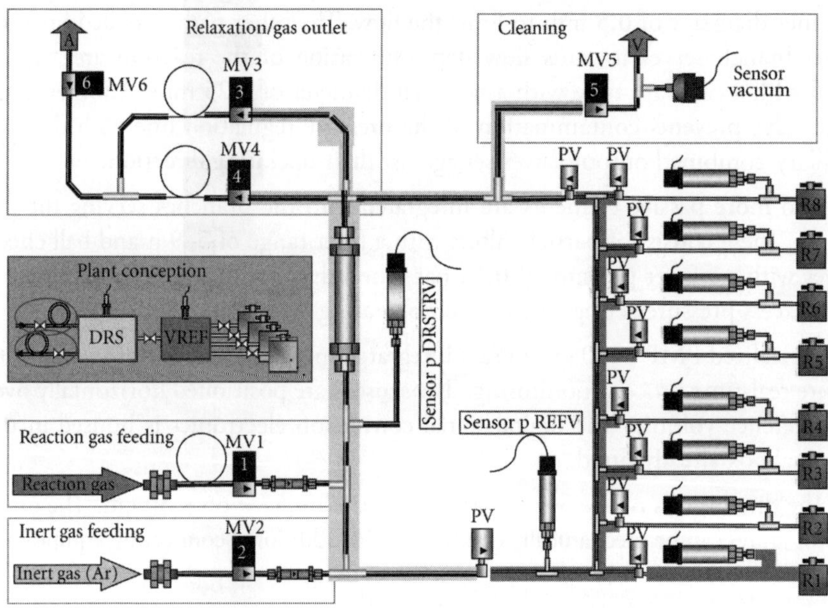

Figure 5. Detailed reaction system equipment structure.

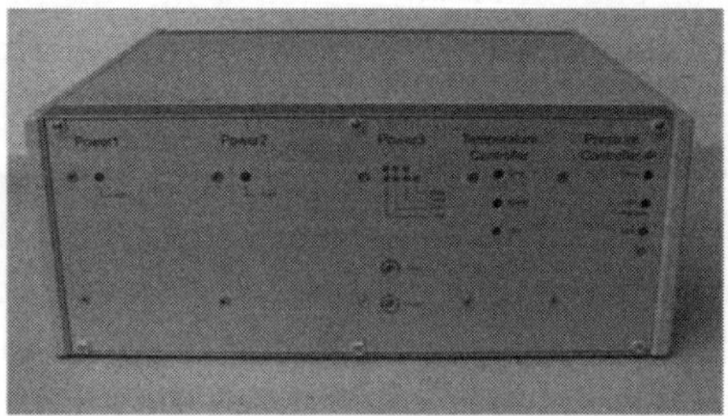

Figure 6. Electronic equipment rack.

The two other software modules presenting the local control system use the PC as their platform. It is important to distinguish the server used as the system service, which is mainly responsible for data transfer, and the client, which handles visualization, operation, and protocol logging.

The server runs on the computer that is connected to a maximum of three electronic equipment rack units. The three client applications responsible for visualizing the individual 8-fold parallel reactor blocks can be connected to this computer. Data transfer between the reactorc ontrol and the local control system uses the CAN field bus. On the other hand, data is transferred between the client and server via the network using an IP protocol.

The actual user interface is provided using complex control software enabling control of the three autoclave arrays—or 24 individual autoclaves.

Local Control System

The local control system handles planning, control, and monitoring of automated reaction processes in an 8-fold auto clave array.

The reaction can be halted manually at any time. The control system reacts to any errors that have occurred by halting the reaction.

The field bus is used for communication with the auto clave array using a proprietary protocol.

An LIM system can be used for planning and starting the reaction.

The user interface ensures user-friendly chemical reaction operation. This requires data transfer with the server usingaTCP/IP-based interprocess communication protocol. The reactor data are logged to a protocol file for later review. An export function enables subsequent processing and analysis of the log data using, for example, MS Excel.

The functional emphasis of the system mainly lies in the following areas:

(i)   visualization andoperation;

(ii)   process management and control;

(iii)   archiving andprotocollogging;

(iv)   process communication.

The control software is Windows-based and has a two-part design corresponding to the client-server model. The clients are responsible for the visualization, process management, and logging function while the server manages time-critical communication with the distributed units at process level. In order to ensure at least close to real-time conditions, this part ofthe software runs as a high-priority system service [12,13].

Communication between the two software parts is based on a TCP/IP protocol to ensure remote communication between the server and client. The client is responsible for visualization and runs a typical Win32 MDI (multiple-document interface) application that can handle a number of documents at the same time.

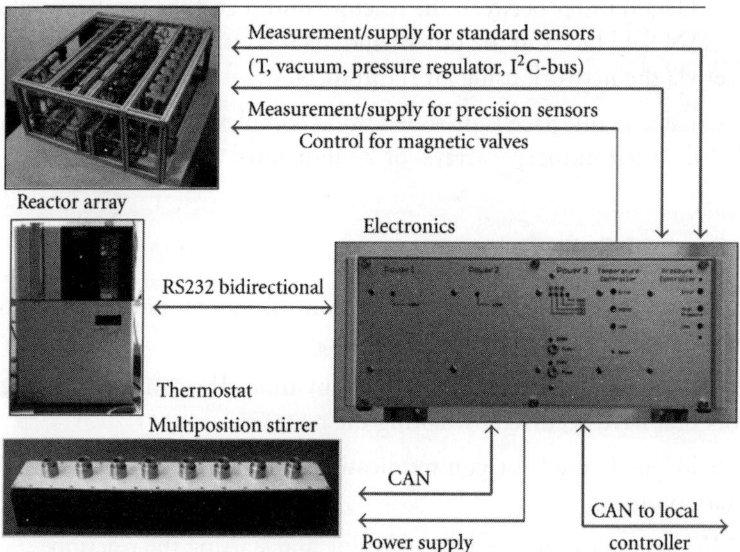

Figure 7. Local control system.

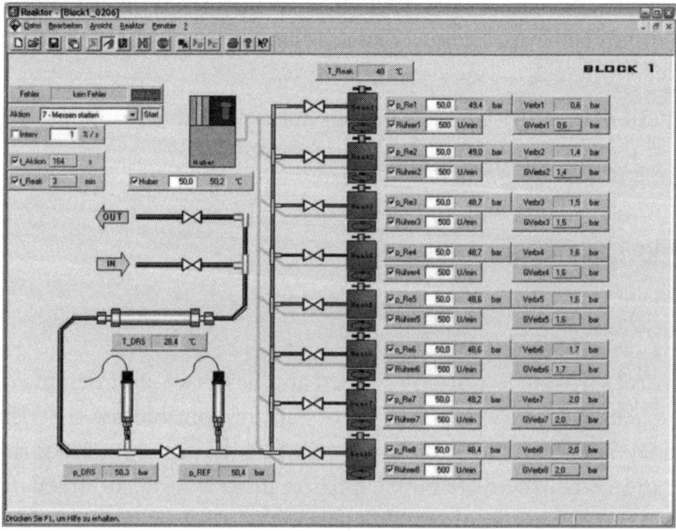

Figure 8. Control software and reactor control document layout.

Visualization and Operation Function

There are two document types. One of their functions is to visualize the state of there actor system and its operation, and the other is to edit the command lists for automatic experiment control.

The document for operating the reactor system is run as an OLE container, and its appearance can be altered according to the preferences of the end user. The layout consists of a background (an image or graphic) and a number of visualization and data-entry elements that can be placed as required. Figure 8 shows the default configuration for operating a reactor array, which is modeled on the systematic structure of the actual system.

Global operation takes place using the control sat the top left-hand part. Control commands can be sent to the pressure management unit using thedrop-down menu ("Aktion" control).The system status visualization is also shown here.

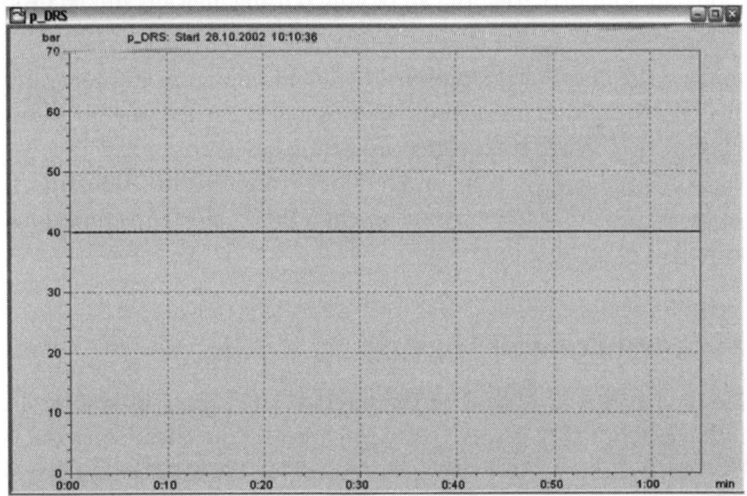

Figure 9. Diagram representation.

Figure 10. Example of an exported file.

The target values can be entered into the corresponding entry fields. The respective actual values are shown as numbers. There is also the possibility of displaying the progress of each value of any control element in a diagram, allowing the user to follow the reaction trends in real time. Figure 9 shows an example of this diagram in which both the time and value axes are scalable.

Process Management and Control

Process management and control can take place on two levels. In the first, the user can influence the reaction manually using a mouse and keyboard on the layout interface described above. In the second level, the system allows automated reaction management using an editable command list, which can implement processes based on time orevent. This sets the basis for carrying out reproducible experiments.

The list can be shown and edited using the second document type that the control software supports. The list can be edited using logically linked drop-down menus.

Command list creation is simplified by an additional macro recording function. This allows the user to record the commands carried out on the software control layout, which are then stored as a time-based command list. Note that only time-based processes can be programmed using this functionality. The user will have to use the drop-down menus to enter limit value and conditional statements as well as branching instructions.

Archiving and Protocol Logging

The values and events displayed on the interface are logged by the local control system. This functionality is specifically provided by the client software. All data created is recorded every 250 ms into a protocol log file with the rpf extension. A new file is created everyday and carries the current date and the computer name of the server used in its file name.

The client has an export function that can export the rpf file into a format that can be read by spread sheet software. The export process is controlled using menus on the client, and allows the user to select certain data, time intervals, and periods (as multiples of the basic interval). The result is a tab-separated text file in ASCII format that can be read by anyDOS-based system, and can be opened and processed by standard data processing packages such as MS Excel or MATLAB (Figure 10). Therefore, the process data logged can also be used after the experiment to analyse the chemical reaction course.

Process Communication

The distributed microcontroller units and the higher-level control system structures have to be able to communicate.

The control system is connected to the process level units using the server application and a CAN interface. The latter is managed using a software library supplied by the manufacturer that gives the system the corresponding driver function [12]. Status information on each process unit is read at 250 ms intervals.

The higher-level control level is connected using the client software and a TCP/IP protocol. A script interface allows the user to enter target values, read actual values, and execute command lists. This enables the integration of the reaction equipment into the whole automation system with other dosing and analysis function units.

Operating Modes

The reactor can be operated in two operating modes—isochoric and isobar modes.

Isochoricmode

The sensors that directly measure the innerpressure are read at set intervals, thus delivering current information on the actual reaction process. Due to the characteristics of the multichannel analog-digital converter and the input circuit, the operator has to wait for a constant value reading after switching channels. A measuring interval of 200 ms has been selected that leads to a total running time of 1.6 s taking array dimension into account.

The measuring process takes place throughout the entire reaction period and is directly connected to the filling system, which is made possible by the integration of high-pressure sensors in the compressed gas system for each individual autoclave.

The decreasing pressure during the reaction gives information on the gas consumption in the reaction.

Isobar Mode

The advantages of isobar mode operation are the constant pressureconditions during the reaction. The user has to define a maximum permissible pressure decrease in the reactors. If the pressure falls below a certain threshold value, a sequence is triggered that adjusts the pressure back to the target pressure via the supply system.

Chemical Application System Test: Carbonylation

Introduction to the Reaction

The system was tested with catalytic carbonylation reactions using a carbon monoxide supply. The reaction selected for determining precision and screening reaction parameters was carbonylation of iodobenzene to methylbenzoate using a palladium catalyst (see Figure 11) as described in the literature [14].

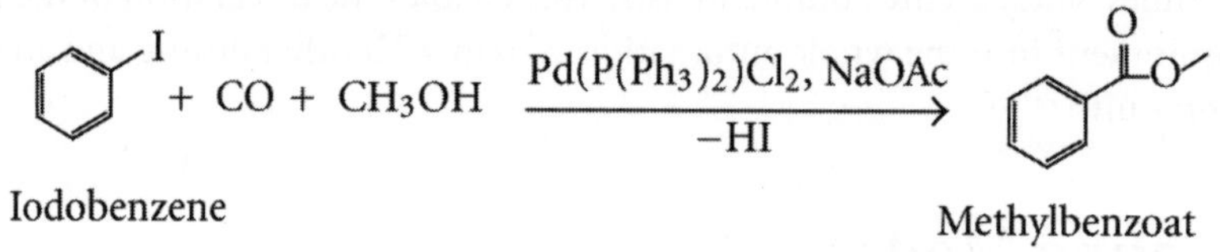

Iodobenzene

Methylbenzoat

Figure 11. Reaction diagram forc arbonylation of iodobenzene to methylbenzoate.

Table 1. Example of an exported file.

Reactor	6 h at 60°C	1 h at 80°C
	Rate in %	Rate in %
1	79.40	75.77
2	78.14	74.05
3	74.87	79.45
4	80.22	76.78
5	79.59	77.49
6	81.03	75.35
7	80.37	77.49
8	80.10	83.74
Average	79.216	77.516
Standard deviation, abs.	1.951	2.993
Standard deviation, rel.	2.463	3.862

Examination of Precision Under Comparison Conditions

To examine the precision under comparison conditions, all eight reactors were filled with identical reaction solutions, and the contents of each reactor were exposed to the same reactor conditions (60°C temperature and 6 h reaction time, or 80°C temperature, 1hreactiontime, H2 starting pressure: 50 bar, stirrer speed: 300 rpm). The results are summarized in Table 1.

In contrast to the reports from the literature stating that are action temperature of 60°C at a pressure of 200 psi (14 bar) and a reaction time of 48h should yield a rate of 80%, this rate was already reached after 6 h at are agent ga spressure of 50 bar, and 60 min after increasing the temperature to 60 min. At relative average standard deviations of 2.46% and 3.86%, the values were reached under comparison conditions were present.

Screening Examinations

Screening experiments with variations in the following reaction parameters were carried out to demonstrate the power of the octuple parallel reactor system:

(i)   temperature: 60–90°C;

(ii) CO starting pressure: 10–40 bar;

(iii) stirrer speed: 100–900 rpm;

(iv) reaction time: 1–3 h.

Figure 12 shows a graphical representation of values reading using double determination by varying reagent gas pressure and reaction temperature at a reaction period of 3 h and a stirrer speed of 300 rpm.

The graphic shows that the maximum of the range examined is already reached at a temperature of 80°C and reagent pressures between 30 and 40 bar.

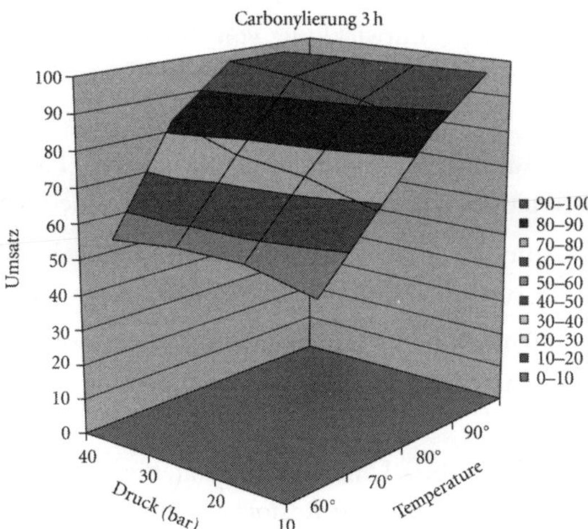

Figure 12. Summary of results for a 3 h reaction period, 300 rpm stirrer speed.

Summary

As the results show, the octuple reactor is a parallelized system that can be used for various reactions with gases under high pressure at volumes ranging from 1.5 to 2 ml. With inert gas conditions ensured, air-sensitive substances can also be used and several reaction parameters can be optimized at the same time.

Development of various septum adapters will allow substance injection at higher pressures into the reactor.

References

1. B.T. Haag, "Automated and unattended parallel synthesis integrating work-up and analysis," Chimia, vol. 54, no.4, pp. 163– 164, 2000.

2. Frost & Sullivan (Hrs.), Combinatorial Chemistry: Powerful New Tool for Discovery and Product Development, Marktreport, New York, NY, USA, 2nd edition, 2002.

3. Y. K. Yun and J. Labadie, "Parallele katalytische Hydrierung fokussierter Verbindungen," GIT Laborzeitschrift, vol. 44, no. 12, pp. 1456–1458, 2000.

4. G. Jung, Combinatorial Chemistry, VCH Verlagsgesellschaft mbH, Weinheim, Germany, 1999.

5. F. A. Rampf and W. A. Herrmann, "High-Throughput-Katalysator screening unterHochdruckbedingungen," Chemie Ingenieur Technik, vol. 73, no.1-2, pp. 97–99, 2001.

6. K. Thurow, A. Allwardt, C. Wendler, and N. Stoll, Kombinatorische Methoden zur effizienteren Entwicklung von Katalysatoren -Teilprojekt Automation, Schlussbericht zum Projekt 03C0304B an dasBMBF, Universitat Rostock 2003.

7. R. Lemke, "Automation eines miniaturisierten Hochdruck reaktions systems,"Dissertation, Universitat Rostock, Rostock, Germany, 2003.

8. A.G.Degussa, B. E. Bosch, T. Riermeier, etal., Array of Autoclaves, DE10049078A 1und WO2002030557A2, 2002.

9. Firmenvero offentlichung, Betriebsanleitung alzthermo-Umwstat Unistat, HuberKoltemaschinenbau GmbH, Betriebsanleitung, Offenburg, Germany, 2000.

10. J. Kozik, Anbindung eines Umlaufthermostaten and as Controller-Area-Network, Strudienarbeit, Universitat Rostock, 1999.

11. A. Meinel, "Entwicklung einerverschleißfreie Mehrpositions Ruruhreinrichtung f'chemische Reaktoren auf Basis einer Mikrocontrollersteuerung mit CAN-Anbindung," Diplomarbeit, Universitat Rostock, Rostock, Germany, 1999.

12. H. Dahl, "Entwicklung einer programmierbaren SoftwareoberflurWindowsNT4.0 zurzeit-und ereignisbasierten achef Ablaufsteuerung eines chemischen Reaktors," Diplomarbeit, Universitat Rostock, Rostock, Germany, 1997.

13. H. Dahl, M. Krohn, S.Junginger, T. Roddelkopf, and N. Stoll, Prozessleitsystem unterWindowsNT4.0furvernetzte intelligente Komponenten einesdezentralenMess-undSteuerungssystems, 2. Wismarer Auto matisierungs symposium, Wismar, Germany, 1999.

14. J. K. Stille and P. K. Wong, "Carboalkoxylation of aryl and benzyl halides catalyzed by dichlorobis (triphenylphosphine) palladium(II)," Journal of Organic Chemistry, vol. 40, no.4, pp. 532–534, 1975.

Assessing the Reliability and Credibility of Industry Science and Scientists

Craig S. Barrow and James W. Conrad Jr.

ABSTRACT

The chemical industry extensively researches and tests its products to implement product stewardship commitments and to ensure compliance with governmental requirements. In this commentary we argue that a wide variety of mechanisms enable policymakers and the public to assure themselves that studies performed or funded by industry are identified as such, meet high scientific standards, and are not suppressed when their findings are adverse to industry's interests. The more a given study follows these practices and standards, the more confidence one can place in it. No federal laws, rules, or policies express a presumption that scientific work should be ignored or given lesser weight because of the source of its funding. To the contrary, Congress has consistently mandated that agencies allow interested or affected parties to provide information to them and fairly consider that information. All participants in scientific review panels should disclose sources of potential biases and conflicts of interest. The former should be considered in seeking a balanced panel rather than being used as a basis for disqualification. Conflicts of interest generally do require disqualification, except where outweighed by the need for a person's services. Within these constraints, chemical industry scientists can serve important and legitimate functions on scientific advisory panels and should not be unjustifiably prevented from contributing to their work. Key words: bias, chemical industry, conflict of interest, industry science, industry scientists.

Introduction

The business of chemistry takes very seriously its responsibility to protect people and the environment throughout the entire life cycle of its products. This commitment is embodied in the product stewardship aspects of Responsible Care [American Chemistry Council (ACC) 2005a], the ACC's initiative to continuously improve its members' environmental, health, safety, and security performance, as well as the Long-Range Research Initiative (ACC 2005b), ACC's voluntary initiative to fund research to increase understanding of the potential impacts chemicals may have on human health and the environment. Support for both programs is a condition of ACC membership. Also, a comprehensive set of U.S. and foreign government rules strictly regulate how chemicals are developed, manufactured, distributed, and used. The chemical industry conducts extensive research and testing on its products to implement product stewardship commitments and to ensure compliance with those governmental requirements. This work is conducted both directly by companies and indirectly through contracts with and grants to external scientists and research institutions. Consequently, the products of chemistry are among the most thoroughly evaluated and regulated in commerce. The research and testing conducted by the chemical industry are absolutely necessary, as they provide information on the potential health and environmental risks of substances that manufacturers, users, and government agencies all rely on to determine the conditions under which chemical products can be safely used.

In recent years some have questioned the reliability and credibility of public health and environmental research conducted or funded by the chemical industry, suggesting that industry research is fundamentally conflicted and hence unreliable (Devine 2001; Sass et al. 2005). These critics also challenge the legitimacy of allowing scientists employed or funded by industry to participate on scientific review panels (Center for Science in the Public Interest and Environmental Working Group 2004; Devine 2001; Greer and Steinzor 2002; Sass et al. 2005). Both contentions are mistaken and counterproductive to societal interests.

Chemical Industry Science

Scientific studies conducted or funded by the chemical industry have long been acknowledged by government agencies, nongovernmental organizations, and the scientific community at large as necessary and valuable contributions to the understanding of potential public health and environmental effects related to the manufacture and use of chemicals [Environmental Defense 1998; Health and Environmental Sciences Institute 2005; U.S. Environmental Protection Agency (EPA) 2005d]. These same groups generally also recognize that, historically and for the foreseeable future, the costs associated with conducting chemical product testing have been and will be borne largely by industry, not the public sector, reflecting both the free enterprise view that those who benefit from an activity should bear the costs of that activity as well, and chronic resource limitations on the public sector.

A wide variety of mechanisms exists by which policymakers and the public they serve can assure themselves that studies performed by or funded by industry are identified as such, meet high scientific standards, and are not suppressed when their findings are adverse to industry's interests. These practices and standards include the following:

- The ability of test sponsors to contractually authorize investigators—regardless of the results—to submit the investigators' findings for publication in the peer-reviewed scientific literature without sponsor approval. This is the policy of ACC's Long-Range Research Initiative.

- The practice of Environmental Health Perspectives and virtually all other scientific journals to require disclosure of funding sources.

- Peer review, which both government agencies and private entities may conduct or fund.

- The U.S. EPA requirement that all studies required to be submitted in connection with chemical regulation and pesticide statutes be conducted in accordance with U.S. EPA-approved guidelines for test protocols and Good Laboratory Practice (GLP) regulations (U.S. EPA 2005b, 2005c), which entail full availability to government authorities of the raw, quality-assured data files for review and audit. Research required to be submitted to regulatory agencies in member countries of the Organisation for Economic Cooperation and Development (OECD) must also follow OECD GLP principles (OECD 2005), which serve to substantiate the high quality and validity used for determining the safety of chemicals and chemical products in those countries.

- Information Quality Act (2000) guidelines issued by all federal agencies. These guidelines require scientific data to meet applicable standards for accuracy, reliability, and lack of bias and that apply to privately generated information when agencies rely on it for regulatory purposes (Office of Management and Budget 2002; U.S. EPA 2002a).

- U.S. EPA requirements under chemical and pesticide regulatory statutes (Federal Insecticide, Fungicide and Rodenticide Act 1972; Toxic Substances Control Act 1976), and similar global equivalents, that chemical manufacturers and pesticide registrants provide the U.S. EPA and its equivalents with timely notification of any adverse effects findings.

- The prospect of tort liability for suppression of adverse research findings.

- Finally, and most fundamentally, the scientific process itself through which different investigators attempt to reproduce the findings of others—a process that has led to the retraction of papers for which results could not be reproduced (McLachlan 1997).

The more a given study follows the above practices and standards, the more confidence one can place in it. In particular, the use of U.S. EPA-approved test protocols helps ensure that high standards of quality are being followed. Such protocols have been validated by the U.S. EPA and chosen after extensive and careful review to determine that the results would provide reproducible information suitable for regulatory decision making. Use of such protocols helps to produce a degree of certainty regarding the reliability and relevance of test results, which in turn provides the confidence necessary for making safety and regulatory determinations. Similarly, when research studies adhere to GLPs, as is the norm for industry health and environmental studies, reviewers and those acting on the science may have a high degree of confidence that the experimenters adhered to the specific and detailed experimental protocol employed, took all of the steps and measurements claimed to be taken during conduct of the study itself, and accurately reported the test results (Anderson et al. 2001).

No federal laws, rules, or policies express a presumption that scientific work should be ignored or given lesser weight because of the source of its funding. To the contrary, an entire body of federal law embodies a congressional mandate that agencies allow interested or affected parties to provide information to them and fairly consider that information. Primary among these is the Administrative Procedure Act (1946); others include the Information Quality Act (2000), the Federal Advisory Committee Act (1972), the Federal Register Act (1935), the Regulatory Flexibility Act (1980), and the Paperwork Reduction Act (1980).

Ironically, the proliferation of articles and presentations impugning the merits of industry science is having the effect of subjecting that science to much greater public and agency scrutiny than is applied to science conducted or funded by government or nonprofit entities (Miller 2005). Such scrutiny can only increase the likelihood that any flaws in the work will be identified.

Ultimately, all scientific research must stand or fall on its merits. Researchers should disclose their sources of funding because in cases where those sources have a potential interest in a question addressed by the research, others may want to scrutinize that research with heightened care to determine the extent to which it followed the practices and standards discussed above. However, it is unscientific (Society of Toxicology 1997), as well as unfair, to disregard or discount a study based solely on which investigator or institution conducted or funded it.

Chemical Industry Scientists

The chemical industry's commitment to scientific research and product testing includes engaging the highest quality scientists. Our scientists have national and international stature in the scientific community, as reflected by their inclusion on such authoritative bodies as the National Academies' Board on Environmental Studies and Toxicology (National Academies 2005) and the U.S. EPA Science Advisory Board (U.S. EPA 2005a). These scientists have expert knowledge of the chemicals their employers manufacture and fully appreciate the value of their contributions—as objective, trained scientific experts—to the development and interpretation of the science needed to evaluate the health and environmental effects of their products. As members of professional associations such as the Society of Toxicology, industry scientists adhere to both personal and professional commitments to act in accordance with the codes of ethics of their professions (Society of Toxicology 1985).

As noted above, some have argued that scientists employed or funded partially by industry should not be permitted to serve on private or governmental review panels or similar bodies due to an asserted conflict of interest. Any discussion of this issue must carefully distinguish between conflict of interest and bias.

Federal rules issued under the Ethics in Government Act (Office of Government Ethics 1997) provide that true conflicts of interest are limited to instances where a person has a concrete financial interest in the subject being addressed. A conflict might occur, for example, in the case of an employee of a business that generates significant revenues from a product, if that employee is tapped to review a government assessment of that product. A conflict of interest could also occur in the case of a "public interest" representative, particularly if that representative is also serving as an expert witness in connection with ongoing litigation over the same subject matter. It is important to note that these government ethics rules still allow a person with a financial interest to serve where "the need for the individual's services outweighs the potential for a conflict of interest."

By contrast, bias (or "partiality," under government ethics rules) is both unavoidable and unobjectionable. As the National Academies explain, bias derives from

> points of view or positions that are largely intellectually motivated or that arise from the close identification or association of an individual with a point of view of a particular group. (National Academies 2001)

Similarly, a U.S. EPA Science Advisory Board committee has stated that

> [a]lthough it is possible to avoid conflict of interest, avoidance of bias is probably not possible. All scientists carry bias due, for example, to discipline, affiliation and experience. (U.S. EPA 2000)

Fifteen past presidents of the Society of Toxicology have written in Risk Policy Report (2002) that

> [o]f course, all scientists have biases; acknowledging this, we as a society must be aware of those biases and seek to ensure balance in the scientific panels whose task is to provide the best possible technical review of complex, important issues.

Finally, any evaluation of the role of industry scientists in governmental processes must consider the federal laws referenced above that empower interested persons to have input into those processes. Particularly relevant in this context is the Federal Advisory Committee Act (1972), which requires advisory committees to be "balanced" and thus should prohibit both exclusion of, as well as domination by, any interest (Office of Government Ethics 1997).

Expertise is the touchstone that guides the procedures followed by both the National Academies (2001) and the U.S. EPA Science Advisory Board (U.S. EPA 2002b). The National Academies' current policy (National Academies 2003) is a particularly useful and appropriate statement of the relevant issues, for the following reasons:

- It emphasizes that knowledge, training, and experience are the foremost considerations and that no one should be appointed to a panel to represent a particular point of view or special interest.

- It clarifies that "[f]or some studies … it may be important to have an 'industrial' perspective or an 'environmental' perspective," not because these "sides" need to be represented, but because such individuals, through their particular knowledge and experience, are often vital to achieving an informed, comprehensive, and authoritative understanding and analysis of the specific problems and potential solutions to be considered by the committee.

- It notes that "conflict of interest" ordinarily refers to "financial interests" and that these can arise from any quarter, including regulated entities, the government, and private organizations.

- It explains that biases should not be disqualifying—even when a person works for a company with "a general business interest in" the subject of the panel— unless the person

is totally committed to a particular point of view and unwilling, or reasonably perceived to be unwilling, to consider other perspectives or relevant evidence to the contrary.

Such a case of bias would seem to be present in the report of Sass et al. (2005), for example, which argues that peer-review panels involving industry scientists are not "scientifically credible." The article acknowledges funding from the Beldon Fund, which awarded the authors' employer (the Natural Resources Defense Council) a 3-year, $210,000 grant

[t]o implement [Natural Resources Defense Council's] Public Interest Service Initiative, a campaign to remove industry-funded scientists from EPA advisory boards and to appoint scientists dedicated to protecting human health and the environment. (Beldon Fund 2001)

The Office of Management and Budget's recent peer-review guidelines agree with the National Academies and the U.S. EPA that

the most important factor in selecting reviewers is expertise: ensuring that the selected reviewer has the knowledge, experience and skills necessary to perform the review. (Office of Management and Budget 2005)

Consistent with these authoritative sources, scientists employed or funded by industry should be eligible to participate in peer-review panels and similar bodies just like any other scientists, based on the knowledge, training, and experience they bring to the body. All participants in such bodies should disclose sources of potential biases and conflicts. Potential biases should be considered in seeking a balanced panel rather than being used as a basis for disqualifying an expert. True conflicts of interest generally do require disqualification, except when outweighed by the need for a person's services.

Conclusion

The chemical industry takes seriously its central responsibility to conduct or fund research and testing of chemicals for use in the risk assessment processes. The scientific work it conducts and funds has an important and appropriate role in the development of health and environmental information. Its scientists can serve important and legitimate functions on scientific advisory panels. Frequently, they can provide unique knowledge and insight concerning the chemical in question and therefore should not be unjustifiably prevented from contributing to the work of such panels.

References

1. ACC. 2005a. Responsible Care. Washington, DC:American Chemistry Council. Available: http://www.responsiblecare.com [accessed 3 October 2005].

2. ACC. 2005b. Long-Range Research Initiative. Washington, DC:American Chemistry Council. Available: http://www.uslri.com [accessed 3 October 2005].

3. Administrative Procedure Act. 1946. Public Law 79-404.

4. Anderson W, Parsons B, Rennie D. 2001. Daubert's backwash: litigation-generated science. Univ Mich J Law Reform 34:619-682.

5. Beldon Fund. 2001. Beldon Fund 2000 Grants. Available: http://www.beldon.org/grants2000_07.html [accessed 26 February 2001].

6. Center for Science in the Public Interest and Environmental Working Group. 2004. CSPI, Environmental Working Group Challenge Two Scientists on EPA Panel. Available: http://cspinet.org/integrity/press/200412091.html [accessed 3 October 2005].

7. Devine J. 2001. Has there been a corporate takeover of EPA science? Risk Policy Rep 8:35-38.

8. Environmental Defense. 1998. Joint Announcement of Cooperative Program for High Production Volume U.S. Industrial Chemicals. Available: http://www.environmentaldefense.org/article.cfm?ContentID=661 [accessed 3 October 2005].

9. Federal Advisory Committee Act. 1972. Public Law 92-463.

10. Federal Insecticide, Fungicide and Rodenticide Act. 1972. Public Law 92-516.

11. Federal Register Act. 1935. Public Law 74-220.

12. Greer L, Steinzor R. 2002. Bad science. Environ Forum 19:28-43.

13. Health and Environment Sciences Institute. 2005. Homepage. Washington DC:International Life Sciences Institute, Health and Environmental Sciences Institute. Available: http://www.hesiglobal.org [accessed 3 October 2005].

14. Information Quality Act. 2000. Public Law 106-554.

15. McLachlan J. 1997. Synergistic effect of environmental estrogens: report withdrawn. Science 277:459-463. [CrossRef].

16. Miller H. 2005. Some activist groups exhibit a "pathological science" stance; pursued agenda is often not the protection of human health or the environment. Genet Engin News 25:68,70. Available: http://www.genengnews.com/current/article.aspx?cat=Point%20Of%20View&id=323 [accessed 3 October 2005].

17. National Academies. 2001. The National Academies Study Process. Washington, DC:National Academies. Available: http://www4.nationalacademies.org/news.nsf/isbn/07302001?OpenDocument [accessed 3 October 2005].

18. National Academies. 2003. Policy on Committee Composition and Balance and Conflicts of Interest. Washington, DC:National Academies. Available: http://www.nationalacademies.org/coi/BI-COI_FORM-0.pdf [accessed 3 October 2005].

19. National Academies, Board on Environmental Studies and Toxicology. 2005. Roster of National Academies Board on Environmental Studies and Toxicology. National Academies. Washington, DC:National Academies. Available: http://dels.nas.edu/best/members.shtml [accessed 3 October 2005].

20. OECD. 2005. Principles of Good Laboratory Practice. Paris:Organisation for Economic Cooperation and Development. Available: http://www.oecd.org/document/63/0,2340,en_2649_34381_2346175_1_1_1_1,00.html [accessed 3 October 2005].

21. Office of Government Ethics. 1997. Standards of Ethical Conduct for Employees of the Executive Branch. 5 CFR 2635.402. Washington, DC:U.S. Office of Government Ethics.

22. Office of Management and Budget. 2002. Guidelines for ensuring and maximizing the quality, objectivity, utility, and integrity of information disseminated by federal agencies; notice; republication. Fed Reg 67:8452-8460.

23. Office of Management and Budget. 2005. Final information quality bulletin for peer review. Fed Reg 70:2664-2677.

24. Paperwork Reduction Act. 1980. Public Law 96-511.

25. Regulatory Flexibility Act. 1980. Public Law 96-354.

26. Risk Policy Report. 2002. Letter to the Editor. Available: http://www.iwpnews. com/050603_enviro_iwp.php [accessed 21 December 2005].

27. Sass J, Castleman B, Wallinga D. 2005. Vinyl chloride: a case study of data suppression and misrepresentation. Environ Health Perspect 113:809-812.

28. Society of Toxicology. 1985. Code of Ethics. Available: http://www.toxicology. org/AI/ASOT/ethics.asp [accessed 3 October 2005].

29. Society of Toxicology. 1997. Principles for Research Priorities in Toxicology. Available: http://209.183.221.234/ai/asot/principles.asp [accessed 3 October 2005].

30. Toxic Substances Control Act. 1976. Public Law 94-469.

31. U.S. EPA. 2000. Re: Review of the Draft Report to the Congress, "Characterization of Data Uncertainty and Variability in IRIS Assessments, Pre-Pilot vs Pilot/post-Pilot" [Letter]. EPA-SAB-EHC-LTR-00-007. Washington, DC:U.S. Environmental Protection Agency, Science Advisory Board. Available: http:// www.epa.gov/sab/pdf/ehcl007.pdf [accessed 3 October 2005].

32. U.S. EPA. 2002a. Guidelines for Ensuring and Maximizing the Quality, Objectivity, Utility, and Integrity of Information Disseminated by the Environmental Protection Agency. 260/R-02-008. Washington, DC:U.S. Environmental Protection Agency. Available: http://www.epa.gov/quality/informationguidelines/ documents/EPA_InfoQualityGuidelines.pdf [accessed 3 October 2005].

33. U.S. EPA. 2002b. Overview of the Panel Formation Process at the Environmental Protection Agency Science Advisory Board. EPA-SAB-EC-02-010. Washington, DC:U.S. Environmental Protection Agency, Science Advisory Board.

34. U.S. EPA. 2005a. Chartered Science Advisory Board Members Fiscal Year 2005. Washington, DC:U.S. Environmental Protection Agency, Science Advisory Board. Available: http://www.epa.gov/sab/boardmem05.htm [accessed 3 October 2005].

35. U.S. EPA. 2005b. Good Laboratory Practice Standards. 40 CFR 160. Washington, DC:U.S. Environmental Protection Agency.

36. U.S. EPA. 2005c. Good Laboratory Practice Standards. 40 CFR 792. Washington, DC:U.S. Environmental Protection Agency.

37. U.S. EPA. 2005d. High Production Volume (HPV) Challenge Program-- Sponsoring Organizations. Washington, DC:U.S. Environmental Protection Agency. Available: http://www.epa.gov/chemrtk/spncomp.htm [accessed 3 October 2005].

Copyrights

15. Copyright © 2006 Norbert Stoll et al. This is an open access article distributed under the Creative Commons Attribution License, which permits unrestricted use, distribution, and reproduction in any medium, provided the original work is properly cited.

16. Public Domain

Index

Differential Optical Absorption Spectroscopy (DOAS), 59, 60, 67, 77
dimethyl ether (DME), 249
di-n-butyl phthalate (DBP), 127
dioxins, 187
distributed control system, 295
DME. *See* dimethyl ether (DME)
DOAS. *See* Differential Optical Absorption Spectroscopy (DOAS)
dry matter production, 18, 19
Dudhia shortwave scheme, 82
Duke Forest vegetative flux measurements, 28–29

E

early life exposures, to xenobiotic chemicals, 182
ECDs. *See* electron capture detectors (ECDs)
EC 3.2.1.73 enzyme category, 114
ECHA. *See* European Union Chemicals Agency (ECHA)
EDX. *See* energy-dispersive x-ray (EDX)
effluent discharge, 13
EISA. *See* Energy Independence and Security Act (EISA)
electrochemical cells, 269
electron capture detectors (ECDs), 26
electrophoresis, 13, 118, 120
electrostatic precipitators, 143
Emergency Planning and Community Right-to-Know Act, 179
emissions from refinery and power plants, models for studying
    meteorological measurements, 60–61
        pilot balloons, 61
        radiosondes, 61
    mini-DOAS instruments, 59–60
        simulation models, 61–63
Emissions Inventory Improvement Program, 66
empirical model, for design of control system for RO unit, 50–55
energy-dispersive x-ray (EDX), 145

Energy Independence and Security Act (EISA), 244
entrapment, 203
Environmental Defense Fund, 177, 191
environmental health sciences, implications for, 190–91
    data gap, 191
    safety gap, 191–92
    technology gap, 192
enzyme
    activity, 111
    assay, 118–19
    half-lives of, 115
    production, 111–12
    purification, 113–14
    specificity, 114
EPS. *See* extracellular polymeric substances (EPS)
Escherichia coli, 218, 228
ethanol production, 217–19
Ethics in Government Act, 310
ethyl alcohol, 126
Eulerian modelling, of Popocatepetl emissions, 79–80
Eulerian pollutant transport, 83
European Commission Seventh Framework Program, 137
European Union Chemicals Agency (ECHA), 180
European Union, new chemical and product laws in, 192–93
exopolysaccharide alginate, 207
expanded granular sludge bed (EGSB) reactors, 204, 206, 209, 213, 229
extracellular polymeric substances (EPS), 206
ExxonMobil, 244

F

FACE. *See* Free Atmospheric Carbon Enrichment (FACE)
Faroes Statement of the International Conference on Fetal Programming and Developmental Toxicity, 182

G

fuel evaporation, 38–39
influence of various sources on toluene levels at, 33–38
toluene-to-benzene ratio in, 32
toluene *vs.* i-pentane mixing ratios at, 35
vegetative flux measurements, 27–28
warm season toluene enhancements at, 29–32
time-of-flight neutron diffraction, 284
tolerable daily intake (TDI), 127
toluene
biogenic and anthropogenic emissions in New England, 41
^{13}C labeled, 25
effects on human central nervous system, 24
emission rates for sunflowers and pine trees, 25
fluxes, 35, 37, 40
fuel evaporation emissions of, 34
influence of various sources on toluene levels at Thompson Farm, 33–38
vs. i-pentane mixing ratios at Thompson Farm, 35
methods for measuring emission of
Duke Forest vegetative flux measurements, 28–29
Thompson Farm ambient VOC measurements, 26–27
Thompson Farm vegetative flux measurements, 27–28
as pollutant and respiratory irritant, 24
seasonal cycle in urban anthropogenic emissions of, 25
sources of, 25, 34
warm season enhancements at Thompson Farm, 29–32
toluene-to-benzene ratios, 25, 29
observed in air masses, 32
seasonal pattern of, 31
at Thompson Farm, 32
as tracers of urban anthropogenic influence, 30

toxicological evaluation of alcohols
materials and methods used for
chemicals and reagents, 127–28
GC/MS method, 128–29
glassware and reagent control, 128
optimization and validation studies, 130–31
samples, 129–30
statistics, 131
results
method validation and sample measurement, 133–35
parameter optimization for LLE method, 131–33
Toxics Release Inventory, 37, 179
Toxic Substances Control Act (TSCA), 177–79, 181, 182, 184–90, 192, 309
trade secrets, and industrial chemical policy, 187–88
triboelectrostatic method, 158
1,1,2-trichlorotrifluoroethane, 130
Trichoderma harzianum, 116
Trichoderma sp., 111
trickling bed reactors (TBR), 209, 210
trickling filters, 212, 216
triethylamine (TEA), 264, 268, 271
triethylenediamine (TEDA), 266
2,2,4-trimethylpentane, 25
TSCA. *See* Toxic Substances Control Act (TSCA)
Tula industrial complex, 57, 58, 79
comparisons between measured and simulated emissions from, 67–68
NO_2 measurements of, 66
SO_2 and NO_x emission inventories of, 66
Tula–Vito–Apasco industrial corridor, 57
TVC. *See* thermal vapor compression (TVC)

U

ultraviolet (UV) detector, for elemental mercury, 154
ultraviolet fluorescence detector, 145